THE ANT AND THE PEACOCK

THE ANT
AND THE PEACOCK

ALTRUISM AND SEXUAL SELECTION
FROM DARWIN TO TODAY

Helena Cronin

CAMBRIDGE
UNIVERSITY PRESS

Published by the Press Syndicate of the University of Cambridge
The Pitt Building, Trumpington Street, Cambridge CB2 1RP
40 West 20th Street, New York, NY 10011-4211, USA
10 Stamford Road, Oakleigh, Melbourne 3166, Australia

© Cambridge University Press 1991

First published 1991
Reprinted 1992 (three times)
First paperback edition 1993

Printed in the United States of America

Library of Congress Cataloging in Publication Data available.

A catalogue record for this book is available from the British Library.

ISBN 0-521-32937-X hardback
ISBN 0-521-45765-3 paperback

CONTENTS

FOREWORD

There is something distinctly odd about the history of the idea of sexual selection. For Darwin, the idea was an important part of his theory of evolution: he needed it to explain the many examples of elaborate and apparently non-adaptive sexual ornamentation. Yet Wallace, the co-discoverer of natural selection, had little liking for the idea, and it was largely ignored in discussions of evolution for a hundred years. Initially, the main difficulty lay in the concept of 'choice', which did not fit well with the attempts of behaviourists to interpret behaviour in mechanistic terms. In 1930, Fisher offered an explanation for the evolution of choice, but his idea had little impact at the time. During the period of the 'modern synthesis' in the 1940s and 1950s, female choice was accepted, but solely as a process ensuring that a female would mate with a member of her own species. This was perhaps natural, since at that time *the* evolutionary problem was seen as the nature and origin of species, but it did lead to an unfortunate neglect of Darwin's theory of sexual selection.

When, in 1956, I published a paper showing, at least to my own satisfaction, how female fruit flies choose males, I do not remember receiving a single reprint request. My ethological contemporaries who were studying under Tinbergen at Oxford refused to accept my explanation that some males failed to mate because they could not keep up with females during the courtship dance. Although ethologists had introduced the concept of motivation into animal behaviour, they could not understand that an animal did not mate because it could not: the idea that the spirit might be willing but the flesh weak was not then an acceptable one.

This neglect of sexual selection turned to enthusiasm during the 1970s and 1980s. It is tempting to ascribe this change of attitude to the influence of the women's movement. It is certainly not the case that the new research, theoretical and empirical, has been carried out by ardent feminists, but I think it may have been influenced, even if unconsciously, by attitudes towards female choice in our own species. But I suspect that a more important reason lies in the kind of explanation that scientists were seeking. A major characteristic of evolutionary biology since 1960 has been the attempt to give a consistently Darwinian explanation of characteristics that at first sight seem anomalous – for example, sex, ageing, conventional behaviour, sexual ornamentation, and, most important of all, cooperation. I do not know why these matters have seemed so important to us since 1960, whereas, despite the fact that both Fisher and Haldane were interested in cooperation, they were largely ignored before that date. George Williams once told me that he was provoked into writing his influential 1966 book, *Adaptation and Natural Selection*, by hearing a lecture by Emerson on the superorganism: my own interest was similarly provoked by reading Darlington and Wynne-Edwards. But although this illustrates the importance of erroneous ideas in science, it does not explain why an interest in providing a functional explanation for behaviour blossomed in the 1960s, and not 30 years earlier.

Helena Cronin has tackled these questions head on, taking sexual selection and the evolution of cooperation as her central themes. There is a recent fashion in the history of science to throw away the baby and keep the bathwater – to ignore the science, but to describe in sordid detail the political tactics of the scientists. Helena Cronin, I am happy to say, does not belong to this school. She has told me much that I did not know about the ideas of Darwin and Wallace, and the disagreements between them. She has also understood the modern research on the same topics. For Darwin, the ant and the peacock symbolised two major difficulties for his theory – the existence of cooperation, and of apparently maladaptive ornament. I think he would enjoy reading her account of what has happened since.

John Maynard Smith FRS

PREFACE

An awesome gulf divides the pre-Darwinian world from ours. Awesome is not too strong a word to describe the achievements of Charles Darwin and Alfred Russel Wallace. The theory of natural selection revolutionised our understanding of living things, furnishing us with a comprehension of our existence where previously science had stood silent. *The Ant and the Peacock* celebrates that towering theory and its thriving modern descendant.

Towering – but not monolithic. Wallace has been fancifully called the moon to Darwin's sun. But his was no pale, reflected light. He held firmly to his own, distinctive conception of their joint theory. In his eyes, even Darwin could be revisionist – to the extent that Wallace felt moved to declare himself 'more Darwinian than Darwin'. Their disputes were not merely personal preferences, history's marginalia. They amounted to telling divergences in emphasis and interpretation, divergences so significant – and this is one thesis of this book – that they have persisted right up to the present day. Indeed, in recent years, they have been played out with renewed intensity.

For these are exciting times. Since the 1960s, Darwinian theory has been swept along in a transformation. In its wake have come new preoccupations – so new that they may seem remote from those of Darwin and Wallace. And yet, look more closely, and they reveal themselves to be part of that continuing controversy that stretches all the way back to Darwinism's beginnings. In today's disputes we hear echoing exchanges in the voices of the founding fathers themselves. As one leading Darwinian remarked to me on reading a chapter of mine in draft: 'I hadn't realised all these years that I was really a Wallacean!' *The Ant and the Peacock*, then, is in some ways a history of Darwin versus Wallace.

This book traces the careers of two controversial issues: sexual selection and altruism. It was over sexual selection that Darwinism's co-discoverers most notoriously parted company. The theory's initial blaze was followed by long years in the wilderness of fashion. Now, however, sexual selection has once again become hotly debated. Altruism, too, has been much aired of late. On this question, also, the responses of Darwin and Wallace were typically divergent. But here the interest of the history lies less in the continuities and more in the contrasts between then and now. Whereas in the nineteenth century altruism was barely seen as a difficulty, in recent years it has been seized on as an obvious anomaly – and triumphantly solved.

The timing of this book, then, has turned out to be propitious. It has proved the more so because sexual selection and altruism have been surprisingly untouched by that prodigious outpouring of historical research that has come to be known as the 'Darwin industry'. *The Descent of Man*, which spells out the theory of sexual selection, was second only to *The Origin* among Darwin's works. For him, sexual selection was not just a frill, a minor variant of natural selection, but a distinctive and pervasive force in the living world. Yet historians have largely overlooked both book and theory. Equally, little attention has been paid to the history of altruism – above all, to the vague, good-for-the-species view that was rampant from about the 1920s to the 1960s. And yet, with today's fresh understanding, the twists and turns of its fate are ripe for historical scrutiny.

This book is opportune, too, because the story it tells is not confined to the past: it is also a gentle guide to the sometimes bewildering ferment of new ideas. What is the current standing of the latest views and what are the connections between them? *The Ant and the Peacock* sets out the notions that are currently being aired in conferences and corridors.

Indeed, this mapping of the new Darwinian world was one of the starting points of this book. My initial interest in Darwinian theory was roused by philosophers' criticisms – not because I thought that they were right but because I was convinced that they must be seriously wrong. Methodologists had long given Darwinism a bad press: 'untestable', 'circular', 'idle metaphysics', 'an empty tautology'. Even with Darwinism's recent flowering, the judgements were scarcely less harsh. The discrepancy between Darwin's and Wallace's magnificent legacy and these ungenerous appraisals left me dissatisfied. I decided to explore more of the new territory that Darwinian theory was opening up. This book, in part, records my personal journey through this new-found-land.

The Ant and the Peacock is not, however, a book of science, nor of history, nor philosophy, though it combines something of all three. It can certainly be read without any expert knowledge. But I hope, too, that readers from these specialised backgrounds will, in their different ways, find new things in it – in particular from the light that history, science and philosophy can shed on one another.

This book grew out of joint work with Allison Quick; she cooperated closely with me in the early stages; I am much indebted to her and miss our discussions. John Watkins read an early draft with care; any philosopher knows how to criticise but he also knows how to encourage. John Maynard Smith's enthusiasm sustained me through all the usual misgivings. Peter Milne was a merciless critic, but a constructive one.

I am deeply grateful to Sir Richard Southwood for arranging a position for me in the Department of Zoology in Oxford. This proved to be an ideal working environment, with its excellent institution of tea and coffee times as well as more formal seminars. (No sooner had I written that than I was firmly told, at coffee, that it is a departmental platitude – which only goes to reinforce my point.) I am indebted to many people in the department, particularly to Alan Grafen, W. D. Hamilton, Paul Harvey, Andrew Pomiankowski and Andrew Read.

Several other people have generously read draft sections, criticised, discussed, advised and encouraged. Among them are Aubrey Sheiham, Nicholas Maxwell, Michael Ruse, Nils Roll-Hansen, Amanda C. de C. Williams, Michael Joffe, David Ruben, Peter Urbach, John Worrall, N. H. Barton, J. S. Jones, John Durant, Peter Bell, K. E. L. Simmons and Carl Jay Bajema. Alan Crowden at Cambridge University Press has been a committed and amusing editor.

My brother, David Cronin, thought of the title *The Ant and the Peacock*. Richard Dawkins thought of the title of Chapter 1: *Walking archives*.

I should also like to express my thanks to The British Academy, The Leverhulme Trust, The Nuffield Foundation and The Royal Society, all of whom have generously supported my research.

Finally, my especial thanks and appreciation to Richard Dawkins.

Helena Cronin

Picture acknowledgments

Chapter 2 A world without Darwin

The photograph of Darwin is reproduced by permission of the Syndics of Cambridge University Library. Dr Rachel Garden gave permission to reproduce the drawing of Wallace from Williams-Ellis, A. (1966) *Darwin's Moon: A biography of Alfred Russel Wallace*, Blackie, London. Alwyne Wheeler gave permission to reproduce the drawing of the sandsole by Valerie DuHeaume from Wheeler, A. (1969) *The Fishes of the British Isles and North-West Europe*, Macmillan, London. Melissa Bateson drew the diagram of Lamarckism, Weismannism and the extended phenotype.

Chapter 3 Darwinism old and new

Melissa Bateson drew the green woodpecker. T. and A. D. Poyser gave permission to reproduce the woodpecker tongue and hyoid bone diagrams from Campbell, B. and Lack, E. (eds.) (1985) *A Dictionary of Birds*, T. and A. D. Poyser, Calton, Staffordshire. The scanning electron micrograph of the dandelion is reproduced courtesy of the British Museum (Natural History). The photograph of R. A. Fisher is reproduced courtesy of the Department of Genetics, University of Cambridge. John Maynard Smith supplied the photographs of J. B. S. Haldane and of himself. W. D. Hamilton supplied the photograph of himself. *Scientific American* gave permission to reproduce the bower birds' bowers from Borgia, G. (1986) 'Sexual selection in bowerbirds', *Scientific American 254 (6)* (Copyright 1986 by *Scientific American*, Inc. All rights reserved).

Chapter 4 Demarcations of design

Melissa Bateson drew the extended pleiotropy diagram. The diagram reproduced in the 'honest advertisement' illustration first appeared in 1978 in *New Scientist* magazine, London, the weekly review of science and technology.

Chapter 6 Nothing but natural selection?

Joshua R. Ginsberg supplied the photograph of the plains zebras. The picture of the three African plovers is reproduced by permission of the Syndics of Cambridge University Library.

Chapter 7 Can females shape males?

Priscilla Barrett drew the proboscis monkey. The picture of the Argus pheasant is reproduced by permission of the Syndics of Cambridge University Library.

Chapter 8 Do sensible females prefer sexy males?

Melissa Bateson drew the male ornamentation diagram. Blackwell Scientific Publications gave permission to reproduce the drawing of the pelican from Brown, L. H. and Urban, E. K. (1969) 'The breeding biology of the Great White Pelican *Pelecanus onocrotalus roseus* at Lake Shala, Ethiopia', *Ibis 111*.

Chapter 14 Make dove, not war: Conventional forces

Lea MacNally supplied the photograph of the roe deer skulls. Melissa Bateson drew the three stages of combat in red deer.

Chapter 15 Human altruism: A natural kind?

The graph of homicides is adapted from Daly, M. and Wilson, M. (1990) 'Killing the competition', *Human Nature 1*, fig. 1.

My thanks to all those mentioned above and additional thanks to Mark Boyce, T. H. Clutton-Brock, George McGavin, Charles Munn, Amotz Zahavi and especially Melissa Bateson, Euan Dunn, Sean Neill and Priscilla Barrett.

PART ONE

Darwinism, Its Rivals and Its Renegades

1

Walking archives

We are walking archives of ancestral wisdom. Our bodies and minds are live monuments to our forebears' rare successes. This Darwin has taught us. The human eye, the brain, our instincts, are legacies of natural selection's victories, embodiments of the cumulative experience of the past. And this biological inheritance has enabled us to build a new inheritance: a cultural ascent, the collective endowment of generations. Science is part of this legacy, and this book is about one of its foremost achievements: Darwinian theory itself. The story is a success story: a tale of two puzzles that had stubbornly resisted explanation, and of how Darwinism finally resolved them. One puzzle is the problem of altruism, epitomised by the ant of this book's title; the other is the problem of sexual selection, the peacock.

Ants – and other social insects – have long been held up as models of rectitude, as sharing, caring, community-minded creatures who will act for the good of others even at extreme cost to themselves. Such saintly self-abnegation is by no means confined to insects. Many animals put themselves in danger to warn of predators, forgo reproduction to help rear the offspring of others, share food that could alleviate their own hunger. But how could natural selection have led to characteristics that are so obviously disadvantageous to their bearers? How can self-sacrifice, especially reproductive self-sacrifice, which places others at an advantage, possibly be passed on to subsequent generations? How can selection favour you if you insist on putting others first? Natural selection surely prefers the quickest, the boldest, the slyest, not those that renounce tooth and claw for the public-spirited ways of the commune.

With our other eponymous hero, the peacock, the difficulty lies in his splendid tail. It flies in the face of natural selection. Far from being efficient, utilitarian and beneficial, it is flamboyant, ornamental and a burden to its bearer. And 'peacocks' tails' – ornaments, colours, songs, dances – abound throughout the animal kingdom, from insects to fish to mammals.

At first sight the glory of the peacock's plumage or the splendour of the stag's antlers may seem to have little to do with the risks of acting as a sentinel or foraging for others; self-indulgent narcissism may seem to be at opposite poles from self-sacrificial altruism. But to a Darwinian such characteristics pose a common difficulty. Aren't they downright disadvantageous to the bearer? Wouldn't natural selection be expected to eliminate, rather than favour, them?

For over a century these problems, when not neglected, were 'solved' in quite erroneous ways. During this time Darwinian theory was spectacularly successful at accounting for the eye, the spider's web, the woodpecker's beak, a plumed seed – characteristics that were obviously adaptive. Nevertheless, its power to explain the unselfishness of a bird's warning call or the glory of the peacock's tail was overlooked or misunderstood. But in the last few decades Darwinism has undergone a revolutionary change. And, in the wake of this transformation, the obstinate anomalies of altruism and sexual selection are anomalies no more.

An understanding of the present can illuminate the past. The recent revolution in neo-Darwinism provides a powerful tool for sharpening our understanding of earlier Darwinian thinking. In the light of the new ideas we can return to the nineteenth century and take a fresh look at evolutionary theory at that time, at how the problems of altruism and sexual selection were viewed and why they remained unsolved. In return, history can illuminate the present. Continuities with the past can throw an unexpected light on current controversies. More, history can help us to elucidate the status of Darwinism. Despite the theory's obvious success, some philosophers have compared it with their favourite triumphs of science – Newton and Einstein – and found it wanting. Historical insight can help us to see exactly why it is that, such philosophers notwithstanding, Darwinism really does explain so much so well.

Historians often dismiss this retrospective style of viewing history. They are understandably anxious to reject the blinkered complacency of 'Whig' history, for which the past is no more than an inevitable progress towards the triumphs of the present. But when the history concerns science, there surely is reason to expect that the latest really is the best. Any upward trend will have its sawtooth reversals but, give or take the odd blind alley, the scientific knowledge of its time usually incorporates the best attempts to date. Of course, improvement is not guaranteed. But science shows it more reliably than most human activities. So I shall be optimistically retrospective. The vantage point of the present, far from diminishing our empathy with the problems and solutions of the past, helps us to appreciate how they could be reasonable even though wrong; it helps us to value old ideas even when they have disappeared from the textbooks of today. Here is the biologist John Maynard Smith rumbling one of those ritual disavowals of Whiggishness that nowadays tend to precede histories of biology:

> He [it happens to be Ernst Mayr] remarks on the need to avoid writing a Whig history of science, but that is the kind of history he has written. To be fair, I cannot imagine how a man who has striven all his life to understand nature, and who has fought to persuade others of the correctness of his understanding, could write any other kind of history ... After all, if Victorian England really had been the highest peak of civilization yet reached, and if it really had held in itself the guarantee of continued progress, Macaulay's method of writing history would have much to recommend it.
>
> Unfashionable as it may be to say so, we really do have a better grasp of biology today than any generation before us, and if further progress is to be made it will have to start from where we now stand. So the story of how we got here is surely worth telling.   (Maynard Smith 1982a, pp. 41–2)

For better or worse, then, my policy throughout this book has been to begin at the end, with the best of what we know today. In some chapters this is overt; in others it lurks behind the scenes. And while on policy, I should add that my history will also be 'internalist' – confining itself to the scientific content of theories and other matters internal to science. Darwinian historians have recently leaned towards 'externalism' – concentrating on political, economic, social, psychological and other non-scientific influences on individual scientists and their discoveries. With no disparagement towards meticulous scholarship in the archives of Victorian society, there is also a place for scientific history that concentrates on the scientists' science.

Sexual selection and altruism occupy Parts 2 and 3 of this book. Part 1 sets out several themes that have woven their way through the history of Darwinism and, in particular, through the history of theories about sexual selection and altruism. It discusses the success of Darwinism and the failure of its rivals; the key features that distinguish the theory of Darwin and Wallace from its lineal descendant of today; and Darwinian alternatives to adaptive explanations. But for the reader more interested in how the peacock got his tail and the ant her sociable ways, Parts 2 and 3 can be treated as self-contained.

2

A world without Darwin

1859

Imagine a world without Darwin. Imagine a world in which Charles Darwin and Alfred Russel Wallace had not transformed our understanding of living things. What, that is now comprehensible to us, would become baffling and puzzling? What would we see as in urgent need of explanation?

The answer is: practically everything about living things – about all of life on earth and for the whole of its history (and, probably, as we'll see, about life elsewhere, too). But there are two aspects of organisms that had baffled and puzzled people more than any others before Darwin and Wallace came up with their triumphant and elegant solution in the 1850s.

The first is design. Wasps and leopards and orchids and humans and slime moulds have a designed appearance about them; and so do eyes and kidneys and wings and pollen sacs; and so do colonies of ants, and flowers attracting bees to pollinate them, and a mother hen caring for her chicks. All this is in sharp contrast to rocks and stars and atoms and fire. Living things are beautifully and intricately adapted, and in myriad ways, to their inorganic surroundings, to other living things (not least to those most like themselves), and as superbly functioning wholes. They have an air of purpose about them, a highly organised complexity, a precision and efficiency. Darwin aptly referred to it as 'that perfection of structure and co-adaptation which most justly excites our admiration' (Darwin 1859, p. 3). How has it come about?

The second puzzle is 'likeness in diversity' – the strikingly hierarchical relationships that can be found throughout the organic world, the differences and yet obvious similarities among groups of organisms, above all the links that bind the serried multitudes of species. By the mid-nineteenth century, these fundamental patterns had emerged from a range of biological disciplines. The fossil record was witness to continuity in time; geographical distribution to continuity in space; classification systems were built on what was called unity of type; morphology and embryology (particularly comparative studies) on so-called mutual affinities; and all these subjects revealed a remarkable abundance of further regularities and ever-more diversity. How could these relationships be accounted for? And whence such profligate speciation?

In the light of Darwinian theory, the answers to both questions, and to a host of other questions about the organic world, fall into place. Darwin and Wallace assumed that living things had evolved. Their problem was to find the mechanism by which this evolution had occurred, a mechanism that could account for both adaptation and diversity. Natural selection was their solution. Individuals vary and some of their variations are heritable. These heritable variations arise randomly – that is, independently of their effects on the survival and reproduction of the organism. But they are perpetuated differentially, depending on the adaptive advantage they confer. Thus, over time, populations will come to consist of the better adapted organisms. And, as circumstances change, different adaptations become advantageous, gradually giving rise to divergent forms of life.

The key to all of this – to how natural selection is able to produce its wondrous results – is the power of many, many small but cumulative changes (Dawkins 1986, particularly pp. 1–18, 43 –74). Natural selection cannot jump from the primaeval soup to orchids and ants all in one go, at a single stroke. But it can get there through millions of small changes, each not very different from what went before but amounting over very long periods of time to a dramatic transformation. These changes arise randomly – without relation to whether they'll be good, bad or indifferent. So if they happen to be of advantage that's just a matter of chance. But it's not a grossly improbable chance, because the change is very small, from an organism that's not much like an exquisitely-fashioned orchid to one that's ever-so-slightly more like it. So what would otherwise be a vast dollop of luck is smeared out into acceptably probable portions. And natural selection not only seizes on each of these chance advantages but also preserves them cumulatively, conserving them one after another throughout a vast series, until they gradually build up into the intricacy and diversity of adaptation that can move us to awed admiration. Natural selection's power, then, lies in randomly generated diversity that is pulled into line and shaped over vast periods of time by a selective force that is both opportunistic and conserving.

The rival explanations of the same evidence (see e.g. Bowler 1984; Rehbock 1983, pp. 15–114; Ruse 1979a) were unimpressive in the extreme (leaving aside Lamarckism and post-1859 rivals for the moment). When we see how grossly inadequate these theories were at doing their explanatory work, and when we reflect that they were nevertheless the principal explanations that were accepted by eminent thinkers for centuries, then we

Charles Darwin (1809–82)

can only too well imagine what our world would be like without Darwinian theory and what an impoverished world that would be.

Darwin and Wallace first put forward their theory in 1858, in a joint paper to the Linnean Society (theirs was one of those odd cases of near-simultaneous discovery in science); and then in 1859 came Darwin's *Origin*

Wallace in the Brazilian jungle in 1849, aged 26

of Species. Before 1859 much of natural history was closely wedded to natural theology (see e.g. Gillespie 1979; Gillispie 1951). With God at its side, natural history came up with the inevitable answer to where all this apparently conscious design came from: that it actually was designed, the work of the supreme designer. And natural theology used this evidence of supposedly deliberate design in nature to prove the existence of God. This may have made neat theology but it made bad science. And the bad science was not confined to anti-evolutionists. By the middle of the nineteenth century, the idea of evolution was gradually coming to be accepted, after having been almost universally rejected at the beginning of the century. However, evolutionists, too, when pressed for a mechanism, resorted to conscious design (or to vagueness).

But theories that are of no direct scientific value can nevertheless be of scientific interest in other ways – in this case, for the light that they shed on Darwinian theory. From this point of view, these pre-Darwinian deliberate-design theories fall into two distinct groups, depending on which of the two major explanatory problems – adaptation or diversity – they took to be paramount. For some, deliberate design was manifested in the adaptive details of individual organisms; for others, it lay in the grand sweep of nature's total plan.

Take, first, the tradition that saw purpose in the coil of a shell, the sweep of a wing, the shape of a petal, in the minute details of adaptation in every living thing. These naturalists saw their prime task as demonstrating how each part of an organism was of use to it, however small, however apparently insignificant. This natural history movement paralleled a school of natural theology whose central tenet was the so-called utilitarian argument from design. This argument appealed to organic adaptation, to its usefulness and function as evidence of providential design: nature's pervasive purpose was God's purpose. Here is David Hume's spoof of the utilitarian argument, from his *Dialogues Concerning Natural Religion*, published in 1779, three years after his death. (The word 'natural', by the way, is used to distinguish such religion or theology from its so-called revealed counterpart, the idea being that it is founded not on a leap of faith or revelation but on evidence and reason about the natural world – just like natural history, now called biology, or natural philosophy, now called physics.) The *Dialogues* are witheringly critical of natural religion, which is why Hume withheld their publication in his lifetime. But Hume's parody is characteristically sharper and more succinct than many a pious original. This passage likens organisms to machines:

Look round the world, contemplate the whole and every part of it: you will find ...
[that all] these various machines, and even their most minute parts, are adjusted to
each other with an accuracy which ravishes into admiration all men who have ever
contemplated them. The curious adapting of means to ends, throughout all nature,
resembles exactly, though it much exceeds, the production of human contrivance –
of human design, thought, wisdom, and intelligence. Since therefore the effects
resemble each other, we are led to infer, by all the rules of analogy, that the causes
also resemble, and that the Author of nature is somewhat similar to the mind of man,
though possessed of much larger faculties, proportioned to the grandeur of the work
which he has executed.   (Hume 1779, p. 17)

One can immediately see how natural history could reverse this theological
argument: not design in nature as proof of the existence of God, but the
manifest existence of God as an explanation of nature's adaptive design, its
apparent contrivance, its improbable complexity.

At the beginning of the nineteenth century, the utilitarian views were
gathered together, systematised and popularised by Archdeacon William
Paley. His *Natural Theology* (1802) – a work with which Darwin was
thoroughly acquainted – became a classic text. He made famous a particular
version of the utilitarian argument. Just look at an instrument as intricately-
wrought as a watch, he said, and you can see immediately that it must have a
watchmaker; in the same way, an object as complex and well-adapted as an
organism must have a designer. Paley's text was superseded in the 1830s by a
highly ambitious project, the *Bridgewater Treatises* (so called because the
Earl of Bridgewater commissioned them in his will) (see e.g. Gillispie 1951,
pp. 209–16). This was a series of publications from eight contributors in all.
They were asked to demonstrate no less than

the Power, Wisdom, and Goodness of God, as manifested in the Creation;
illustrating such work by all reasonable arguments – as for instance the variety and
formation of God's creatures in the animal, vegetable, and mineral kingdoms; the
effect of digestion, and thereby of conversion; the construction of the hand of man,
and an infinite variety of other arguments; as also by discoveries, ancient and
modern, in arts, sciences, and the whole extent of literature.   (Chalmers 1835, p. 9)

A grandiose scheme indeed: not only the utilitarian argument but evidence of
providential design in every aspect of the world, animate or inanimate, natural
or man-made. And this the contributors supplied, with remorseless
thoroughness, not omitting to mention even the providential proximity of
Britain's iron ore to the coal needed to smelt it, nor our good fortune in
possessing an instinct for property ownership on which we could base our
moral code. But it is a sign of the powerful hold that the utilitarian argument
had over people's imaginations that organic adaptation was by far the most

popular evidence. Even the contribution that was ostensibly about astronomy, written by William Whewell, celebrated historian and philosopher of science, managed to linger long on the wondrous designs of organisms.

This utilitarian way of thinking about natural history and theology gained a firm hold in Britain during the first half of the nineteenth century. In its most popular version the utilitarian argument was combined with special creationism – the theory that God, rather than having created all organic forms in one fell swoop, still intervened in the natural world from time to time to introduce new ones. This 'utilitarian-creationism' even became something of an establishment position during that period (see Gillespie 1979, pp. 172–3, n6 for a list of classic works of this school).

Now to the other tradition of natural history, the tradition that set its sights not on the minutiae of adaptation in individual organisms but on the grand sweep of nature's whole array. It is sometimes called the 'idealist' or 'transcendentalist' view (not to be confused with philosophical idealism or transcendentalism, which are related but distinct ideas). For idealists, deliberate design was to be found above all in likeness-in-diversity. All living things, they claimed, are built on a very few basic structural plans, the divine blueprints for creation. Organic forms are dictated primarily by these blueprints: what organisms manifest above all is ideal patterns. Adaptive modification, the utilitarian-creationists' pride and joy, was viewed as subordinate to these far-reaching designs. Idealists saw their main task as revealing the unifying grand plan that lay behind the diverse appearances of living things. Idealism, then, like utilitarian-creationism, was permeated with the idea of intentional design. But it was design of a different kind. It showed itself not in adaptive detail, in function and utility, but in the symmetry and order of organisms and of the organic world as a whole, in the structural relationships among different species and the so-called unity of plan beneath their diversity. This was a symmetry and order so impressive, so perfect, it was argued, that it could not be accidental. This outlook is epitomised by the theory of archetypes that was developed by Richard Owen, the eminent comparative anatomist (the same Owen who is now best remembered for purportedly having primed Bishop Wilberforce before his notorious oration at the British Association meeting of 1860). Owen held that archetypes were the groundplans of the major groups of organisms, groundplans that existed in God's mind; the successive fossil forms within each of these groups were the result of divine intervention gradually modifying the original forms.

In the eighteenth century this idealist outlook, although influential on the Continent, had not found favour in Britain. But during the first decades of the nineteenth century an idealist school grew up in Scotland under the influential anatomist Robert Knox (later notorious for his unwitting involvement in the

Burke and Hare murders) and his student Edward Forbes (who came to be so highly regarded that he was Darwin's second choice of editor for his manuscripts in the event of his death). During the 1840s this view gained ground in Britain, largely through the work of Forbes and Owen, and from the early 1850s it became a flourishing movement. By 1859 idealism had ousted utilitarian-creationism from first place in natural theology. When Baden Powell, a well-known mathematician and controversial writer on religion, reviewed the literature of natural theology in the 1850s, he was able to record with satisfaction that at least some leading writers had 'grasped clearly the idea of *order* as the true indication of supreme intelligence'; they were not relying merely on 'subserviency of means to an end' (that is, adaptation) for proof of providential design (Powell 1857, p. 170). And this influence was reflected, albeit less strongly, in natural history. The eminent naturalist William Carpenter, for example, claimed that idealism was clearly supported by the evidence and he posed a challenge to utilitarian-creationists:

> if such persons will go to Nature, and interrogate her by a careful and candid scrutiny of the various forms and combinations which she presents, with the real desire to ascertain whether there be a guiding plan, a unity of design, throughout the whole, or whether each organism is built up for itself alone without reference to the rest, – we are confident that they will find the former doctrine to be irresistibly forced upon them ...   ([Carpenter] 1847, pp. 489–90)

Idealists often looked down on utilitarian-creationists as touting empty teleology. This is because utilitarian-creationists explained adaptations in terms of final causes. (An explanation of a characteristic was thought to have reached a final cause, which required no further explanation, if the characteristic had been shown to have been expressly designed for a particular adaptive purpose.) Knox even disparagingly renamed the utilitarian-creationists' classic work '*The Bilgewater Treatises*' (Blake 1871, p. 334) because he regarded an appeal to final causes as vulgar and naive. But idealists had no grounds for such smugness. Admittedly, they avoided talk of a grand designer fussing with adaptive details. But instead they appealed with exquisite vagueness to powers exerted by ideal patterns (more like formal causes, for those who find Aristotelean distinctions enlightening). Admittedly, too, some idealists claimed that they anyway didn't attempt to provide explanations but merely categorised the phenomena using ideal types. Nevertheless, idealism, no less than utilitarian-creationism, depended on a scientifically unacceptable assumption about conscious design. Apart from which, why be satisfied about not even having attempted to explain anything?

In 1859, then, there were two well-established, distinctive ways of interpreting nature. Utilitarian-creationists were preoccupied with the complexity and craftmanship of adaptations, with their ingenious utility, with the careful fit between an animal or plant and its environment. Organisms were studied more or less in isolation without much attention to relationships between species. Idealists, however, were uninterested in what they saw as finicky details. They were preoccupied with the grand plan of creation as a whole, with the patterns that unify nature's diversity. Of course, these outlooks were not so entirely opposed either conceptually or in practice as to preclude eclecticism. Peter Mark Roget (of *Thesaurus* fame), in his contribution to the utilitarian-creationist *Bridgewater Treatises*, seized on unity of plan as evidence of deliberate design; conversely, even the arch-archetypalist Owen was not above exploiting functional adaptation for the same purpose. But, whatever their differences and compromises, on one principle these two schools of thought converged: to look at nature was to see deliberate design. It was against this background that Darwinism offered its alternative interpretation. Let's now look at how Darwinian theory dealt with the two classes of evidence. We'll begin with adaptation.

Adaptations that most justly excite

It was the evidence of adaptation that was the greater challenge – and so the greater triumph – for Darwinism. As Darwin pointed out, the other class of facts, the pattern of diversity, can to some extent be explained merely by positing evolution; but the major problem is to find a mechanism for evolution that will explain the complexity of adaptive design:

> In considering the Origin of Species, it is quite conceivable that a naturalist, reflecting on the mutual affinities of organic beings, on their embryological relations, their geographical distribution, geological succession, and other such facts, might come to the conclusion that each species had not been independently created, but had descended, like varieties, from other species. Nevertheless, such a conclusion, even if well founded, would be unsatisfactory, until it could be shown how the innumerable species inhabiting this world have been modified, so as to acquire that perfection of structure and coadaptation which most justly excites our admiration. (Darwin 1859, p. 3)

Darwin was forcefully impressed by this during his *Beagle* voyage when, on the South American Pampas, he discovered striking resemblances between fossil and modern forms, and continuities in the geographical distribution of the modern flora and fauna. He realised that to explain all this merely by evolution was not adequate unless the mechanism of evolution could also explain adaptation:

Continuities in time

The *mataco* or three-banded armadillo (from Darwin's *Journal of Researches*): In La Plata in South America, Darwin was much struck by the close resemblance between the armadillos now living there and the fossilised armour buried beneath their home; the close relationship between the modern species and the gigantic extinct forms was, he said in the *Origin*, 'manifest, even to an uneducated eye'. But he must also have been struck by the contrast between those ancient giants and the shy little animals that he encountered:

"The pichy [Dasypus minutus] ... often tries to escape notice, by squatting close to the ground ... The instant one was perceived, it was necessary, in order to catch it, almost to tumble off one's horse; for in soft soil the animal burrowed so quickly, that its hinder quarters would almost disappear before one could alight. It seems almost a pity to kill such nice little animals, for as a Gaucho said, while sharpening his knife on the back of one, 'Son tan mansos' (they are so quiet)." (Darwin: Journal of Researches)

It was evident that such facts as these ... could only be explained on the supposition that species gradually become modified ... But it was equally evident that [one needed to] ... account for the innumerable cases in which organisms of every kind are beautifully adapted to their habits of life ... I had always been much struck by such adaptations, and until these could be explained it seemed to me almost useless to endeavour to prove by indirect evidence that species have been modified. (Darwin, F. 1892, p. 42)

We have seen that Darwinism explains adaptation by cumulative selection: small, undirected variations that are channelled by selective pressures, resulting, after long periods of time, in vast, complex, diverse and, above all, adaptive changes. One can think of adaptation as the successful incorporation of information about the world (Young 1957, pp. 19–21). The small changes that provide the raw materials for adaptation are undirected, random relative to the organism's environment. But the selective forces that shape these variations into adaptations carry vital information, often exquisitely detailed information, about that environment. So an organism inherits from its parents a model of aspects of its world, a match with its environment (or, rather, of their world and that of more distant ancestors). 'The adult organism can ... be considered as containing a representation of the environment, transmitted to it from the genes' (Young 1957, p. 21). The control that natural selection exercises over random variations is somewhat like an engineer's idea of negative feedback: constant comparison between the representation of the world and new information coming in from it, and constant adjustment and readjustment in the light of that comparison (Young 1957, pp. 23–7). The end result, adaptation, simulates deliberate, conscious design.

So Darwin and Wallace made pioneering use of what is now recognised as a standard solution to the problem of explaining design-without-a-designer. We can appreciate their success all the more when we see the two ways in which the rival theories had to rely on deliberate design – neither of which, in in spite of misconceptions that we'll look at in a moment, can be found in Darwinian theory.

First, with natural selection, the raw materials – the changes, the differences, the mutations, out of which evolution is built – are not subject to design at their source, when they arise; they are random, blind. 'Random' in this context doesn't mean events that appear random to us merely because of our lack of knowledge (in philosopher's language, it shouldn't be understood as an epistemological notion). The small changes that natural selection works on may indeed appear this way to us. But random is intended as a description of the state of the world rather than our perception of it (what philosophers call an ontological description). It is not, however, intended to mean 'lawless' or even necessarily 'not deterministic'; there's nothing lawless or particularly non-deterministic about, say, cosmic rays causing mutations. Rather, it means 'not pre-selected', random with respect to adaptive value. Darwin uses the following analogy to illustrate his view that there are laws (which he holds are deterministic) but that their existence is compatible with randomness in this sense:

Let an architect be compelled to build an edifice with uncut stones, fallen from a precipice. The shape of each fragment may be called accidental; yet the shape of each has been determined by the force of gravity, the nature of the rock, and the slope of the precipice, – events and circumstances, all of which depend on natural laws; but there is no relation between these laws and the purpose for which each fragment is used by the builder. In the same manner the variations of each creature are determined by fixed and immutable laws; but these bear no relation to the living structure which is slowly built up through the power of selection ...

(Darwin 1868, ii, pp. 248–9)

Contrast this randomness in the theory of natural selection with its counterpart in the theories of so-called evolutionary teleology that became regrettably popular for some time after 1859. Consider, for example, the 'improvements' that Asa Gray, the distinguished American botanist, proposed for Darwinian theory (Gray 1876). Gray considered himself to be a Darwinian and was a leading champion of Darwinism in the United States. But he couldn't bring himself to jettison divine intervention. So he devised what the philosopher John Dewey scathingly called 'design on the installment plan' (Dewey 1909, p. 12): the theory that God provides a pool of pre-eminently suitable variations for selection to work on. Gray reintroduced conscious design in the source of variation, whilst still leaving a role for selection. Or, at least, purportedly so. Obviously, if unnatural selection does too much pre-selecting, there won't be anything left for natural selection to do. Darwin thought that, even given Gray's own objectives, he had inadvertently made natural selection entirely redundant (Darwin 1868, ii, p. 526). Incidentally, he objected to Gray's theory not only because it reintroduced intentional design but also on the empirical grounds that there was overwhelming evidence, particularly from domestic selection, that variations really were undirected (e.g. Darwin, F. 1887, i, p. 314, ii, pp. 373, 378; Darwin, F. and Seward 1903, i, pp. 191–3).

It may seem odd today that anyone, particularly an apparently enthusiastic Darwinian, could go to such desperate lengths to retain a designer. But Gray was by no means alone in his view – nor, by the way, in his theological motivation for it. Several leading scientists of the time, among others, adopted an evolutionary theory of directed variation. They included Charles Lyell, leading geologist and one of Darwin's mentors, and John Herschel, a celebrated astronomer, like his father William Herschel (see Darwin, F. 1887, ii, p. 241; Darwin, F. and Seward 1903, i, pp. 190–2, n2, pp. 330–1, n1, n2; Herschel 1861, p. 12). And, to mention a less elevated figure, the Duke of Argyll had great success with a popular book along these lines, *Reign of Law* (1867). Perhaps it was because Darwinism was silent on the question of the origin of the variations that natural selection worked on that this seemed a heaven-sent opportunity for bringing a designer back in. I hasten to say that Darwinism is, of course, a perfectly adequate theory even though it offers no explanation of the origin of variations. Nevertheless, this silence has troubled some Darwinians. Even so distinguished a biologist as Peter Medawar lamented that 'The main weakness of modern evolutionary theory is its lack of a fully worked out theory of variation, that is, of *candidature* for evolution, of the forms in which genetic variants are proferred for selection' (Medawar 1967, p. 104).

The alternative theories also depended on a designer for the selective process, for the elimination and retention of variations. On the theory of natural selection, however, this process takes its course without benefit of a selector. There is no deliberation, no planning, no 'mind', nothing incorporating ends or goals to direct the selection. It is achieved through nothing more forward-looking than pressures of the environment. Remember the engineer's negative feedback; as Wallace put it: 'The action of this principle is exactly like that of the centrifugal governor of the steam engine, which checks and corrects any irregularities almost before they become evident' (Darwin and Wallace 1858, pp. 106–7).

The notion of selection without conscious design is independent of the notion of randomness precluding design. This can be seen from the familiar example of domestic selection. As Darwin pointed out, the source of variation is the same as in the wild (although he believed that it was 'writ large' under domestication) but the other agent of modification – selection – is different: 'when man is the selecting agent, we clearly see that the two elements of change are distinct; variability is in some manner excited, but it is the will of man which accumulates the variations in certain directions; and it is this latter agency which answers to the survival of the fittest under nature' (Peckham 1959, pp. 279–80). Darwin's and Wallace's theory, then, achieved what none before had managed: it showed that variation and selection alone could be a prodigious creative force, although the variation was undirected and the selection had no deliberate selector.

Astonishingly, in spite of the elegant simplicity and immense explanatory power of that solution, Darwinism has, throughout its history, been dogged by a vocal minority of critics who have denied that the theory really does solve the problem of design-without-a-designer. These critics fall into two broad categories – and, tellingly, their views are mutually contradictory. Some accuse Darwinian theory of relying on blind chance, pointing out that chance is highly unlikely to come up with adaptations. They are, of course, right that it is vanishingly improbable for complex, well-functioning entities to arise without guidance at a stroke; but they are, of course, utterly misguided to imagine that Darwinism makes any such assumption. Other critics have voiced the opposite complaint: that, far from relying on chance alone, Darwinism covertly retains a designer, that it fails to exorcise deliberate design. Again, the complaints are hopelessly off-target; selection is a

The power of adaptation

Humming-bird and humming-bird hawk-moth (from Bates's *The Naturalist on the River Amazons*)

"Several times I shot by mistake a humming-bird hawk-moth instead of a bird ... It was only after many days' experience that I learnt to distinguish one from the other when on the wing. This resemblance has attracted the notice of the natives, all of whom, even educated whites, firmly believe that one is transmutable into the other. They have observed the metamorphosis of caterpillars into butterflies, and think it not at all more wonderful that a moth should change into a humming-bird ... The negroes and Indians tried to convince me that the two were of the same species. 'Look at their feathers', they said; 'their eyes are the same, and so are their tails'. This belief is so deeply rooted that it was useless to reason with them on the subject. The Macroglossa moths are found in most countries, and have everywhere the same habits; one well-known species is found in England. Mr Gould relates that he once had a stormy altercation with an English gentleman, who affirmed that humming-birds were found in England, for he had seen one flying in Devonshire, meaning thereby the moth Macroglossa stellatarum." (Bates: The Naturalist on the River Amazons)

powerful shaping force but it has no eye on the future. The first group, then, argues that Darwinism's conclusion (adaptive complexity) does not follow from its premises. The second group argues that it does, but only because a designer has been smuggled in by the back door.

Two criticisms with but a single fallacy: the assumption that there is no third path between vast leaps of blind, untutored, unchannelled chance on the one hand and the fine discrimination of deliberate, directed design on the other. Historically, there's a long tradition behind this assumption. On one side, there has been a minority view, stretching right back to the Epicureans in the third century BC, that has invoked chance in order to fend off a designer. That is surely a resort of desperation. Intuitively, it seems more satisfactory to leave the question in abeyance than to insist that chance could give rise to the wealth of highly improbable organic adaptations. And, indeed, on the other side, the majority did view the idea that blind chance alone could give rise to complex, functional order as implausible, so implausible that they used it as a *reductio ad absurdum* to support the design-therefore-designer view: 'There must be a designer because otherwise chance alone would be responsible for design – which is manifestly absurd!' These criticisms of the Epicurean view were revived in the seventeenth century, when providential design was felt to be in need of defence against the dangerous atheism that atomistic theory was thought to be fostering. By 1859 these arguments were well entrenched within natural theology. So when Darwinism came on the scene, the opposition was thoroughly, albeit mistakenly, geared up to wield the 'if-not-design-then-blind-chance' dichotomy against it. Some of these arguments came from the popular press (Ellegård 1958, pp. 115–16); but they were also loudly voiced by highly eminent critics, who apparently failed to notice how grossly inappropriate the dichotomy had become now that natural selection offered a genuine alternative to it.

John Herschel, for example, was one of the critics who seemed to be under the impression that natural selection amounted to nothing more than blind chance. He drew a triumphant analogy with Swift's sardonic account in *Gulliver's Travels* (Swift 1726, pp. 227–30) of the Laputan practice of writing books by combining words randomly: 'We can no more accept the principle of arbitrary and casual variation and natural selection as a sufficient account, *per se*, of the past and present organic world, than we can receive the Laputan method of composing books (pushed *à l'outrance*) as a sufficient one of Shakespeare and the *Principia*' (Herschel 1861, p. 12). No less a figure than Lord Kelvin, the eminent physicist, found Herschel's criticisms to be 'most valuable and instructive' (Thomson 1872, p. cv); as late as 1871 he was airing them approvingly in his Presidential Address to the British Association. And the celebrated German embryologist Karl Ernst von Baer also used Gulliver's ironic tale as a *reductio ad absurdum* of what he took to be Darwinism:

For a long time the author of these Laputan reports was taken to be joking, because it is self-evident that nothing useful and significant could ever result from chance events ... Now we must acknowledge this philosopher as a deep thinker since he foresaw the present triumphs of science!

Accidents!? ... These countless accidents would have to be in marvellous harmony if anything orderly were ever to result. (Baer 1873, pp. 419–25)

This remained a popular line of argument throughout the nineteenth century. It is epitomised in an influential book that was published the year before Darwin died, written by the widely read and highly respected author William Graham. The 'most important issue raised ... by Darwin', he announced, is 'whether chance or purpose governs the world' (Graham 1881, p. 50). Chance (Darwinism), he decided, was inadequate to explain evolution: 'we must use the notion of design, because the only alternative, chance, is still wider away from the facts ... [If] design be denied, chance must be offered as the explanation' (Graham 1881, p. 345). Echoes of these voices resound even now in popular debates (there are no such debates within science) about the purported death of Darwinism (e.g. Hoyle and Wickramasinghe 1981, pp. 13–20; Koestler 1978, pp. 166–8, 173–7; Ridley 1985a criticises several other examples).

That's one group of misinformed critics. Attacking from the opposite direction, the other group have held that the theory of natural selection, far from depending on blind chance, covertly introduces a selector, a designer. Many nineteenth-century commentators were of this opinion. It seemed to them that Darwin's theory depended either on an analogy with domestic selection or on a personified Nature. As Wallace wrote to Darwin:

> I have been ... repeatedly struck by the utter inability of numbers of intelligent persons to see clearly, or at all, the self-acting and necessary effects of Natural Selection ... [A recent article] concludes with a charge of something like blindness, in your not seeing that Natural Selection requires the constant watching of an intelligent 'chooser', like man's selection to which you so often compare it ... [and another] considers your weak point to be that you do not see that 'thought and direction are essential to the action of Natural Selection'. (Marchant 1916, i, p. 170)

Historians are inclined to view this nineteenth-century position as a remnant of the grand designer bias of natural theology (e.g. Gillespie 1979, p. 83). Yet even today popular writings are littered with commentators who labour under the same misconception (Ridley 1985a cites examples). And some historians of science (e.g. Manier 1980; Young 1971) have taken at best an ambiguous position; they maintain that the metaphor of a selector aided the acceptance of Darwinism but they fail to make clear that Darwinism-plus-a-selector is no Darwinism at all. It seems that even those who have (presumably) not been clinging to a great-selector-in-the-sky have faltered at the idea that pressures of the environment take the place of the domestic breeder. But, of course, the analogy with domestic selection (which Darwin appeals to in the *Origin*) is not essential to Darwinian theory. Indeed, Wallace explicitly rejected this comparison in his contribution to the joint Linnean Society paper in which they first publicly presented their theory (Darwin and Wallace 1858, pp. 104–6). And, anyway, how on earth could an analogy be essential if the theory is to be interpreted realistically?

It may seem unjust – even rash – to accuse these critics, many of them eminent, of misunderstanding. After all, take what might seem to be a similar case: one would not accuse Einstein of misunderstanding quantum mechanics because he claimed that it was an unsatisfactory theory. A scientist might accept that if the premisses of a theory were true then the conclusion would follow but might nevertheless refuse to accept the premisses. This was indeed Einstein's position: 'God does not play dice'. But this has not been the standard position of Darwin's critics. They seem to have misunderstood, to an embarrassing extent, what is involved in the premisses. According to the Darwinism-is-blind-chance view, the deduction to adaptation cannot be made; according to the Darwinism-involves-design view, Darwinism explains 'design' by positing a designer. Unlike the case of Einstein, these do appear to be sheer misunderstandings.

It is surprising that misunderstandings so fundamental should be so widespread and so persistent, even among scientists highly distinguished in their own fields. After all, as Howard Gruber points out, already in Darwin's time analogous systems could be found in other disciplines: the development of self-regulating machines was well under way, and the idea of uncoordinated chance elements giving rise to order at a higher level was familiar from economists and moral philosophers, epitomised by Adam Smith's 'hidden hand' (Gruber 1974, p. 13). Admittedly these analogies are very inexact, particularly in the social sphere. Nevertheless, as we have seen, Wallace found the image of the governor of the steam engine to be apt, and it is likely that Darwin found the social theories suggestive (Schweber 1977).

Adaptations, in Hume's delightful phrase, 'ravish into admiration all men who have ever contemplated them'. But how ravishing, how perfect should we expect them to be? We shall see later that this has long been a point of contention among Darwinians. For now, we'll find that a comparison with the two pre-Darwinian schools of thought reveals a great deal about Darwinism's outlook. When utilitarian-creationists dealt with adaptation, they saw perfection. When idealists dealt with the same evidence, they saw structures that were only imperfectly adapted to their functions. Darwinism steered a course between them: the power of selection can achieve the marvels of adaptation so beloved of the creationists; but, because the starting point is random variations imposed on solutions appropriate to previous generations, the results bear the tell-tale marks of doing the best with what is to hand rather than displaying the faultless imprints of an untrammelled designer; nevertheless, these imperfections are not the purposeless structures of ideal

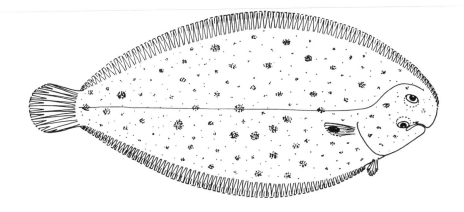

The stamp of history

The distorted eyes of the sand sole (*Pegusa lascaris*) tell a tale of adaptive 'imperfection'
– 'If I were you, I wouldn't start from here'

"The Pleuronectidae, or Flat-fish are remarkable for their asymmetrical bodies. They rest on one side ... But the eyes offer the most remarkable peculiarity; for they are both placed on the upper side of the head. During early youth, however, they stand opposite to each other, and the whole body is then symmetrical ... Soon the eye proper to the lower side begins to glide slowly round the head to the upper side; but does not pass right through the skull, as was formerly thought to be the case. It is obvious that unless the lower eye did thus travel round, it could not be used by the fish whilst lying in its habitual position on one side. The lower eye would, also, have been liable to be abraded by the sandy bottom." (Darwin: Origin)

forms but good solutions within constraints. We'll take first the contrast between Darwinism and the utilitarian-creationists, and then move on to the contrast with idealism.

The reason why utilitarian-creationists anticipate perfection where natural selection expects imperfection is to do with the role of history. For utilitarian-creationists, an eye, a wing, a fin is adapted in the sense that it was designed from the outset for its end. Each species was created readymade and has remained unchanged since its creation. So the adaptations of organisms are not constrained by those of their ancestors. The Darwinian understanding of adaptation, however, involves the historical starting point as well as the final state of adaptation. The utilitarian-creationist designer, looking upon natural selection's way of going about things, would be moved to echo the Irish advice to the lost traveller: 'If I were you, I wouldn't start from here'. A wing is appropriate to a bird's environment not because it has been created to fit that environment but because the lineage of ancestors leading to the bird has adapted to past environments. So 'imperfection' is to be expected: the legacies of ancestral adaptations may act as constraints on present perfection. Of course, historical theories, too, may lead one to expect perfection – theories that envisage the unfolding of a developmental plan, for example, with the adaptations at each stage perfectly fitted for their function. But Darwinian history is without foresight.

Darwin had no trouble in finding imperfections. Vestigial and rudimentary organs, for example, could prove a source of considerable embarrassment to utilitarian-creationists:

> Organs or parts in this strange condition, bearing the stamp of inutility, are extremely common throughout nature ... In reflecting on them, everyone must be struck with astonishment: for the same reasoning power which tells us plainly that most parts and organs are exquisitely adapted for certain purposes, tells us with equal plainness that these rudimentary or atrophied organs, are imperfect and useless.   (Darwin 1859, pp. 450–3)

Half the human race was a testimony to imperfection: 'If anything is designed, certainly man must be ... yet I cannot admit that man's rudimentary mammae ... were designed' (Darwin, F. 1887, ii, p. 382) ('man', for once, really meaning 'man'). But such structures present no difficulties for his own theory:

> On my view of descent with modification, the origin of rudimentary organs is simple ... They may be compared with the letters in a word, still retained in the spelling, but become useless in the pronunciation, but which serve as a clue in seeking for its derivation ... [O]rgans in a rudimentary, imperfect, and useless condition, or quite aborted, far from presenting a strange difficulty, as they assuredly do on the ordinary doctrine of creation, might even have been anticipated, and can be accounted for by the laws of inheritance.   (Darwin 1859, pp. 454–6)

History also leaves a legacy of changes of function. Parts are adapted to new uses for which they were clearly not 'created', new adaptations recycled from old: 'in certain fish the swim-bladder seems to be rudimentary for its proper function of giving buoyancy, but has become converted into a nascent breathing organ or lung' (Darwin 1859, p. 452); (modern zoologists think that the recycling went the other way, the primitive lung being pressed into service as a swimbladder). This is not the hallmark of an efficient Creator. Adaptations, then, look more like the work of a 'skilful tinkerer' than of a 'divine artificer' (Jacob 1977; see also Ghiselin 1969, pp. 134–7; Gould 1980, pp. 19–44, 1983, pp. 46–65, 147–57).

Darwin's work on orchids (1862) is tellingly entitled *On the Various Contrivances by which British and Foreign Orchids are Fertilised by Insects.* Michael Ghiselin, in his appraisal of Darwin's method, has pointed out that it is a delightful example of Darwin showing up the 'contrivances' so beloved by utilitarian-creationists for the 'contraptions' that they really are (Ghiselin 1969, pp. 134–7). Darwin's book is a persuasive demonstration of how 'pre-existing structures and capacities are utilised for new purposes' (Darwin 1862, p. 214; see also e.g. pp. 348–51), the purpose in the case of orchids being cross-fertilisation. The title was undoubtedly an ironic comment on the utilitarian-creationist insistence on perfect design; 'contrivance' was one of Paley's favourite concepts. (Just in case you are wondering why an omnipotent God would need to resort to contrivance, the answer is that, according to Paleyite theology, it was God's clue to us that he exists – as opposed to such clues as symmetry, unity or plenitude (Manier 1978, p. 72).) When the book was first published Darwin wrote these revealing words to Asa Gray:

> I should like to hear what you think about what I say in the last chapter of the orchid book on the meaning and cause of the endless diversity of means for the same general purpose. It bears on design, that endless question ... [N]o one else has perceived that my chief interest in my orchid book has been that it was a 'flank movement' on the enemy.
>
> (Darwin, F. and Seward 1903, i, pp. 203, 202; see also Darwin, F. 1887, iii, p. 266)

Given Gray's own strong leanings towards teleology it's not surprising that he was quick to detect a 'flank movement' on the defenders of deliberate design. But Darwin's irony was quite lost on many of his less Darwinian contemporaries. The Duke of Argyll, for example, triumphantly concluded that Darwin was unable to do away with designer-teleology:

> It is curious to observe the language which this most advanced disciple of pure naturalism instinctively uses when he has to describe the complicated structure of this curious order of plants. 'Caution in ascribing intention to nature' does not seem to occur to him as possible. Intention is the one thing which he does see, and which, when he does not see, he seeks for diligently until he finds it ... 'Contrivance', – 'curious contrivance' – 'beautiful contrivance' – these are expressions which recur over and over again. ([Argyll] 1862, p. 392; see also Darwin, F. 1887, iii, pp. 274–5)

Turning now to the idealists, we find a very different view of perfection in adaptation. For them, purposeful design was manifested in the total pattern of creation, not in particular organisms – so much so that adaptive efficiency in a particular species might well be sacrificed to maintain the grand plan of all species. Thus, far from stressing perfection, idealists even went out of their way to emphasise imperfection (Bowler 1977; Cain 1964; Ospovat 1978, 1980; Yeo 1979), thereby, according to the distinguished zoologist Arthur Cain, making the first major break in a tradition, stretching back to Aristotle, that explained structure by its impressive fit to function (Cain 1964, pp. 37–8, 46). If form takes priority over function, then inutility, redundancy and maladaptation are to be expected. The useless vestiges that the utilitarian-creationist sweeps with embarrassment under the Grand Designer's carpet are seized on, given a new twist and set on a pedestal by idealists. What better proof that structure rather than function is paramount than, say, homologies (similarities of structure across species) that make no functional sense but fit neatly into a scheme of archetypes?

So we find Owen (1849; see also Cain 1964), for example, comparing the limbs of vertebrates and asking why there is homology in the 'fin' of the dugong, the wing of the bat, the leg of the horse and the human limb. If, as utilitarian-creationists assume, their design criterion was function, why is there so little correlation between structure and use? Why are structures for doing such different things nevertheless so alike? After all, instruments built by humans for different purposes are structurally diverse. Utilitarian-creationists should expect to find as much diversity in nature's instruments: 'the same direct and purposive adaptation of the limb to its office as in the machine' (Owen 1849, p. 10). But, on the contrary, there is 'a much greater amount of conformity in the construction of the natural instruments ... of ... different animals' (Owen 1849, p. 10). These homologies are thus inexplicable on the utilitarian-creationist view. They are, however, exactly what one would expect if all vertebrates are built on the same essential plan. Just as Darwin was later to do, Owen emphasised that, far from being perfect machines, organisms exhibit incongruities and redundancies. The utilitarian-creationist, who expects a machine-like match between structure and function, cannot explain such anomalies: 'The fallacy perhaps lies in judging of created organs by the analogy of made machines' (Owen 1849, p. 85; see also e.g. Knox 1831, p. 486).

Some of this was obviously congenial to Darwin. Owen's evidence of the inutility manifested by homologies, for instance, could be used both against utilitarian-creationism and, at the same time (archetypes having been replaced by phylogeny), in favour of evolution. Here, for example, are undoubted echoes of Owen:

> What can be more curious than that the hand of a man, formed for grasping, that of a mole for digging, the leg of the horse, the paddle of the porpoise, and the wing of the bat, should all be constructed on the same pattern, and should include the same bones, in the same relative positions? ... Nothing can be more hopeless than to attempt to explain this similarity of pattern in members of the same class, by utility or by the doctrine of final causes. The hopelessness of the attempt has been expressly admitted by Owen in his most interesting work on the 'Nature of Limbs'.
>
> (Darwin 1859, pp. 434–5)

(Owen wasn't, of course, forced to 'admit' this point; on the contrary, he was drawing attention to it as evidence against the utilitarian-creationist view and in favour of his own.) In the same spirit, Darwin took an example of apparently functionless serial homology (the homology of different parts in the same organism):

> Most physiologists believe that the bones of the skull are homologous with ... the elemental parts of a certain number of vertebrae ... How inexplicable are these facts on the ordinary view of creation! Why should the brain be enclosed in a box composed of such numerous and such extraordinarily shaped pieces of bone? As Owen has remarked, the benefit derived from the yielding of the separate pieces in the act of parturition of mammals, will by no means explain the same construction in the skulls of birds.   (Darwin 1859, pp. 436–7)

Darwin made great play of such 'inutilities'. Indeed, according to Arthur Cain, 'he and Owen (and the "Naturphilosophen" [idealists]) assessed the imperfection of animals, the degree to which they are not adapted for their mode of life, as far higher than anyone else had done for centuries' (Cain 1964, p. 46).

Too much imperfection, however, would not serve Darwin's purpose any more than would too much perfection. After all, natural selection can achieve astonishing feats of design. So, against the idealists' imperfectionism, Darwin had to stress a more 'perfectionist' view or else show that the 'imperfections' were anyway of the kind that his theory anticipated. Again, he draws both these lessons from his work on orchids. When idealists assume that the Designer's templates are paramount they should look more closely:

> Some naturalists believe that numberless structures have been created for the sake of mere variety and beauty, – much as a workman would make a set of different patterns. I, for one, have often and often doubted whether this or that detail of structure could be of any service; yet, if of no good, these structures could not have been modelled by the natural preservation of useful variations ...
>
> (Darwin 1862, p. 352)

And 'imperfections' are not elements of a grand pattern but the legacy of phylogeny and natural selection:

It is interesting to look at one of the magnificent exotic species ... and observe how profoundly it has been modified ... Can we, in truth, feel satisfied by saying that each Orchid was created, exactly as we now see it, on a certain 'ideal type'; that the Omnipotent Creator, having fixed on one plan for the whole Order, did not please to depart from this plan; that He, therefore, made the same organ to perform diverse functions – often of trifling importance compared with their proper function – converted other organs into mere purposeless rudiments, and arranged all as if they had to stand separate, and then made them cohere? Is it not a more simple and intelligible view that all Orchids owe what they have in common to descent ... and that the now wonderfully changed structure of the flower is due to a long course of slow modification ... ?   (Darwin 1862, pp. 305–7)

Darwin's careful reworkings of utilitarian-creationist perfection and idealist imperfection are not mere triumphalism. They are to do with how the already existing knowledge supports his theory. How is any new theory corroborated by old evidence, by evidence that corroborated its predecessors? Consider, to take one way, Karl Popper's example of how Newton dealt with the legacy of Galileo and Kepler. Newtonian theory did not simply take over the work of these predecessors wholesale. Far from it. The new theory not only explained but also corrected the old: 'far from being a mere conjunction of these two theories ... *it corrects them while explaining them*' (Popper 1957, p. 202). And it corrects them using only the fundamental assumptions of Newtonian theory, without need of subsidiary aid. As J. W. N. Watkins puts it, Newtonian theory 'is corroborated by those two, seemingly disjoint, sets of earlier results because it explains close approximations of both from one and the same set of fundamental assumptions; and its revisions of them, far from being mere tinkerings, are induced *systematically* by those fundamental assumptions' (Watkins 1984, p. 302). So old knowledge was transformed into new, providing powerful corroboration for Newtonian theory. Now, in the case of Darwinism and the existing theories we have nothing so exact as the observations that Newton used – certainly nothing numerically expressed. But it is clear that in a qualitative way something in the spirit of Popper's idea is going on: a telling correction of empirical implications of the previous theories, a significant new look to old data. And this reinterpretation flows naturally from the basic assumptions of Darwinian theory alone, from the assumptions of variation, heredity and selection, without resort to any of the (largely historical) subsidiary assumptions.

Likeness in diversity

Intricate, adaptive complexity was one major problem that Darwin and Wallace solved. The prodigious diversity combined with striking similarity among groups of organisms was the other. Today, we can hardly look at this evidence without seeing the solution: evolution. Evolution is descent with modification. Descent results in likeness. Modification results in diversity. But likeness is preserved in spite of modification because the change is gradual, incremental, based on small differences, not on great leaps to radically new forms. Broadly speaking, one can think of the two parts of Darwinian theory – a general theory of evolution and the particular mechanism of natural selection – as accounting for the two classes of evidence respectively. Nature's grand scheme displays fundamental likenesses; this results from a history of descent. Adaptation displays a diversity imposed on this likeness; this results from the modifications made by natural selection, the mode of evolution. That was Darwin's own view of the relationship between his theory and the two classes of evidence:

> It is generally acknowledged that all organic beings have been formed on *two great laws – Unity of Type, and the Conditions of Existence*. By unity of type is meant that fundamental agreement in structure, which we see in organic beings of the same class, and which is quite independent of their habits of life. On my theory, *unity of type is explained by unity of descent*. The expression of *conditions of existence ... is fully embraced by the principle of natural selection*. For natural selection acts by either now adapting the varying parts of each being to its organic and inorganic conditions of life; or by having adapted them during long-past periods of time ...   (Darwin 1859, p. 206; my emphasis)

It was with this second class of evidence that idealism came into its own. The search for resemblances was at the very core of the idealist programme. By contrast, utilitarian-creationism had little to say about the details of this aspect of organic design. But, although idealism fared better than utilitarian-creationism, by the time of the publication of the *Origin* neither school of thought was faring well.

As the century had progressed so had natural history and by 1859 this evidence was far richer and more comprehensive than it had been even a decade or two before. And as the evidence grew, the alternatives to Darwinism found it increasingly difficult to account convincingly for the facts. Such judgements cannot be very precise; an idea like special creationism is so essentially vague that it is impossible to say quite when it succeeds or fails. Nevertheless, it is clear that by this time all these theories were running aground in some respect or other. Remember that their principal tenet was an appeal to intentional design. Certainly the evidence was revealing some patterns in nature. But these patterns were becoming too arbitrary, too lacking in apparent plan to be plausibly explained as the work of a well-organised creator. Take the field of classification. The results were beginning to look more like the work of an inept do-it-yourself enthusiast than the elegant craftmanship of an omniscient Designer. So, for instance, naturalists were constantly having to revise the status accorded to groups of organisms – promoting a species to a genus here or demoting an order to a family there – not because some intrinsic hierarchy was revealing itself but because intrepid discoverers were uncovering unpredictable and apparently arbitrary pockets of creation (Darwin 1859, p. 419). It was a similar story with geographical distribution. Of course, God was free to put any organic forms anywhere. But consider, say, island species (Darwin 1859, pp. 388–406). Why are there typically fewer species on islands than on the mainland? Why are whole classes (such as that of frogs, toads and newts) absent from islands although they thrive if introduced? Why do remote islands seem to favour bats but exclude all other mammals? Why do species on neighbouring islands resemble one another more than they resemble species on distant islands? Why – to take a question traditionally, although perhaps wrongly (Sulloway 1982), associated with Darwin – did God see fit to furnish each of the Galapagos Islands with its own species of finch and tortoise?

Utilitarian-creationism and idealism were being let down by their ideas of deliberate design. Such ideas could be fruitful for suggesting patterns in nature but they provided no means for dealing with phenomena that had an unplanned air about them. Idealists soon found that nature did not always conform neatly to the supposed transcendental plan; increasingly, modifications to the purportedly a priori patterns had to be 'read off' retrospectively from the data. Utilitarian-creationism suffered the worst of both worlds in this respect. On the one hand it was too weak even to suggest what patterns would be found in this class of evidence; on the other hand, being wedded to providential design, it was clearly embarrassed by growing evidence of apparently imperfect planning.

For Darwinian theory, the evidence of likeness in diversity was easily explained by descent with modification. Nature's large-scale patterns could be explained by descent; nature's vagaries were to be expected because of the adaptive modifications wrought by natural selection. But how exactly did this class of facts corroborate Darwinian theory? Our usual idea of corroboration involves the successful prediction of unexpected new facts, like Eddington's triumphant corroboration of Einstein's startling prediction that light rays would bend in strong gravitational fields. Such corroboration depends on predictions that are temporally novel in the sense that no facts of that kind are already recorded in the background knowledge (Popper 1957). But, clearly, when it came to classification, geographical distribution and so on, Darwinian theory was not strong on temporally novel predictions. For the most part, this evidence was already well known, thoroughly documented by pre-Darwinian natural history.

Evidence can be novel, however, without being temporally novel (Zahar 1973). The point is that one can't rig up a theory to encompass known evidence and then parade predictions that are based on this evidence as corroboration. Predictions of that kind cannot count as novel. But suppose that the theory is not constructed on the basis of the evidence and suppose that a successful prediction can nevertheless be derived from it, a prediction that falls out of the theory 'without contrivance' (Watkins 1984, p. 300). That kind of prediction can count as novel and provide corroboration for the theory. This idea 'embodies the simple rule that one can't use the same fact twice: once in the construction of a theory and then again in its support. But any fact which the theory explains but which it was not in this way pre-arranged to explain supports the theory *whether or not the fact was known prior to the theory's proposal*' (Worrall 1978, pp. 48–9). It may appear that in order to judge whether a fact is novel in this sense one has to take into account the way in which the theory was built – in particular whether its construction was guided by the heuristics of the research programme (Worrall 1978; Zahar 1973). But novelty can be decided without investigating how the theory was constructed (Watkins 1984, pp. 300–4). The important conditions are whether the fundamental assumptions of the theory play a key role in the derivation of the predictions and whether these fundamental assumptions at the same time endow the theory with a greater predictive and explanatory power than that of its rivals. For in that case, its ability to predict and explain a particular fact that was already known, far from resulting from some *ad hoc* adjustment to it, indicates the superiority of its fundamental assumptions. As long as the theory was not the product of mere *ad hoc* tinkering in the light of the evidence then, however familiar the evidence and whatever role it played in the construction of the theory, it still confirms the theory. And that is why the evidence of likeness in diversity, although well known long before the days of Darwinism, could nevertheless constitute impressive corroboration for Darwinian theory.

The 'grand facts', as Darwin called them, that made up nature's likeness in diversity, covered a wide and disparate range of evidence, from resemblances between the embryos of frogs and humans, to the patchy distribution of freshwater fish, to the similarities between the long-extinct and more modern gigantic birds of New Zealand. It is a sign of the comprehensiveness of Darwinian theory that it could encompass it all, that it could deal collectively with this whole body of evidence, as well as with individual aspects. Pre-Darwinian theories had concentrated on just a few areas of the evidence and barely touched on others. For idealism, the unity of type that emerged from classification and the mutual affinities that emerged from morphology and embryology were, of course, central to its programme. Nevertheless, it was less concerned with geological succession, except as a clue to unravelling the transcendental plan. And geographical distribution was largely ignored by both schools of thought; idealism hardly touched on it and utilitarian-creationism merely muttered vaguely about 'centres of creation'. In the hands of such theories, then, these hitherto unrelated groups of phenomena remained unrelated. Darwin himself seized on this comprehensiveness of his theory as a point in its favour. Let's give him the last word. He says of natural selection:

> this hypothesis may be tested, – and this seems to me the only fair and legitimate manner of considering the whole question, – by trying whether it explains several large and independent classes of facts; such as the geological succession of organic beings, their distribution in past and present times, and their mutual affinities and homologies. If the principle of natural selection does explain these and other large bodies of facts, it ought to be received.
>
> (Darwin 1868, i, p. 9; see also Darwin, F. and Seward 1903, i, p. 455)

'It is no scientific explanation'

It is the ability to explain not only these 'large and independent classes of facts' but to explain also the evidence of adaptation, and to explain both of them as a natural consequence of the theory, that provides the most impressive demonstration of Darwinism's unity and explanatory power. Let us give Darwin the last word, too, on how this achievement compared with the pre-Darwinian schools of thought.

Take his reaction to the utilitarian-creationists' attempts to deal with the anomaly (for them) of rudimentary organs. They generally responded by briskly abandoning the idea of exact adaptation and retreating instead to some kind of total harmony, such as the principle of plenitude or symmetry (an escape clause that owed more to idealism than to their own concept of design). Darwin was contemptuous of all such empty appeals:

> In works on natural history rudimentary organs are generally said to have been created 'for the sake of symmetry', or in order 'to complete the scheme of nature': but this seems to me no explanation, merely a restatement of the fact. Would it be thought sufficient to say that because planets revolve in elliptical courses round the sun, satellites follow the same course round the planets, for the sake of symmetry, and to complete the scheme of nature? (Darwin 1859, p. 453)

And this is from his book on orchids:

> every detail of structure which characterises the male pollen-masses is represented in the female plant in a useless condition ... At a period not far distant, naturalists will hear with surprise, perhaps with derision, that grave and learned men

formerly maintained that such useless organs were not remnants retained by inheritance, but were specially created and arranged in their proper places like dishes on a table (this is the simile of a distinguished botanist) by an Omnipotent hand 'to complete the scheme of nature'.   (Darwin 1862, 2nd edn., pp. 202–3)

Another utilitarian-creationist tactic, even more implausible, was flatly to deny the inutility of rudimentary organs. After all, if only they were taken from the right perspective, didn't they exhibit economy? This is Darwin's account:

There was ... a new explanation ... of rudimentary organs, namely, that economy of labour and material was a great guiding principle with God (ignoring waste of seed and of young monsters, etc.) and that making a new plan for the structure of animals was thought, and thought was labour, and therefore God kept to a uniform plan, and left rudiments. This is no exaggeration.   (Darwin, F. 1887, iii, pp. 61–2)

Darwin also dismissed, as unscientific, the pre-Darwinian attempts to explain homologies. Of utilitarian-creationism he said: 'On the ordinary view of the independent creation of each being, we can only say that so it is; – that it has pleased the Creator to construct all the animals and plants in each great class on a uniformly regulated plan; but this is not a scientific explanation' (Peckham 1959, pp. 677–8). And of idealism: 'homological construction ... is intelligible, if we admit ... descent ... together with ... subsequent adaptation ... On any other view the similarity ... is utterly inexplicable. It is no scientific explanation to assert that they have all been formed on the same ideal plan' (Darwin 1871, pp. 31–2).

Similarly, talking of classification, he says: 'many naturalists think that ... the Natural System ... reveals the plan of the Creator; but unless it be specified whether order in time or space, or what else is meant by the plan of the Creator, it seems to me that nothing is thus added to our knowledge' (Darwin 1859, p. 413). These criticisms echo a comment that Darwin jotted down in some early notes (probably 1838):

The explanation of types of structure in classes – as resulting from the *will* of the deity, to create animals on certain plans, – is no explanation – *it has not the character of a physical law* / & is therefore utterly useless. – it foretells nothing / because we know nothing of the will of the Deity, how it acts & whether constant or inconstant like that of man. – the cause given we know not the effect ...
                                                          (Gruber 1974, pp. 417–18)

In the same notes he also dismissed final causes as 'barren Virgins' (Gruber 1974, p. 419).

Last, an example from idealism. It is easy to underestimate just how deeply idealists were steeped in the idea that knowledge of nature's grand patterns, its fundamental laws, could be gained without serious recourse to experience

– and, as a consequence, just how cavalier they could be in dealing with the anomalies posed by vulgar facts, especially with adaptations that didn't fit into this grand design:

> particular phenomena which appeared to follow these laws were not, strictly speaking, to be taken as evidence in support of the laws, but rather as illustrations of what was regarded as known a priori. And equally important, phenomena which appeared to violate the laws were not of great concern, since their inconsistency could be caused by inadequate interpretation, or by the incomplete state of the science.   (Rehbock 1983, p. 21)

One has to bear this in mind to understand how idealism could pay so little attention to adaptation. The following comment from one of Darwin's contemporaries, writing in the *Gardeners' Chronicle* of the 1870s, testifies to the stultifying impact of idealism on the study of adaptation and the immense contribution made by Darwin:

> Most of us remember the use that Paley made of the watch as an evidence of design, and of necessity of a designer. Twenty or thirty years ago ... a new school arose ... Modifications in form were set down as variations from an ideal pattern or type, and adaptations to special ends, though admitted in some cases, were discredited in others. Not the least service which Mr Darwin has rendered to science has been the demonstration that many adaptations formerly supposed either to be of trifling moment or purposeless illustrations of a particular preordained pattern, are really adaptations to special purposes ...   (Barrett 1977, ii, p. 187)

(although admittedly this commentator spoils things by welcoming Darwin's adaptationism as support for utilitarian-creationist natural theology!).

Darwin's comments on his rivals provide a useful corrective to a trend that has recently dominated Darwinian historiography. Many historians bend over backwards in order to understand Darwin's rivals from a nineteenth-century point of view. In this position, they lose sight of how far superior Darwin's contribution was. In spite of his nineteenth-century vantage point, Darwin knew better than these twentieth-century Darwinians.

Rivals and follies: 1859 and beyond

So far, Darwinism has emerged as having no serious rival. But then we have compared it only with manifestly inadequate theories, permeated by natural theology and not even in the realm of science. What about Lamarckism? And what about other alternatives, after the mid-nineteenth century – surely more scientific than those early rivals? Admittedly they turned out in the end not to be true. But at least they were candidates for filling the same explanatory gap as Darwinian theory.

Or were they? It is precisely that assumption that the rest of this chapter will challenge. We shall look at some arguments to the effect that the apparently serious rivals to Darwinism are actually incapable, not only in fact but in principle, of doing the job. These arguments are not based on empirical evidence about how the alternative theories fit – or, rather, don't fit – the actual world (a posteriori arguments); they are more like pure-reason, first-principles arguments, a priori arguments, about what a theory needs in order to explain any world, any possible world, in which adaptive complexity is found – and about why the alternatives don't meet those requirements whereas Darwinism does. This is not to say that there couldn't possibly be a rival to Darwinism. But nobody has as yet come up with anything even remotely like one.

I'm going to concentrate on Lamarckism because historically it has been the most serious alternative to Darwinian theory. Lamarckian theory can be summed up in the phrase *use-inheritance*. The idea of *use* is that an organism's activity shapes the organism appropriately for that activity. The more a giraffe stretches its neck, the longer its neck becomes; the more a blacksmith uses his biceps, the larger they grow; the more we do aerobic exercise, the more our lung capacity increases. And, conversely, there is disuse: the less an ostrich uses its wings, the less it will be capable of flight. This idea of use and disuse is coupled with a particular idea of *inheritance* – the inheritance of acquired characteristics. This is a theory that the characteristics that are acquired during an organism's lifetime through use and disuse will be passed on to the organism's offspring.

What I've just set out as Lamarckian theory may not be the authentic Lamarck. The question of what he actually said is hazy. But I have described Lamarckism as it was understood and taken up at least in Britain; it's the Lamarckism that made an impact on the history of Darwinism (Bowler 1983, pp. 58–140). Jean-Baptiste Antoine de Monet, known to history as Lamarck, set out his theory in his famous *Philosophie Zoologique* in 1809. He had few followers during his lifetime and died, in 1829, in some obscurity. But in Britain, by the second half of the nineteenth century, most Darwinians (including Darwin himself – but not Wallace) accepted use-inheritance as a subsidiary agent in evolution. They thought that natural selection was by far the predominant force but they welcomed a little help from other mechanisms. It was August Weismann, the distinguished German biologist and ardent Darwinian, who led the attack on use-inheritance – or, more generally, on the inheritance of any acquired characteristics. His efforts met with a sadly mixed response (see e.g. Bowler 1984, pp. 237–9). On the one hand, it was through his work that, beginning in about the late 1880s, Lamarckism lost favour with Darwinians as a supplementary mechanism. On the other hand, by sharpening up the differences between the two theories, Weismann stimulated naturalists who were ill-disposed to Darwinism into looking again at Lamarckism as an alternative comprehensive theory of evolution. The result was that the theory of use-inheritance underwent a major revival in Britain, under the name of neo-Lamarckism. This was during the long period when Darwinism was most widely rejected – which stretched from soon after Darwin's death in the 1880s right up until about the 1940s. It's a phase in Darwinian history that has been called the eclipse of Darwinism (see Bowler 1983). The zenith of the neo-Lamarckian alternative was from about 1890 until around the turn of the century; by the 1920s it was in decline.

In spite of the end of that story, there are two important reasons why the status of Lamarckism is still a serious issue for Darwinians. First, claims of purportedly Lamarckian discoveries crop up at times even now, and are greeted by many non-biologists and even by some biologists with hopeful interest. The second reason is what concerns us here: that once we properly understand what Lamarckism lacks, we can see what is required for any theory of evolution – and why Darwinism, unlike Lamarckism, satisfies those requirements. We know that the living things on our earth haven't got here by Lamarckian means. But is that just a contingent fact about our planet or are there more fundamental reasons why we shouldn't expect ever to meet Lamarckian life anywhere in the universe, life that has evolved in a Lamarckian way unaided by natural selection? We'll see that there are indeed such reasons.

I find it puzzling that Lamarckism has had such an enduring attraction. I'll list some of the (no doubt overlapping) reasons that Lamarckian sympathizers have given. Having examined the status of the theory, one can see that these arguments, far from being persuasive, are spurious – and indeed that the very issues they raise, apparently in Lamarckism's favour, actually make us all the more indebted to Darwin and Wallace.

One reason why some Darwinians have been attracted to Lamarckism is that they feel Darwinian theory to be incomplete because it doesn't explain the origin of the variations on which natural selection works – a point that, as we have seen, apparently worried even Peter Medawar. This seems to be one of the great strengths of Lamarckism. It appears to explain the source of adaptations: they arise as a response to needs, to challenges posed by the environment.

Another reason is the hope, a hope that has been voiced often throughout Darwinian history, that Lamarckism would lend an order to evolution, a guiding hand that Darwinism has been felt to lack. This motivated, among others, E. J. Steele, an immunologist who created a temporary stir in the early

1980s with his claim that immunological tolerance acquired by a parent in its lifetime could be passed on to its offspring (Steele 1979) (a claim that was not sustained on further investigation (see e.g. Howard 1981, pp. 104–5; see also Dawkins 1982, pp. 164–73)). According to Steele, Darwinism

> provides no satisfactory explanation for our intuitive belief that there appears to be an element of 'directional' progress in the complexity and sophistication of adapted living forms ... [We should therefore consider] the possibility that in many ... organisms ... there is an undercurrent of Lamarckian modes of inheritance which ... provide a continuous 'sense of direction' to the evolution of biological complexity.   (Steele 1979, p. 1)

A further reason, again commonly offered, was expressed by the nineteenth-century novelist and polemicist Samuel Butler. Butler blew hot and cold over Darwinism for a while but ended up very cold (e.g. Butler 1879). He felt that it excluded mind, will, intention from any serious role in nature. By contrast, he was attracted to the Lamarckian mechanism of organisms responding appropriately and creatively to their environments. This, he felt, gave purpose the central place that was its due.

A final lure of Lamarckism has perhaps been as much political and social as scientific. In his book *The Case of the Midwife Toad* (1971), Arthur Koestler tells the story of Paul Kammerer, an experimental biologist who worked in Vienna around the time of the First World War. Koestler had an enduring commitment to Lamarckism and a perverse blindspot on Darwinism. (By the way, one of Koestler's books inspired E. J. Steele with 'a penetrating sense of awe' (Steele 1979, Preface) and Koestler financed some of Steele's work.) Kammerer was Koestler's earlier counterpart (though at that time with more scientific justification). Here is Kammerer expressing one reason for his faith in Lamarckian inheritance – the help that it can offer us in attaining a better future:

> on the hypothesis of the inheritability of acquired characters ... the individual's efforts are not wasted; they are not limited by his own lifespan, but enter into the life-sap of generations ... By teaching our children and pupils how to prevail in the struggles of life and attain to ever higher perfection, we give them more than short benefits for their own lifetime – because an extract of it will penetrate that substance which is the truly immortal part of man. Out of the treasure of potentialities contained in the hereditary substance transmitted to us from the past, we form and transform, according to our choice and fancy, a new and better one for the future.   (Koestler 1971, p. 17)

So the improvements that we attain through struggle and effort in our lifetime need not die with us: the blacksmith's son will be born with bulging biceps (and so, presumably, will his daughter); the linguist's offspring will be born, if not multilingual, at least with a greater facility for learning languages; and, above all, the political lessons that we learn with such cost at the barricades and in the trenches need not be so painfully relearned by each generation. That, at least, is how committed Lamarckians have generally seen it. Equally, of course, Lamarckian inheritance could ensure that we are all born conservatives, bereft of the blessed ability to make a radically new start in each generation. It could ensure that our legacy from a history of victimisation, colonisation, deprivation would be the genes of the victimised, colonised and deprived. Indeed, Lamarckism could look just as attractive to reactionaries: 'Lamarckism is now [the 1930s] being used to support reaction. A British biologist who holds this view thinks that it is no good offering self-government to peoples whose ancestors have long been oppressed, or education to the descendants of many generations of "illiterates"' (Haldane 1939, p. 115).

We don't need to discuss the empirical arguments against Lamarckism. Everybody knows that the theory was eventually abandoned because acquired characteristics turned out not to be inherited. But I'd like to mention one point about the most famous and influential experiments that illustrated this, performed by Weismann (largely between 1875 and 1880), which involved cutting off the tails of mice to see whether the mutilation was inherited. Why on earth did anyone need to perform such experiments? Didn't everyone know, from common, everyday experience, that hereditary change isn't generally induced by such injuries, even injuries repeated over generations? If all acquired characteristics are inherited, why do Jewish male babies need to be circumcised, when they are born of fathers who were circumcised, themselves the sons of circumcised fathers, and so on, for many a pious generation? But did everyone know this? Here's John Maynard Smith's characteristically unforgettable answer to that:

As a boy, I acquired, from reading the preface to Shaw's *Back to Methuselah*, a picture of Weismann as a cruel and ignorant German pedant who cut the tails off mice to see if their offspring had tails. What a ridiculous experiment! Since the mice did not actively suppress their tails as an adaptation to their environment, no Lamarckist would expect the loss to be inherited. Much later, I discovered that Weismann was not as I had imagined him. His experiment on mice was performed only because, when he first put forward his theory, he was met with the objection that, (as was, it was claimed, well known) if a dog's tail is docked, its children are often tailless – an early use of what J. B. S. Haldane once called Aunt Jobisca's [sic] theorem, 'It's a fact the whole world knows'. (Maynard Smith 1982c, p. 2)

Lamarckians did try to deal with many of the obstinate cases of non-inheritance by stressing that it is adaptations (changes resulting from use and disuse), not just any changes, that are inherited. And, indeed, this distinction is vital to the theory, given the unhappy fact that many changes that individuals undergo in their lifetimes are really rather unpleasant – disease, injury, ageing. But Lamarckians have never been able to explain how the body manages to distinguish useful acquired characteristics from these less felicitous ones, to distinguish adaptations from the heart-ache and the thousand natural shocks that flesh is heir to. Why should the blacksmith's son inherit his highly developed muscles but not his badly injured back or his burn-scarred hands? Lamarckians have traditionally appealed vaguely to the notion that the hereditary system accepts only those changes that are responses to 'striving' or 'need'. We shall see that this answer smuggles in assumptions that are inherently unLamarckian, and indeed Darwinian.

It wasn't through torturing innocent mice that Weismann came to reject Lamarckism. It was because he had an alternative theory, a theory for which he coined an alluring image: the immortal river of the germ plasm (see e.g. Maynard Smith 1958, pp. 64–8, 1982c, pp. 2, 4, 1986, pp. 9–10). According to Weismann, the hereditary units (the parts of organisms that are passed down from generation to generation) are, at least by and large, both inviolable and immortal. What he meant by that was the following. Think of any organism as divided into two parts. On the one hand there are the cells that make up its body (the soma, the somatic cells). On the other hand, there are its reproductive cells (the germ plasm), the materials from which its offspring – new bodies – arise. Weismann's theory states that the interaction between somatic and germ cells is strictly one-way. Germ cells give rise to bodies, they determine what the offspring will be like – male or female, tiger or snail, large or small. But bodies have no influence whatsoever over germ cells. They merely carry the germ plasm within them down the generations, its passive keepers. Germ cells are shaped by other germ cells, not by the bodies in which they reside. Germ cells are inviolable, then, because they are not changed by somatic changes. And they are potentially immortal because identical replicas of them pass down from generation to generation. On Weismann's theory, the inheritance of acquired characteristics is impossible because there is no flow of information from an individual's body to its germ cells about bodily changes that have occurred during that individual's lifetime. The immortal river of the germ plasm flows on indifferent to which individuals it happens to be flowing through.

What has not been appreciated until recently is that there are also, as I mentioned, more fundamental objections to Lamarckism. This insight we owe to the zoologist Richard Dawkins (Dawkins 1982, pp. 174–6, 1982a, pp. 130–2, 1983, 1986, pp. 287–318, particularly pp. 288–303). He has developed three powerful arguments that completely demolish Lamarckism's

Lamarck

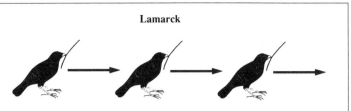

Bodies are copied directly from parental bodies. A bowerbird replicates its parents, acquired characteristics and all.

Weismann

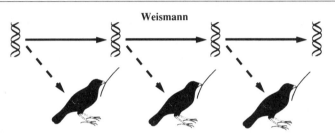

Only genes are copied. Bowerbirds are just genes' way of making more genes. The immortal river of germ plasm is unaffected by bodily changes ...

Extended phenotype

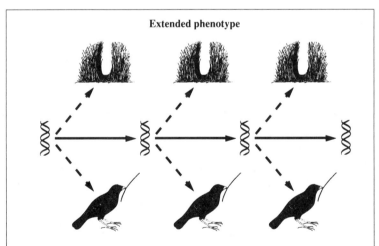

... but, as we shall see in the next chapter, genes can also have phenotypic effects beyond the bodies that house them: bowers as well as bowerbirds.

What gets replicated? And how?

pretensions to be a comprehensive rival to Darwinism, or, indeed, to add anything to it other than a slight reinforcement of what natural selection is getting on with anyway. Lamarckism maintains that characteristics acquired through use and disuse are inherited. The first of Dawkins's arguments is about use and disuse. The second is about the acquiring of characteristics. The third is about the embryology that use-inheritance would need.

First, could use and disuse be appropriate instruments for bringing about adaptation? Lamarckism maintains that organisms typically improve themselves, refine their adaptations, by exercising the capacities they have so that they stretch them further and further (or run them down by disuse). Adaptations arise because the act of doing something improves the organism's ability to do the same thing in the future. The blacksmith's bulging biceps and the giraffe's super-stretched neck come instantly to mind. And in these stock examples one can see intuitively how Lamarckism might work. Perhaps improved muscle power and greater height could arise in this way. But what about improved eyesight? The adaptive complexity of the eye could never be brought about from primitive beginnings merely by exercising more of the same. On Lamarckian theory, adaptive improvement has to be linked to use and disuse. But generally it isn't – and nor is it all likely to be for any highly complex adaptations, any adaptations of sophistication and intricacy. Natural selection has no such difficulty. It will seize on any advantageous characteristic, of whatever kind, however small and insignificant, however deeply buried, however rarely exercised, however indirect. So there is an automatic fit between what is of advantage and what evolves. Adaptations don't have to pass through the highly restrictive requirement of use and disuse.

The second of the arguments is that Lamarckism cannot explain why organisms respond adaptively. We take it for granted that a hungry giraffe will stretch up to the high branches when the lower ones are overgrazed; but why doesn't it stoop – and starve? We take it for granted that its neck will lengthen rather than shorten, that its muscles will stretch rather than contract, that it will try to eat rather than, say, kick its heels. But we shouldn't take these things for granted. Why is an animal's behaviour appropriate to its needs, and its physiology appropriate to its behaviour? Why, when it learns, does it learn to do the right thing rather than plumping for the multitude of wrong ones? Its behaviour, learning, and so on are all adaptive responses. But however many ways there are of responding adaptively, there are many, many more ways of responding non-adaptively (including not responding at all). So we should ask how the adaptive response comes about. To this question, Lamarckism and Darwinism have fundamentally different answers.

A useful distinction makes this difference clear. It is the distinction between *instructive* and *selective* models of the origins of adaptations. Picture a locksmith who has to make a key to fit a particular lock. The instructive way is to take a wax impression of the lock and make up a custom-fitted key, taking instruction from the environment about the exact design that is required. The selective way is to take a random bunch of keys and try them until one of them works. Lamarckism is an instructive theory; Darwinism is a selective theory. Darwinism can easily explain how the giraffe originally came to do the appropriate, adaptive thing. It is descended from a long line of giraffes who happened – out of the random pool of possible genetic changes – to hit on changes, however small, that constituted improvements. Note, by the way, that the key-maker analogy is too stringent here. The giraffes don't have to find the one key that will finally open the lock, but merely a key that comes a bit closer, no matter how tiny that bit is. It is the fact that adaptive keys can be chosen incrementally that makes it possible, out of the whole bunch of keys, to hit on one that is satisfactory. By contrast, Lamarckism needs to explain how the giraffe came, in some mysterious way, to be guided into doing the right thing. The theory assumes that an organism responds adaptively because it learns from its environment, gathers information from it, takes 'instruction' from it about what response is needed. But it doesn't account for the organism's capacity to accept such instruction in the first place. After all, the instructive locksmith has to use wax; wood or water or wobbly jelly wouldn't do. How does Lamarckism answer this problem? By relying covertly on Darwinian adaptations, by taking for granted that the giraffe will stretch rather than stoop, that its muscles will give support rather than snap, that it will develop a taste for the nutritious rather than the noxious, that it will try to avoid pain rather than seek it. Lamarckian mechanisms cannot originate adaptations; they can do no more than carry over, into future generations, tendencies to 'acquire' that are originated by Darwinian means. Any instructive theory must ultimately rest on a selective model (or resort to deliberate design). So Lamarckism could never be more than a limited adjunct to Darwinian theory. It could never replace Darwinism as a comprehensive theory of evolution.

It's ironic that Lamarckians have traditionally looked to their theory for the very qualities it cannot deliver. They have pinned their hopes, for example, on learned behaviour being passed down the generations. But it turns out that, ultimately, learning is adaptive for Darwinian, not Lamarckian, reasons. And they have welcomed use-inheritance for its creative, initiating role, for guiding the path of evolution in a way that supposedly blind Darwinian forces, passively steered by chance, are powerless to do. But, again, it turns out that such a lead is precisely what Lamarckism is inherently, in principle, incapable of providing. If Lamarckian mechanisms are to go anywhere, they must ride on the back of Darwinian achievements.

Now to the third of our anti-Lamarckian arguments. Unlike the other two, it applies more to our actual world than to any possible world, for it takes certain principles of embryology as given. It rests on a distinction between recipes and blueprints. Recipes are instructions that are irreversible. A recipe for a cake cannot be reconstructed from the cake itself, nor can a word-processing program be reconstructed from a print-out. In cases like these, the relationship between the end-product and the instructions is so tortuous that there is no simple one–one mapping between them. So the process is not reversible. Blueprints, however, are instructions that can go both ways. The method for making a dolls' house can be discovered by carefully measuring an existing house. In this case, there is a simple – and reversible – one–one correspondence between structure and plan.

In the history of embryological ideas there have been two opposing schools of thought about how single cells get transformed into full-blown organisms: the epigenetic (recipe–cake) view and the preformationist (blueprint–house) view. If we lived in a world of blueprint embryology then the inheritance of acquired characteristics would be possible. Lamarckism requires a flow of information from the body to the genes, so that bodily changes in one generation can be incorporated in the next. If embryology worked by blueprints, the two ends of the embryological process – the body and the gene (or the DNA, the material in which genes carry information about heredity) – would have the same structure. The isomorphism would automatically provide rules, built-in rules, for reversing the process of instruction. In that case the phenotype (what Weismann called the soma – eyes and wings and shells and petals and other ways that genes exert their effects) could be mapped back into the genotype (the organism's genetic constitution, its particular set of genes), and then that information could be read back into the phenotype in the next generation.

But it turns out that embryology is recipe-like (see e.g. Maynard Smith 1986, pp. 99–109). Genes convey information about making bodies and behaviour in the same irreversible way that a recipe conveys information about making a cake, not in the reversible way that blueprints convey information about making buildings. DNA issues instructions about how cells should multiply, die, join with other cells, and so on, in an orderly process, step by step in carefully controlled sequence; each stage in the procedure builds on the previous stages, each development is influenced by previous developments. So the parts of the whole are fundamentally influenced by the history of the procedure, by where they find themselves and when; bits are not preserved as identifiable, as discrete. If there is any mapping at all, it is in the correspondence between the set of instructions and the successive stages of the epigenetic process – just as a recipe, if it maps onto anything at all, maps not onto the parts of the cake but onto the successive stages of assembling ingredients, mixing, baking and so on. All this tells crucially against Lamarckism and for Darwinism. The Lamarckian hope that a discovery is just around the corner, that somewhere, perhaps in some recondite corner of the immune system, acquired characteristics will turn out to be inherited, is in vain. In our world of epigenetic recipe embryology, the inheritance of acquired characteristics is impossible.

That third anti-Lamarckian argument depended on the contingent fact that embryology is recipe-like not blueprint-like. Is this contingency just a matter of complete chance? Or is there, perhaps, something intrinsically unlikely about blueprint embryology? How would lifeforms with such an embryology manage their development, and what, if anything, would they lose? Such speculations, tempting as they are, take us too far from our purpose here. For us, the upshot is that even the most serious alternative to Darwinism, the one on which anti-Darwinians have pinned their highest hopes, is not so serious a candidate after all.

Now a brief look at the other historical contenders for Darwinism's place. We can group them into two camps: orthogenesis or 'straight-line evolution' and mutationism or 'evolution by directed mutation alone' (Bowler 1983, pp. 141–226, 1984, pp. 253–6, 259–65; Dawkins 1983, pp. 412–20, 1986, pp. 230–6, 305–6; for a turn-of-the-century assessment, see Kellogg 1907, pp. 274–373).

Orthogenesis is the theory that evolution proceeds along 'straight lines', driven by forces that are internal to the organism and not the result of environmental pressures. It first became popular through the advocacy of the Swiss-born zoologist Theodor Eimer, writing in Germany in the last three decades of the nineteenth century; the American palaeontologist Henry Fairfield Osborn became an influential proponent in the first few decades of the twentieth. As a more-or-less comprehensive theory of evolution, orthogenesis was influential for several decades during the very end of the nineteenth century and the earlier part of the twentieth. With the final decline of Lamarckism in the 1920s and 1930s it even outstripped that theory as the most serious alternative to Darwinism.

Needless to say, those internal driving forces that were supposedly propelling evolution eluded detection. But orthogenesis was anyway an unlikely candidate to rival Darwinism as a comprehensive explanation of adaptation. How could an internal driving force, unaided by selection, give rise to forms that match their environments? How would that force manage to pick its way down intricate, detailed pathways to adaptive complexity?

Orthogenetic theories are instructive theories, and, like all instructive theories, they just push the problem of design back one stage. Not surprisingly, proponents of orthogenesis, rather than attempting to explain adaptation, tried to argue it away. They pounced on what appeared to be adaptive anomalies and attempted to reinterpret the Darwinian evidence as less utilitarian, less elegant, less economical than had been supposed. Rather than detecting adaptation in nature, they claimed to detect a pattern on the grand scale, a regularity and order that Darwinism was supposedly powerless to explain. They pointed triumphantly to the massive antlers of the extinct 'Irish elk', the ever more involuted, elaborately coiled shells of fossil ammonites, and other such strange escalations, as evidence of a momentum that drove species in fixed directions, regardless of adaptive advantage, even to the point of evolutionary disaster. Palaeontology, it was argued, revealed a wealth of orthogenetic trends that led inexorably to degenerating, deleterious structures and eventually to extinction. Palaeontologists nowadays would argue that these 'trends' were largely artefacts of the orthogenetic imagination (e.g. Simpson 1953, pp. 259–65).

All this stress on nature's grand patterns at the expense of adaptation sounds reminiscent of idealism. And orthogenesis was, indeed, a direct heir to that view: 'In their fascination with the regularity of development at the expense of utilitarian factors, the supporters of orthogenesis reveal the last vestige of the influence of idealism on modern biology' (Bowler 1984, p. 254). The development of orthogenesis in the United States, for example, was fostered in the mid-nineteenth century by the anti-evolutionary idealism of the renowned palaeontologist Louis Agassiz (a Harvard professor but, significantly, with a Continental background – Swiss born, German and French educated). It is in this climate that such an unpromising candidate as orthogenesis managed to make headway as an alternative to Darwinism. It succeeded insofar as it was able to underplay the extent of adaptation and to magnify an apparent fundamental direction or order in evolution.

Now to mutationism. This is a saltationist theory. Saltationism is jumpy evolution, the view that evolution proceeds by the sudden appearance of radically new forms. Saltationism can allow a role for selection. And saltationists may indeed be right that occasionally a random change with large effects could be advantageous and push evolution on in a new direction. A change that involved more-of-the-same (like the change from an unsegmented animal to segmented repetition of the basic design) rather than a radically new complex adaptation (an eye from undifferentiated skin) would not be hopelessly unlikely embryologically; neither would such a change be quite so likely to constitute a gigantic plunge into selective disaster. So macro-mutations could be of some importance in the history of life on earth. The idea of 'hopeful monsters', proposed by the American (originally German) geneticist Richard Goldschmidt in the 1940s, was a theory of this kind.

So much for selectionist saltationism. Its mutationist version is a less respectable position. It does not allow a serious place for selection. According to mutationism, random changes in the hereditary material are sufficient for adaptation without much, or any, selection at all. Mutations just somehow happen to be adaptive, the right changes simply manage to occur. The inadequacies of this view are obvious: either the mutations must be directed by some mysterious force yet to be discovered, which itself will take some explaining, or their adaptive fit depends on a staggeringly generous dose of good luck, far too generous to be taken seriously. Like orthogenesis, mutationism had its heyday at the beginning of this century. And like orthogenesis, it had a predictably non-adaptive bias. Among its proponents (at some time during their careers) were the Dutch botanist Hugo de Vries, the Danish botanist Wilhelm Johannsen, the English biologist William Bateson and the American founder of chromosomal theory Thomas Hunt Morgan.

A century ago Weismann wrote: 'We must assume natural selection ... because all other explanations fail us and it is inconceivable' ('improbable' would have been nearer the mark) 'that there could be yet another capable of explaining the adaptation of organisms *without assuming the help of a principle of design*' (Weismann 1893, p. 328). We can now understand why Weismann's intuition was likely to have been right.

Goodbye to all that

In the library of the Zoology Department in Oxford is a copy of a book, published in 1907, called *Darwinism To-day*. Its author was Vernon L. Kellogg, a zoologist and at that time a professor at Stanford University. Kellogg gives us a survey of the standing of Darwinian theory and its alternatives at the beginning of this century. And a full and judicious survey it is, carefully sifting majority from minority views, and largely unbiased by the author's own declared leanings towards Lamarckism. Kellogg was perhaps particularly well-equipped for such a task, for he had worked for a few years in Leipzig and Paris and so was in touch with Continental, as well as North American, thinking. The book is a fascinating glimpse into the state of evolutionary ideas at a time when Darwinism was at a low ebb.

And yet, however low that ebb, reading Kellogg's undoubtedly non-partisan and careful accounts of orthogenesis and mutationism, I still found it hard to believe that these theories could be considered as serious comprehensive alternatives to Darwinism, or even as useful supplements to it.

Surely someone who had fully understood Darwin's contribution and appreciated its immense explanatory power would feel that these contenders were not just wrong (or largely so) but not even in the same league as Darwinism, often barely within the realms of science at all.

My hopes were confirmed at least as far as one biologist of the time was concerned. The Zoology Department copy of *Darwinism To-day* bears this sad inscription: 'Bequeathed by Captain Geoffrey Watkins Smith, Fellow and Tutor of New College, Lecturer and Demonstrator in Zoology from 1905–1914, who fell in action in France July 10, 1916'. Across the generations, a spontaneous interview emerged from those pages. Up and down the margins, in what I established was Geoffrey Smith's hand, are pencilled comments. And the appraisals – 'Rot! ... Bunkum' among them (pp. 141, 306) – have the value of honest, heartfelt reactions, unfettered by the politenesses of publication.

Smith was deeply unimpressed by orthogenesis and mutationism. Take, for example, the 'evidence' for orthogenesis (the scare quotes are his) involving parallel evolution in different branches of the same large group, such as 'the reduction of the hind toes among the Artiodactyls in several genera (giraffe, camel, llama) up to a complete disappearance'; to orthogenesists, this represents 'a definite or determinate direction of modification' (Kellogg 1907, pp. 279–80). Smith disagrees: 'Isn't selection enough for them?' he asks rhetorically. To the evidence that the 'constitution, or actual chemical composition of the body permits, in many cases, changes only in few directions' (Kellogg 1907, p. 280), Smith objects: 'This is no particular evidence for orthogenesis. The direction of selection is of course limited by the nature of the varying organism'. As for the argument that palaeontology seems 'to prove the existence of orthogenetic evolution ... [because] we always see a limited number of lines of development' (Kellogg 1907, p. 281), Smith exclaims: 'Who would expect anything else?' He finds mutationism equally unpersuasive. When Kellogg states that its 'principal criticisms' of Darwinism are that natural selection cannot explain either 'forthright development along fixed lines not apparently advantageous' (Smith underlined 'apparently') or, worse, 'ultra-development ... even ... to death and extinction' (Kellogg 1907, pp. 274–5), a laconic note in the margin reads: 'These two don't come to much'. Incidentally, the respected Cambridge zoologist Sir Arthur Everett Shipley wrote when Smith was killed: 'He was a zoologist of the most extraordinary ability, and one who did not lose his head like so many of the Mendelian enthusiasts have' (Anon 1917, p. 36) – other Mendelian heads having been lost, presumably, to some version of mutationism (Schuster and Shipley 1917, p. 278).

Kellogg declares unequivocally that Darwinism does 'not satisfy present-day biologists' (Kellogg 1907, p. 375). That was undoubtedly true taking all biologists world-wide. And where Darwinism was thought to fail, orthogenesis and mutationism flourished. But Kellogg adds a footnote: 'However, there still exist, especially in England, thorough-going Darwinians who see nothing serious in all this criticism of their great compatriot's explanation of the origin of species' (Kellogg 1907, p. 389); and he mentions that 'neo-Darwinism' as it was then called (Darwinism fortified with Weismannism) is 'accepted more or less nearly completely by Wallace and a number of other English biologists, and by a few naturalists of Europe and America' (Kellogg 1907, p. 133). Perhaps in England the direct line of descent from Darwin and Wallace exerted a more powerful influence; and, moreover, idealism had never gained as strong a hold. In Geoffrey Smith's case, his 'thorough-going Darwinism' and hostility to rival theories were certainly not the result merely of insularity. He had worked in the lively, international atmosphere of the Stazione Zoologica in Naples and was clearly familiar with the criticisms of Darwinism and the fashionable alternatives. His position, it seems, was the considered conclusion of an informed scientist of the time.

That voice from the past also teaches us another lesson. We know that if the scientific theories of former times are judged with the hindsight of today's knowledge, then the blind alleys of science, which have left no descendants in the current textbooks, might come to be undervalued. This could tempt us to err on the side of leniency, to curb critical judgements. After all, if scientists of the period took a theory seriously, who are we to take it less seriously, albeit we now know that it happens to be wrong? Such tolerance has its merits. But, as Geoffrey Smith's reactions remind us, it also has its limits. We should not treat the alternatives to Darwinism with undue generosity for fear that it is hindsight alone that reveals their inadequacies. Equipped not with hindsight but with Darwinian understanding, Smith, and no doubt others, rejected the alternatives even at a time when they had a large and influential following. Having seen what orthogenesis and mutationism had to offer, his response was a firm 'goodbye to all that'.

This brings us to an odd irony in that phase of Darwinian history. For it was not Darwinism's rivals so much as Darwinism itself that, typically, was condemned for being unscientific. From the end of the nineteenth century right up until several decades into the twentieth, Darwinism was quite commonly castigated for hindering scientific progress by insisting on asking the wrong questions. These were the days when biology was finding its feet as a respectable science. And respectable was commonly interpreted to mean laboratory based, no-nonsense fact-gathering, experimental in the narrowest sense (as, shamefully, 'scientific' is still all too often taken to mean today). In this light, Darwinian theory was stigmatised as speculative, untestable, inexact and – worst of all, for this put it entirely beyond the pale of science – teleological. Science, it was proclaimed, should not even attempt to ask adaptive questions; a precise description of, say, biochemical or physiological pathways was all that was required.

This spirit is reflected in Erik Nordenskiöld's influential history of biology, written at that time, which was notably hostile to Darwinism:

> It is asked: Why has a cat claws? ... [Darwin says: In] order to enable it to survive in the struggle for existence ... But ... the question ... is absurd ... Biology can only endeavour to find out the conditions under which cats' claws are developed and used, but never anything more; those who question beyond that fail to fulfil Bacon's requirement that we should 'ask nature fair questions'. But Darwin and his contemporaries are constantly putting such wrong questions to nature.
>
> (Nordenskiöld 1929, p. 482)

It was these teleological Darwinian questions, Nordenskiöld complained, that had 'in no small degree contributed towards retarding the development of biology into an exact science' (Nordenskiöld 1929, p. 471). There is no need, for example, Nordenskiöld informs us, to resort to a speculation like Darwin's theory of sexual selection in order to explain why males are gaudy and undiscriminating whereas females are drab and choosy. The answer 'is internal secretion and the connexion of the secondary sexual characters with it; both sexual coloration and mating-play have their explanation in this' (Nordenskiöld 1929, p. 474). In other words, the reason why males and females differ is that they have different hormones, and that is all there is to it. Ask *why* they have different hormones, and that would be dabbling in teleology. The same attitude – indeed the same example – crops up in Emanuel Rádl's history of biology, also from that period and also unsympathetic to Darwinism:

> When Darwin discusses animal beauty he deals almost entirely with ... secondary sexual characters; some biologists, however, believe that these are entirely due to the influence of the primary sex glands. They think that the development of the colour, markings, horns, antlers, and other charactistic 'masculine' adornments are due to secretions of the male glands, while these glands also inhibit the development of the corresponding feminine qualities. The female glands have exactly the opposite effect.   (Rádl 1930, pp. 105–6)

Nordenskiöld contrasts Darwinism with the study of heredity, his model for good science:

> Heredity has been the most popular field of research of the age ... Natural selection is certainly retained in principle by some students of heredity ... but it is

really of no practical importance; the phenomenon cannot be observed and it is therefore not possible to fit it into a subject of research that is based on exact observations ... for the very reason that it [heredity research] has become an exact science it has not been able to follow the old Darwinism in its speculative ranging, but whatever may have been lost in the way of the general conception of life has undoubtedly been won in the way of concentration on facts and reliable results.

(Nordenskiöld 1929, p. 594)

Charles Singer's history of biology, also from that time, displays the same feelings: 'the "chance" element in Darwin's scheme was but a veiled teleology. Natural selection had been elevated to the rank of a "cause", and science has to deal not with causes but with conditions. Darwin was occupying himself with the "might" and the "may be" and not with things seen and proved' (Singer 1931, p. 305; see also p. 548). Singer, incidentally, was advised by Thomas Hunt Morgan (Singer 1931, p. ix) who, although by that time a leading Mendelian and self-proclaimed Darwinian, had not shaken off his early misconceptions about natural selection and teleology: 'it is clear that he was never at ease with the idea of selection ... The concept, perhaps the very term "selection", bothered him; it sounded purposeful, and with his strong dislike of teleological thinking, Morgan reacted against purposefulness in evolutionary theory' (Allen 1978, p. 314; see also pp. 115–16, 314–16) – or, rather, what he took to be purposefulness.

The celebrated ethologist Niko Tinbergen, who was a student at the time, recalled how he was rounded on for having the temerity to ask an adaptive question:

In the post-Darwinian era, a reaction against uncritical acceptance of the selection theory set in, which reached its climax in the great days of Comparative Anatomy, but which still affects many physiologically inclined biologists. It was a reaction against the habit of making uncritical guesses about the survival value, the function, of life processes and structures. This reaction, of course healthy in itself, did not (as one might expect) result in an attempt to improve methods of studying survival value; rather it deteriorated into lack of interest in the problem – one of the most deplorable things that can happen to a science. Worse, it even developed into an attitude of intolerance: even wondering about survival value was considered unscientific. I still remember how perplexed I was upon being told off firmly by one of my Zoology professors when I brought up the question of survival value after he had asked 'Has anyone an idea why so many birds flock more densely when they are attacked by a bird of prey?' (Tinbergen 1963, p. 417)

As we shall see when we look at adaptive explanations, even today the spirit of that professor has not quite been laid to rest.

It is perhaps understandable that scientists and historians should have had some such conception of science at a period when biology, aspiring to reach a more elevated rung, looked up to physics. The subject was glorying not merely in its new status but also in genuine progress into worlds hitherto, in Darwin's phrase, 'quite unknown'. What is less understandable is that the Darwinian enterprise should be so misinterpreted that it was thrown on the scrapheap of non-science. This raises suspicions that it was the dislike of Darwinism that came first; the misguided positivism that purportedly led to its rejection was actually daubed on afterwards as methodological gloss. And, talking of meretricious methodology, Nordenskiöld and others who cited the weighty authority of Bacon in their support were entirely off-target. They took his attack on final causes to embrace Darwinism's adaptive explanations. But Bacon was condemning 'what he considered to be genuine final causes – these being essentially connected with a conscious purpose, either God's purpose or that of a human mind'; such explanations, Bacon objected, were likely to be sterile, as the example of Aristotle showed (Urbach 1987, p. 102; see also pp. 100–2). Bacon's criticism certainly applied to the kind of explanation that, as we have seen, typified pre-Darwinian natural history and sometimes also Darwinism's later rivals. But it leaves Darwinian explanations unscathed – as Bacon would surely have recognised. To invoke Bacon against natural selectionist explanations is so misplaced as to suggest a fundamental misunderstanding of what such explanations are about. It brings to mind a remark made by the distinguished American geneticist H. J. Muller, in a slightly different context, about his former teacher Thomas Hunt Morgan and other anti-Darwinians of the time: 'it seemed to us as if he somehow couldn't understand natural selection. He had a mental block which was so common in those days' (Allen 1978, p. 308).

We have seen why Darwinism was in 1859, and still is, the best explanation of why living things are as they are – not only, it turns out, on this planet, but in any world that resembles ours in several fundamental respects. Since 1859, the legacy of Darwin and Wallace has undergone a number of major transformations. The most recent, and one of the most dramatic, has been the transition from classical to modern Darwinism. And it is to this latest transformation that we shall now turn.

3

Darwinism old and new

Anticipations of things past

During the last few decades Darwinian theory has undergone a revolutionary change. This revolution combines two new ways of thinking. Where once Darwinian explanations focused on the individual organism and made only tacit reference to hereditary units, Darwinian ideas now give place of honour to the gene. And where once Darwinism concentrated on the structures of organisms, there is now a rapidly burgeoning study of their behaviour, above all their social behaviour, and the schemes and stratagems that are part of their evolutionary endowment.

Consider two of Darwin's favourite illustrations of the workings of natural selection: the exact adaptation of the woodpecker's beak and tongue to specialised feeding, and the elegant plumes by which certain seeds transport themselves on even the gentlest breeze (e.g. Darwin 1859, pp. 3, 60–1; Darwin and Wallace 1858, pp. 94, 97; Darwin, F. 1892, p. 42; Peckham 1959, p. 357). Such cases are typical of classical Darwinism – the approach epitomised by the *Origin* and Wallace's *Darwinism*. It deals above all with structures and with those structures that aid the bearer with its own survival or reproduction by conferring benefits on it or on its offspring. By contrast, within modern Darwinism that same woodpecker or seed might find itself viewed as a ruthless blackmailer or a calculating gambler, as a skilled tactician whose behaviour varies depending on that of others, as one who systematically exploits its neighbours or, conversely, puts itself at their service even to the detriment of its own chances of survival or reproduction.

It has been claimed that this Darwinism is a fundamentally different theory from the one it has displaced and that the two are incompatible (e.g. Sahlins 1976). In fact, as we shall see, the transformation in Darwinian thinking is a development of what went before, an extension in scope. Relative to today's Darwinism the classical theory is restricted; nevertheless it anticipates modern views: 'Darwin's theory has certain implicit consequences which

Two adaptations that delighted Darwin:

"The structure ... of the woodpecker, with its feet, tail, beak, and tongue, so admirably adapted to catch insects under the bark of trees." (Darwin: Origin)

The woodpecker's tongue is often sticky-coated and barbed; in *Celeus flavescens*, *Veniliornis olivinus* and *Dryocopus lineatus* it has four to six barbs. The tongue can protrude to remarkable lengths, with the help of elongated, flexible hyoid bones or 'horns'; the horns end at different places in different species, in some cases (as in *Picoides villosus*, *Hemicircus concretus* and *H. canente*) winding their way round the back of the skull and then looping the right eye socket.

woodpeckers and plumed seeds

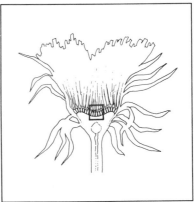

"Seeds ... furnished with wings and plumes, as diversified in shape as elegant in structure, so as to be wafted by every breeze." (Darwin: Origin)

The scanning electron microscope has revealed worlds that Darwin could no more than guess at. This is a 27-times-magnified view of the florets deep inside the flower of a dandelion (*Taraxacum officinale*).

follow logically from it, but which have only recently been noticed' (Dawkins 1978a, p. 710). Nostalgic defences of Darwin's own 'true' Darwinism notwithstanding (e.g. Sahlins 1976, pp. 71–91), if classical Darwinism had failed to adumbrate modern developments, then it would have lacked the impressive explanatory power that it managed to exhibit for over a century.

Mention that there has been a recent revolutionary change in Darwinism to most Darwinians and they will immediately think of the modern synthesis of evolutionary biology, which took place between about 1930 and 1950. That revolution fortified Darwinian theory with Mendelian genetics and showed in detail how this powerful combination could explain the variation and geographical distribution of natural populations, the origin of new species and the history of life on earth as revealed in the fossil record. It was a momentous advance in Darwinian theory. To omit to discuss it is not to depreciate it. But most aspects of this revolution are well recognised and have been extensively documented and analysed. The revolution that concerns us in some respects develops the work of the modern synthesis and in other respects has stronger affinities with earlier Darwinian thinking.

On the one hand, it extends the modern synthesis into neglected areas and makes explicit ideas that it drew on. I have placed this recent revolution as occurring in the last few decades – since about the mid-1960s – because this has been the period of its greatest impact. But this way of thinking was anticipated to a remarkable extent in two of the classics of the modern synthesis, R. A. Fisher's *The Genetical Theory of Natural Selection* (1930) and J. B. S. Haldane's *The Causes of Evolution* (1932), and in the work of Sewall Wright, eventually published as *Evolution and the Genetics of Populations* (1968–78). Why this particular aspect of their work was not more widely taken up, I don't know. We shall see that Darwinian theory could have been saved several unrewarding detours if their contribution had been better appreciated. In fact, it was left to later generations to realise the potential of their ideas.

On the other hand, we shall find that there is also a continuity between this latest revolution and classical Darwinism that is independent of the modern synthesis. As far as this progression is concerned, the limitations of classical Darwinism do not arise from lack of mathematical tools or an adequate theory of heredity. Although developments on both fronts were essential to the full power of the modern position, informal gene-centred and strategic thinking do not require either.

Because there is a continuity between recent and classical thinking, the new developments help us to understand the nature of classical Darwinism and, in particular, its limitations. One can use the hindsight afforded by the modern view to pick out the most important elements of the classical approach.

R. A. Fisher in 1952, at his desk calculator at Whittingehame Lodge, the official residence of the Arthur Balfour Professor of Genetics at Cambridge.

Taking the two guiding principles of modern Darwinism – its gene-centredness and its strategic turn of mind – we shall see just how and why classical Darwinism was different. (The main features of modern Darwinism are nicely conveyed in several recent books. George C. Williams's influential *Adaptation and Natural Selection* offers enduring insights (Williams 1966); Richard Dawkins's *The Selfish Gene* and *The Blind Watchmaker* are classics (Dawkins 1976, 1986); Mark Ridley's *The Problems of Evolution* is also a good source (Ridley 1985); for animal behaviour, see the excellent textbook

J. B. S. Haldane in 1948, lecturing at University College, London. John Maynard Smith, who was then an undergraduate, says that the picture was taken by a fellow undergraduate 'at the risk of his life'.

edited by J. R. Krebs and N. B. Davies, *Behavioural Ecology* (Krebs and Davies 1978, 2nd edn.) and Marian Stamp Dawkins's *Unravelling Animal Behaviour* (1986); for the idea of the extended phenotype, which we shall examine later in this chapter, see Dawkins 1976, 2nd edn., pp. 234–66, 1982; for the notions of game theory and evolutionarily stable strategies, which we shall also touch on in this chapter, see Dawkins 1980; Maynard Smith 1978b, 1978c, 1982, 1984; Parker 1984.)

I shall concentrate on the first few decades of Darwinian theory, particularly Darwin's own writings, as representative of the classical approach..This is not to suggest that Darwinism was static for the next half century or more. But we shall note subsequent developments when we deal with sexual selection and altruism. There was, in any case, startlingly little change during that time in those parts of the theory that have recently been so radically transformed.

There is an obvious difficulty in showing that a position is not held. Negative examples are hardly compelling and even raise suspicions that the missing view was expressed, perhaps regularly expressed, elsewhere. One solution I shall adopt is to choose as illustrations cases well known to classical Darwinism that for modern Darwinians are prime candidates for gene-centred or strategic explanation. Another is to explain how classical Darwinism could be so successful in spite of its restrictions. And a third is to bring to light, in the case of strategic thinking, some background reasons that explain why classical Darwinism adopted so different an orientation.

First, however, let me forestall the mutterings of disagreement that can already be heard in the background. By no means all modern Darwinians would accept my characterisation. But I am dealing with the theory, not with how individuals have chosen to interpret it. One must distinguish between the fundamental tenets of a theory (what the theory actually says) and how it is viewed by some practitioners (what is said about it). I am dealing with the former.

Organism to gene

M. Jourdain discovered after forty years that he had been talking prose all that time and hadn't known it. Modern Darwinism discovered after a century that Darwinian theory had been talking about genes – or, at least, hereditary units – all that time and had been equally unaware of the fact. Although classical Darwinism analyses the workings of natural selection from the point of view of individual organisms and their offspring, the idea that natural selection ultimately deals with replicators rather than the organism that houses them obviously lurks somewhere inside this organism-centred view; that, after all, is what reproductive success is really about. Today's Darwinism takes up this vantage point. And, from there, it is discovering a wealth of hitherto unrecognised and highly fruitful implications.

Modern Darwinian theory is about genes and their phenotypic effects. Genes do not present themselves naked to the scrutiny of natural selection. They present tails, fur, muscles, shells; they present the ability to run fast, to be well camouflaged, to attract a mate, to build a good nest. Differences in genes give rise to differences in these phenotypic effects. Natural selection acts on the phenotypic differences and thereby on genes. Thus genes come to be represented in successive generations in proportion to the selective value of their phenotypic effects.

I should stress at this point that an expression like 'a gene for green eyes' is not about a particular gene that maps onto green eyes. It is about differences. The individual words in a cake recipe do not map onto bits of cake. But a one-word difference in recipes – lemon to vanilla – does map onto differences between two kinds of cake. Similarly, particular genes do not map onto a particular bit of the body. But identifiably different genotypes correspond to identifiably different phenotypes. This point is important because expressions like 'genes for green eyes' (which I use throughout this book) have quite commonly been taken as assuming a crude single-gene/single-phenotypic-effect model or as positing the existence of specific genes without a shred of evidence.

A gene may have a number of phenotypic effects, each of which may be of positive, negative or neutral selective value. It is the net selective value of a gene's phenotypic effects that determines the fate of the gene. When we talk of a gene for green eyes we are picking out just one property from its effects. But the differences between genes that bring about green rather than brown eyes might also be bringing about all sorts of other things – thinner toe-nails, longer limbs and a smaller chin. Darwin noted that these 'correlations of growth', as they were called, can be 'quite whimsical: thus cats with blue eyes are invariably deaf; ... [h]airless dogs have imperfect teeth; ... pigeons with feathered feet have skin between their outer toes' (Darwin 1859, pp. 11–12). We tend to notice the whimsical cases. But pleiotropic effects, as they are now called, are the norm. From a gene's-eye-view, phenotypes are not neatly divided into adaptations and their side effects. There are simply several phenotypic effects, and adaptations are the special case in which their total advantage outweighs their total cost. Thus the costs of an adaptation such as an eye are not merely the costs of building protein or laying down pigment or utilising vitamin A; they are also the costs of any accompanying phenotypic effects.

So natural selection acts on genes through feedback from phenotypic effects. Genes are perpetuated insofar as they give rise to phenotypes that have selective advantages over competing phenotypes. Typically, we think of the phenotypic effects as manifesting themselves in the organism that houses the gene. But as Richard Dawkins, one of the leading architects of the recent revolution, has argued, we need not stop there: Darwinism could quite naturally and cogently – and, indeed, very fruitfully – extend the idea of a phenotype.

Consider a bird's nest. We readily accept that the beak with which the bird collected the materials is a phenotypic effect of genes for beak-building. In precisely the same way, the nest could be regarded as a phenotypic effect of genes for nest-building. The difference is only that, with the nest, the phenotypic effects extend beyond the bird's body. So we can think of the nest as a phenotype but an extended phenotype.

Extended phenotypes need not be confined to artefacts – to the bird's nest or spider's web or beaver's dam. We standardly think of genetic control of an organism's behaviour as stemming from the genes in its own body. But, in principle, there is no reason why one organism should not exploit the ready-made nervous system, muscle-power and behavioural potential of another organism in much the same way as it exploits its ready-made protein, vitamins or minerals. Natural selection would favour genes in one organism that could successfully manipulate the behaviour of another organism for their own benefit. Such manipulation of one organism by another could be viewed as the extended phenotypic effect of manipulating genes. Take the example of parasites. They have traditionally been thought of as just gratefully enjoying a free lunch, not as actively concocting the menu. Any effects that the uninvited guests have had on their hosts have been seen as mere unintended side effects of the parasites' depredations. But organisms that are parasitised sometimes behave in a way that is no good for them but very good for the parasite. This telling fact suggests that manipulation may be afoot:

> One of the most familiar literary devices in science fiction is alien parasites that invade a human host, forcing him to do their bidding as they multiply and spread to other hapless earthlings. Yet the notion that a parasite can alter the behaviour of another organism is not mere fiction. The phenomenon is not even rare. One need only look in a lake, a field or a forest to find it.   (Moore 1984, p. 82)

Look into a typical lake and you may see *Gammarus* and other 'fresh-water shrimps' – strictly not shrimps but amphipod crustaceans. The behaviour of these amphipods changes dramatically when they are parasitised. Among

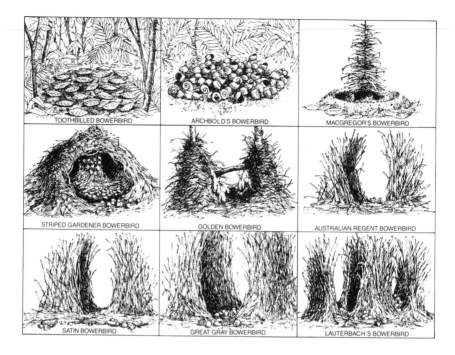

TOOTHBILLED BOWERBIRD | ARCHBOLD'S BOWERBIRD | MACGREGOR'S BOWERBIRD
STRIPED GARDENER BOWERBIRD | GOLDEN BOWERBIRD | AUSTRALIAN REGENT BOWERBIRD
SATIN BOWERBIRD | GREAT GRAY BOWERBIRD | LAUTERBACH'S BOWERBIRD

The architecture of male bowerbirds
Birds' bodies are phenotypes; birds' nests and bowers can be regarded as extended phenotypes. Male bowerbird constructions vary from sparsely ornamented clearings on the ground to elaborately decorated bowers. These edifices are as characteristic of a bird's species as is the bird's own body.

their parasites are three species of acanthocephalans or 'thorny-headed worms': *Polymorphus paradoxus, P. marilis* and *Corynosoma constrictum* (Bethel and Holmes 1973, 1974, 1977; see also Dawkins 1990; Holmes and Bethel 1972; Moore 1984, pp. 82–5, 89, Moore 1984a; Moore and Gotelli 1990). Uninfested amphipods move away from light and avoid the water's surface; if disturbed, they immediately dive down, vanishing into the turbid depths, and burrow into the safety of the mud. When parasitised, however, they become less evasive, more reckless. If infested with *P. paradoxus*, they move up to the surface light and, when touched, lock themselves tenaciously onto vegetation or whatever else disturbed them – or, failing to get a hold, they skim along the water's surface, creating a conspicuous disturbance, until they find some object to clamp onto; all this turns them into easy targets for surface-feeding predators, particularly mallards, beavers and muskrats. Amphipods infested with *P. marilis* move towards the light, but not all the way to the surface; they make themselves more vulnerable to diving ducks, notably lesser scaup. And amphipods that are hosts to *C. constrictum* move towards the surface light, more than half of them diving when disturbed and the others remaining at the surface; they get picked off by both diving and surface-feeding ducks. Why this suicidal behaviour? And why commit suicide in three different ways? There is an ominous clue: these three groups of predators are the next hosts in the life-cycles of our three acanthocephalan species – mallard, beaver and muskrat for *P. paradoxus*, scaup for *P. marilis* and both mallard and scaup for *C. constrictum*. And there is a further clue: the amphipods do not undergo their improvident changes until their thorny-headed invaders are ready to proceed to their next host. It seems, then, that the parasites are manipulating their hosts' bodies for their own adaptive ends. The behaviour of the amphipods can be regarded as the extended phenotypic effect of the worms' manipulative genes.

'Why drag in genes?' you might ask. 'Why not simply say that the acanthocephalans are manipulating the amphipods?' But it is not I who have dragged in genes, it is natural selection. Genes must be involved because we're talking about a Darwinian adaptation. It just happens that this phenotypic effect of the genes shows itself (at least to us) as strikingly in the behaviour of the amphipods as in that of the genes' bearers, the thorny-headed worms.

Extended phenotypic manipulation is by no means the monopoly of parasites. We shall see this when we examine sexual selection and altruism. What matters here is how this perspective differs from that of Darwinism's first hundred years. Classical Darwinism is about how adaptations are of advantage to their bearer or its offspring. Extended phenotypes remind us that modern Darwinism has transformed that maxim: 'An animal's behaviour tends to maximize the survival of the genes "for" that behaviour, whether or not those genes happen to be in the body of the particular animal performing it' (Dawkins 1982, p. 233).

Once we look on natural selection as acting not on individual organisms that are harmonious wholes but on the phenotypic effects of their selfish, manipulative genes, we open up the possibility of conflicts of interest between genes that share a body. Organism-centred Darwinism tacitly took harmony for granted. Questioning this assumption, modern Darwinism has come up with ideas that would once have seemed utterly bizarre. Take the phenomenon of outlaw genes. These are genes that have phenotypic effects that favour their own selection but are deleterious to most of the other genes in the genome (the totality of genes housed in the organism). Consider, for example, a so-called segregation distorter – a gene that has the phenotypic effect of influencing meiosis (cell division during the formation of sex cells) so that it has more than its 50% Mendelian chance of ending up in a sex cell (sperm or egg). A gene that manoeuvres meiosis in this way tends, other things being equal, to be favoured by natural selection. It could also have phenotypic effects detrimental to the rest of the genome. Indeed, it very likely will: most new mutations have several pleiotropic effects and most effects of new mutations are deleterious. In that case, a segregation distorter would be an outlaw; it would spread through the population in spite of its detrimental effect on other genes.

We have travelled far from the organism-centred view of classical Darwinism – from a Darwinism that is about the survival and reproduction of individuals. Indeed, looking back, one wonders how an approach so different from the gene-centred could more or less agree with it over a wide range of explanations. After all, Darwinism no longer even expects the interests of all the genes in an organism to coincide. And gene-centred Darwinism has been highly successful, solving problems that had previously proved intractable, and opening up new issues and fruitful ways of dealing with them. How, then, could an organism-centred theory, which depends on an appeal to the interest of the individual, be so outstandingly successful for so long?

The answer, in short, is that organism-centred Darwinism is a good approximation; even for a selfish gene, a successful strategy is very likely to promote the survival and reproduction of the organism that houses the gene.

A great deal of the work of natural selection comprises individuals striving to keep alive, to have offspring and to care for them. This is far from adequate to characterise the full range of natural selection's activities but it captures more or less well an impressive proportion of them. Individual survival is no negligible matter even on a gene-centred view. After all, even if the organism is regarded as no more than the vehicle of its genes, it has to be roadworthy. So the various genes in the genome will probably all stand to gain from the individual's survival much of the time. What is more, genes are selected not in isolation but against the background of other genes in the gene pool; they are thus to some extent selected for their compatibility with other genes with whom they share a body. So, in spite of saboteurs like outlaw genes, when it comes to individual survival, Darwinism doesn't generally need to take account of such conflicts of interest. As to reproduction, one would expect the classical theory to approximate to the gene-centred view because some kind of replicator-centred notion is obviously implicit in the idea of reproduction.

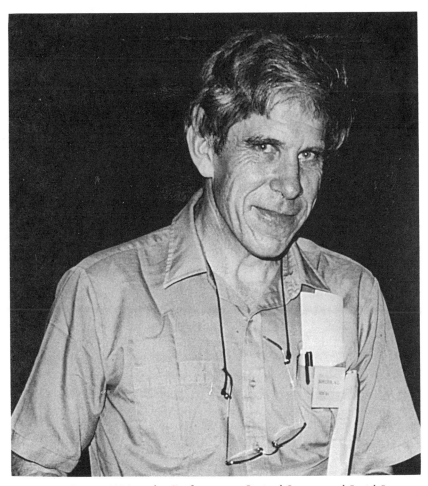

W. D. Hamilton in 1986, at the Conference on Optimal Strategy and Social Structure in Kyoto.

After all, in sexually reproducing populations organisms do not reproduce perfect facsimiles of themselves. More important, no organism, sexual or asexual, does so; how could it unless its offspring inherited its acquired characteristics? Classical Darwinism recognises that natural selection requires the reproduction not of identical individuals but of characteristics – not an identical woodpecker but its well-adapted beak. As for caring for offspring, this is merely a special case of what is now recognised to be the more general principle of kin selection – the principle that natural selection may favour the act of an organism giving help to its relatives. Because parental care is by far the commonest case, classical Darwinism could encompass a great deal of the sphere of kin selection even though its application was so restricted.

More generally, there is good reason to expect harmony to come forth from selfishness, harmonious individuals from selfish genes. All genes have to pass through the isthmus of reproduction. And this engenders a common interest:

> If all replicators 'know' that their only hope of getting into the next generation is via the orthodox bottleneck of individual reproduction, all will have the same 'interests at heart'; survival of the shared body to reproductive age, successful courtship and reproduction of the shared body, and a successful outcome to the parental enterprise of the shared body. Enlightened self-interest discourages outlawry when all replicators have an equal stake in the normal reproduction of the same shared body.   (Dawkins 1982, pp. 134–5)

Much of the time, then, the idea of the individual's interest and that of its offspring will give a good and workable approximation to a gene's-eye-view. That is why, in spite of being restricted to organisms, classical Darwinism could get so impressively far.

Structures to strategists

Two Canadian hikers were startled to find a grizzly bear coming up fast behind them. They immediately started to run, the bear in hot pursuit. Suddenly, one of them stopped, searched frantically in his haversack and pulled out his running shoes. 'You surely don't think that will help you to outrun the bear', panted his astonished companion. 'No. But it will help me to outrun you', was the reply.

That runner has the true spirit of a modern Darwinian strategist. He thinks about behavioural responses, not merely running ability; he weighs up the costs of pausing to change against the benefits of running faster; he considers a range of strategies and chooses from among them; he decides on a strategy in the light not only of what the bear is doing but also what others like himself are doing; and he chooses a strategy that is 'selfish' – one that benefits him, regardless of cost to others.

The development of strategic thinking has involved two major shifts from classical Darwinism: first, a view of adaptations that is more conscious of their costs and less sanguine about their benefits, and, second, a greater emphasis on behaviour, particularly social behaviour. The strategists, of course, are not runners – nor even robins or rats: they are genes.

Let's start with the transformation in how adaptations are viewed. Classical Darwinism is well tuned to spotting the advantages of adaptations but rather poor at taking their costs into account. Modern Darwinism takes a more cost-conscious, less beneficent view of adaptations; it is quicker to appreciate the sacrifices with which adaptations are bought and it adopts a more attenuated, less blithe view of adaptive benefits.

Natural selection is now thought of as if it were scanning a range of possibilities and choosing the option that optimises within the given constraints. The costs are built in as part of the final choice because an adaptation involves a trade-off; it results from a balance among competing demands and has to be paid for. Costs include a drain on materials, energy and time; other changes in the organism; opportunity costs; perhaps a degradation of the environment. So an adaptation intrinsically brings losses in its wake. Classical Darwinism viewed adaptations more as outright benefits than as a compromise between their advantages and disadvantages. It certainly incorporated some aspects of the modern view – in particular, the idea of adaptations having 'unintended consequences'; we've noted Darwin commenting on correlations of growth. Nevertheless, it generally failed to account fully for the price at which adaptations are bought; costs were often underestimated, seen as neutral rather than deleterious, or overlooked. In short, for modern Darwinism, the costs of adaptations are inevitable; for classical Darwinism, they were incidental.

Classical Darwinism's failure to do justice to costs was perhaps a legacy of the utilitarian-creationist way of thinking. Utilitarian-creationists saw nature as essentially benign and concentrated on the beneficial aspect of adaptations. Although, with Darwinism, struggle became paramount and nature was seen as more ruthless, there were still remnants of the utilitarian-creationists' insistence on seeing adaptations as an unalloyed good. From time to time accusations of Panglossianism fly around the Darwinian camp. They are largely misplaced. But on the question of costs they could have been made to stick. Classical Darwinism did not exhibit Dr Pangloss's tendency to see perfection in all things but it did show some of his optimism – his inability to see the bad side of things.

This will emerge in the discussion of sexual selection and altruism. Here, I shall take just one example. It is all the more striking because it appears at first sight to be more of a counterexample. It is classical Darwinism's treatment of 'imperfect' adaptations. We have seen that for early Darwinians there were considerable attractions in emphasising the apparent imperfections of adaptations, as evidence against conscious design and for the makeshift workings of natural selection. Such an emphasis might seem to sit oddly with a failure to appreciate costs. But Darwin and his contemporaries were more preoccupied with how unadaptive such characteristics were than with how natural selection, nevertheless, managed to balance the books. To argue that a trait is adaptive, one must suggest how the benefits might outweigh the costs. But to argue that a trait is imperfect, one merely needs to catalogue the imperfections, without subjecting them to any cost–benefit analysis. An imperfect adaptation was seen as an adaptation that fell short of ideal, not one that incurred costs because of that imperfection.

Consider, for example, how early Darwinians viewed adaptations that involved non-uniform behaviour. Behaviour that is varied in execution or outcome was regarded as a prime candidate for imperfectionist treatment. It was often explained, for instance, as an adaptation for a unified response that had been imperfectly evolved, or as a chance reversion to ancestral habits, or as unimportant individual variation.

Take Darwin's discussion of the apparently erratic egg-laying habits of American 'ostriches' (rheas) and cowbirds (*Molothrus bonariensis*) (Darwin 1859, p. 218; Peckham 1959, pp. 395–6). He is trying to show a gradation in parasitism from the rheas at one end, through to the cowbirds, right up to the well-honed parasitic instincts of the European cuckoo (Peckham 1959, pp. 390–6). This is a standard Darwinian procedure; such gradations provide models for how natural selection could have acted. But it is the 'imperfection' displayed along this gradation that most interests Darwin, for this suggests that deliberate design has not been at work. The cowbirds' habits are 'far from perfect' he says (Peckham 1959, p. 395). They lay their eggs in foster nests in such large numbers that most must be lost; they waste eggs by dropping them on the bare ground; and they sometimes start to build very inadequate nests, which they don't complete or use. Such imperfection, Darwin says delightedly, is enough to turn even a creationist into an evolutionist: 'Mr Hudson is a strong disbeliever in evolution, but he appears to have been so much struck by the imperfect instincts of the Molothrus bonariensis that he quotes my words, and asks, "Must we consider these habits, not as especially endowed or created instincts, but as small consequences of one general law, namely, transition?"' (Peckham 1959, p. 396). Rather than trying to show how natural selection compensates for the damages, Darwin triumphantly writes them off as imperfection unworthy of a designer. The rhea's behaviour is equally untidy and irregular. Several females lay their eggs in a single nest – an adaptation, Darwin says, to cope with laying a large number of eggs and at two to three day intervals between each, which would make incubation and care of young in a single nest difficult. 'This instinct, however, ... as in the case of the Molothrus bonariensis, *has not as yet been perfected*; for a surprising number of eggs lie strewed over the plains, so that in one day's hunting I picked up no less than twenty lost and wasted eggs' (Peckham 1959, p. 396; my emphasis). Again, he is more interested in evidence of imperfection than in explaining how natural selection could allow the rhea to develop such profligate reproductive habits. Incidentally, this whole discussion occurs in the *Origin* under the heading 'Instinct' – a section that one might have expected to yield the richest crop of strategic thinking.

One should not get the impression that varied behaviour or structure were always treated as mere imperfections. Whilst needing to score imperfectionist points against design, Darwinism needed even more to account for variation adaptively, and finding too many imperfect adaptations could come perilously close to scoring an own goal. After all, to lose one egg may be regarded as a misfortune; to lose many looks like carelessness, more carelessness than natural selection would tolerate. So variability was also explained adaptively. We shall see in the discussion of altruism, for example, that Darwin persevered in searching for an adaptive purpose, where previously there was thought to be none, in the dimorphism (the two different forms) displayed by some plants. But there's no doubt that for early Darwinians there was more interest in seizing on the 'imperfection' of varied behaviour than in explaining it adaptively.

In contrast, modern Darwinism, far from explaining variability as natural selection having slipped up, stresses that it is often to be expected. Darwinians have recently taken another look at the ostrich (this time the real African ostrich, *Struthio camelus*, which behaves in the same way) (Bertram 1979, 1979a). The apparently cavalier habits of egg-laying are no longer seen as a single behavioural adaptation imperfectly executed, but as selection for varied behaviour. There is a range of ways in which to care for eggs; both hatching eggs (one's own as well as those of others) and not hatching them have costs and benefits. The eggs of others can be a buffer against predation of one's own, for example; a female without a mate to help her care for her eggs might do best by farming out the incubation to a mated female in spite of the risk that the foster mother will eventually discard the eggs in favour of her own; the costs of building a nest might outweigh the disadvantages of having others incubate the eggs. The result is a mixed fate for the eggs – some hatch, some perish. When classical Darwinism looks at a mix of hatched and spoiled eggs it sees above all an imperfect instinct that points to absence of conscious design. For modern Darwinism, the same mix is the result of selection for mixed tactics.

A second case in which Darwinians latched on to imperfections but underestimated costs was that of the eye. Far from being Panglossians, Darwinians were highly embarrassed by its apparent perfection; Darwin confessed that at one time the thought of it gave him a cold shudder (Darwin, F. 1887, ii, pp. 273, 296). And understandably. Its precision engineering seemed to support the utilitarian-creationist doctrine of a grand optic designer far better than the Darwinian assumption of *ad hoc* tinkering. Darwinians were eager to seize on any evidence that the eye was not a perfect optical instrument. Fortunately, where eyes were concerned, help could hardly have arrived from a more distinguished source than the renowned physiologist and physicist Helmholtz, who came to Darwin's rescue just in time for the second edition of *Descent of Man*:

> We have ... no right to expect absolute perfection ... in a part modified through natural selection ... for instance, in that wondrous organ the human eye. And we know what Helmholtz, the highest authority in Europe on the subject, has said about the human eye; that if an optician had sold him an instrument so carelessly made, he would have thought himself fully justified in returning it.
>
> (Darwin 1871, 2nd edn., pp. 671–2; see also 1859, p. 202)

So Darwinians were quick to see not only the advantages but also the imperfections of the eye. Nevertheless, they were slow to see its possible costs. Take Darwin's explanation of why the eyes of some burrowing creatures like moles are rudimentary or covered by skin and fur (Darwin 1859, p. 137). Darwin acknowledges that the eyes would be liable to frequent inflammation. And he concedes that natural selection might have played some role in developing the protection. Nevertheless, he feels, the injury would not be so severe that natural selection would go to the lengths of covering the eyes. And he concludes that it is the inherited effect of gradual reduction from disuse, a Lamarckian mechanism, that is mainly responsible. In his view the eyes of these creatures are not so much disadvantageous and costly as unused and neutral. Incidentally, it was only when Lamarckians, later in the century, tried to exploit this case that Darwinians were pushed into seeing that the deleterious effects could have been sufficient for natural selection alone to have acted (see e.g. Wallace 1893, pp. 655–6).

As well as being more cost-conscious than classical thinking, modern Darwinism also views adaptations in a less beneficent light. According to classical Darwinism, an adaptation is a characteristic that has been selected because it is good for the bearer or its offspring. Modern Darwinism challenges this charitable assumption.

On a gene-centred view, it is not organisms but genes that adaptations are 'good for'. The bearer of a characteristic, far from being the beneficiary, may be the subject of selfish manipulation by a gene in another organism. Indeed, as we learn from outlaw genes, there may not be any organism at all that is benefiting.

What's more, the very idea of benefit has changed. We shall see this when we examine modern explanations of sexual selection. Here I'll take the example of the evolutionarily stable strategy (ESS). The ESS is a central concept of evolutionary game theory, a theory that has borrowed principles from the mathematical theory of games and applied them with enormous success to evolutionary problems. Imagine a set of strategies available to the individuals in a population; imagine, in other words, that there are 'genes for' a certain specified range of alternative behaviour patterns. There could be, say, a set of decisions on when to escalate a fight and when to give up and slink away. (These are, of course, not conscious decisions; we are talking about genes that have the phenotypic effect of making an organism act as if it had decided to behave that way.) We can consider this strategy set as constituting an evolutionary game. An ESS is a 'solution' to such a game. It is a solution in the sense that it is an uninvadable strategy, a strategy such that, if at any given time the majority of the population adopts it, then natural selection will favour it over any of the other strategies available (those who have adopted it always doing better than those who have not). An intuitively clear (albeit not precise) way to think of an ESS is as a strategy that does well against itself. This is because, over evolutionary time, any strategy that is successful will proliferate in the population. So eventually the strategy that it meets most often will be itself. If, then, it is not to be invaded, it must do well when encountering itself. With the concept of the ESS comes a crucial shift of emphasis. Traditionally, the most important question about an adaptation was 'What benefit does it confer?' But evolutionary game theory attaches equal importance to the question 'Is it evolutionarily stable?' And the theory can go further, wreaking havoc with the very notion of 'benefit'. Consider the hypothetical 'scorpion game' (Dawkins 1980, pp. 336–7). Under the conditions specified for this game, a strategy of attempting to sting one's murderer lethally with one's dying gasp could become the ESS. Yet this behaviour is of no benefit in any sense that classical Darwinism would recognise:

> As far as his survival or genetic success is concerned, retaliation is pointless for the individual retaliator. Once he has been stung he is doomed. Stinging back does him no good at all. Yet retaliation is the dominant strategy ... because it is the ESS. We are breaking down the idea that animal behaviour should necessarily be interpreted in terms of individual benefit. Why do scorpions retaliate? Not because it benefits their inclusive fitness to do so; it does not. Scorpions retaliate because ... [the strategy] retaliator is the ESS.   (Dawkins 1980, p. 336)

Now to the second way in which classical Darwinism is less strategically-minded than its modern counterpart: its concentration on structure and relative neglect of behaviour, particularly social behaviour. This might seem to run counter to a common impression of Darwin's own work. Aren't we often told that he founded not only ecology (e.g. Bowler 1984, pp. 151–2; Coleman 1971, pp. 15, 57; de Beer 1971, p. 571; Ghiselin 1974, p. 26; Kimler

1983, p. 112; Manier 1978, pp. 82–3; Ospovat 1981) but also ethology (e.g. Lorenz 1965, pp. xi–xii; Mayr 1982, p. 120; Ruse 1982, p. 189)?

It is certainly true that Darwin's writings are always alert to the idea that the organic world is generally the most important part of an organism's environment. Darwin repeatedly emphasises that, except in very extreme inorganic environments, other organisms are more significant selective forces than are climate or topography (e.g. Darwin 1859, pp. 68–9, 350, 487–8). So organic beings are not merely adapted to the inorganic environment but coadapted to one another. And certainly he sees the organic world as tightly interlocking, a world in which even a slight change in an organism can have far-reaching consequences. Take the plumed seed once more. It is adapted to the wind – but as a response to other plants:

> the structure of every organic being is related, in the most essential yet often hidden manner, to that of all other organic beings, with which it comes into competition for food or residence, or from which it has to escape, or on which it preys. This is obvious in the structure of the teeth and talons of the tiger; and in that of the legs and claws of the parasite which clings to the hair on the tiger's body. But in the beautifully plumed seed of the dandelion ... the relation seems at first confined to the [element] ... of air ... Yet the advantage of plumed seeds no doubt stands in the closest relation to the land being already thickly clothed by other plants; so that the seeds may be widely distributed and fall on unoccupied ground.
>
> (Darwin 1859, p. 77)

Such insights are not uncommon in Darwin's writings. And they indicate that modern Darwinism is not a radical departure from classical ideas; the elements of modern thinking were there. Indeed, they are to be found to a far greater extent in Darwin's own contribution than in the work of many of his successors. Nevertheless, we shall see that Darwin did not do these insights justice.

One shouldn't underestimate the importance of an obvious and mundane reason why, until very recently, structure was more intensively studied than behaviour: the often formidable practical obstacles to systematically observing and recording what animals (and plants) do. Even now it is not unusual to find ethological questions that have remained unanswered for practical rather than theoretical reasons when questions about the same organism's structure have long been settled.

Bear in mind, too, the evidence that Darwinism in its early stages tried to deal with, a legacy of pre-Darwinian natural history. It was devoted to structural detail and very largely ignored behaviour. Both of the pre-Darwinian schools of thought, utilitarian-creationists and idealists, concentrated on the structure rather than the behaviour of organisms, albeit for different reasons. For utilitarian-creationists, the preoccupation with structure arose from their quest for perfection. The organic world is full of structures that are built to the specifications of a skilled craftsman. Behaviour, however, unless perhaps it is highly stereotyped or results in 'perfect' artefacts such as a web or nest, is at first sight less orderly and less amenable to tidy interpretation. It is no surprise to find Paley dipping into anatomy for his favourite illustrations. Idealists concentrated on structure because, for them, the paramount task was to trace variations on ideal types, variations on fundamental structures. Of course, Darwinism broke free of both these traditions. Nevertheless, they provided the bulk of the evidence that early Darwinian theory dealt with.

What pre-Darwinian views there were on behaviour fell into two sharply divided traditions (Richards 1979, 1982). On one side there were the Cartesian and Aristotelian schools of thought. They took human behaviour to be governed by reason, and the behaviour of all other creatures to be regulated by inflexible instinct. On the other side, there was the sensationalist tradition, stemming from Locke, which played down innateness and emphasised the role of reason and experience in all behaviour, human and non-human alike. Clearly, Darwinism did not slot neatly into either of these views. We shall look in detail at what a Darwinian understanding does amount to – and at misunderstandings, too – when we come to the discussion of human altruism. For now, what is most relevant is how nineteenth-century Darwinians dealt with the question of continuities between humans and other animals. For, even when they studied behaviour, their preoccupation with this issue deflected them from analysing its social aspects.

Darwinism is obviously not obliged to argue for continuities on every front. Indeed, as we shall see when we come to human altruism, there are excellent Darwinian reasons why such a sweeping programme is likely to be naive. Nevertheless, in the early days of Darwinism, the appeal to purported discontinuities between humans and all other living things was a favourite anti-Darwinian ploy. To defeat it on as many fronts as possible would obviously add plausibility to the Darwinian case. So it was that nineteenth-century Darwinism's two pioneering studies of behaviour, for many decades the classic works on the subject, were devoted above all to arguing against the claim that there are major gaps between humans and other animals. These two works were Darwin's *The Descent of Man* (1871) and *The Expression of the Emotions in Man and Animals* (1872). One might expect that an interest in humans, far from shifting the focus away from social behaviour, would bring it to the fore. But we're about to see how it led Darwin instead into concentrating on two areas: the mental states and feelings of other animals and non-adaptive explanations of peculiarly human characteristics.

Take first *Descent*, the work in which Darwin deals most extensively with social behaviour. His main interest in discussing it, as we shall see when we look at human altruism, is to establish a continuity between human 'mental powers' and those of other animals. His particular target is our moral sense, for it was this that was generally held to constitute the greatest gap between humans and all other animals. Darwin is out to demonstrate that the apparently distinctively human moral conscience has its evolutionary roots in animal sociality. So most of his evidence of social behaviour in other animals crops up as part of his comparison between their mental powers and ours (1871, pp. 34–106). And although he treats us to a rich panoply of behavioural anecdotes – deceitful elephants, sulky horses, revengeful monkeys, playful ants, jealous dogs, inquisitive deer, imitative wolves, attentive cats and imaginative birds – what actually concerns him is not the behaviour but the accompanying sentiments. He tells us, for instance, that elephants act as decoys, but his interest is not in the adaptive advantages of their stratagem but in whether they know they are practising deceit (1871, 2nd edn., pp. 104–5). He relates the story of a female baboon who adopted young monkeys of other species and even dogs and cats; that looks like curiously non-adaptive behaviour, but his only comment is about her capacious heart (1871, i, p. 41). In the one section that is specifically devoted to social behaviour (1871, i, pp. 74–84) his interest is again in the accompanying mental faculties associated with it – the 'social instincts' – rather than in the behaviour itself. Love, sympathy and pleasure, then, receive far more attention than warning calls, division of labour or mutual defence. The examination of social behaviour is submerged under a tidal wave of emotional, mental and moral considerations.

Now, that is not to denigrate Darwin's method. Indeed, we shall see in the discussion of human altruism that, as far as humans are concerned, he inadvertently hit on a highly fruitful way to proceed. It was the method of evolutionary psychology, an approach that is only now beginning to be appreciated after a hundred dormant years. But, where humans were not concerned, the study of behaviour suffered.

One might expect *Expression of the Emotions* to have more to say about social adaptations. After all, the uses to be made of expressing emotion – information, deception, manipulation – are certainly to do with fellow creatures. But, once again, Darwin's interest in the non-uniqueness of humans takes him elsewhere. His particular target in this case is the separate creationist view that the means of expression in humans are a 'special provision' (1872, p. 10), created solely for communicating emotion and not to be found in other animals (1871, i, p. 5; Darwin, F. 1887, iii, p. 96). Incidentally, the 'special provision' argument was an application of the utilitarian-creationist procedure of regarding each adaptation as serving one particular purpose (see e.g. Ospovat 1980, pp. 188–9); in this case it had what was for utilitarian-creationists the welcome effect of setting humans apart from other creatures. Darwin takes it as his principal task to show that, although human features – facial muscles, for example – now serve as a means of expression, they originally served quite different functions. With this aim, he assiduously sets about investigating the physiological basis of emotional expression – the raw material at the disposal of natural selection. And this is what the bulk of his argument is devoted to. The next stage, the way in which natural selection has put the material to use, he hardly touches on, apart from a relatively brief discussion of means of expression in animals (1872, pp. 83–145). The book is less about the adaptive expression of the emotions than about blood vessels, nerves and muscles.

What is more, even where Darwin might have discussed such adaptation, he is more intent on arguing that characteristics have not been specially created for their expressive use. So, although he concedes that they do express emotions, he often denies that they have undergone any modification exclusively for the purpose of expression. He concedes, for example, that facial expression enhances the communicative power of language; but he claims that not a single muscle has been adapted especially for this function (1872, p. 354). In some cases he even denies that there is any adaptive function at all. Blushing, for instance, according to creationists, is a special provision for expression; Darwin flatly denies that it has any use whatsoever, even in sexual selection (1872, pp. 336–7). The expression of laughter, too, is 'purposeless' other than for the physiological advantage of expending superfluous nervous energy (1872, pp. 196–9) – although, like modern ethologists (e.g. Charlesworth and Kreutzer 1973, pp. 108–10), Darwin views it as developing in a social context.

So, although Darwin by no means ignored behaviour, even social behaviour, his view and that of his contemporaries was blinkered. The roots of this bias lay in the nineteenth century but its influence carried far. Even as late as the 1960s it could still be claimed that 'the study of behaviour [is a vast area of modern biology] ... in which the application of evolutionary principles is still in the most elementary stage' (Mayr 1963, p. 9).

To turn from this 'elementary stage' to Darwinism today is to find, above all, a world in which organisms are social beings. For classical Darwinism the paradigmatic selective forces, apart from inorganic pressures, were relationships between members of one species and another, such as prey–predator or parasite–host. A modern Darwinian organism inhabits a world in which the success of its behaviour may well depend critically on the relative frequency of its own behavioural type in the population to which it belongs

John Maynard Smith in 1984, at his desk at the University of Sussex.

(its species, say, or sex, foraging party or nest): if success depends on being the rarer of two types, then selection will automatically maintain variability. This is known as frequency-dependent selection. It is above all evolutionary game theory that has provided the means for dealing with frequency-dependence. When success is not significantly affected by the behaviour of others, then an adaptation can be analysed simply as optimisation. When frequency-dependence enters the scene, as it often will with social behaviour, game theory is more likely to be the appropriate tool. It is John Maynard Smith, the distinguished evolutionary geneticist, who, more than anyone else, has been responsible for the development of evolutionary game theory. This is how he compares the conditions under which optimisation theory can be used with conditions that call for game theoretical analysis:

> Evolutionary game theory is a way of thinking about evolution at the phenotypic level when the fitnesses of particular phenotypes depend on their frequencies in the population. Compare, for example, the evolution of wing form in soaring birds and of dispersal behaviour in the same birds. To understand wing form it would be necessary to know about the atmospheric conditions in which the birds live and about the way in which lift and drag forces vary with wing shape. One would also have to take into account the constraints imposed by the fact that birds' wings are made of feathers – the constraints would be different for a bat or a pterosaur. It would not be necessary, however, to allow for the behaviour of other members of the population. In contrast, the evolution of dispersal depends critically on how other conspecifics are behaving, because dispersal is concerned with finding suitable mates, avoiding competition for resources, joint protection against predators, and so on.
>
> In the case of wing form, then, we want to understand why selection has favoured particular phenotypes. The appropriate mathematical tool is optimisation theory. We are faced with the problem of deciding what particular features ... contribute to fitness, but not with the special difficulties which arise when success depends on what others are doing. It is in the latter context that game theory becomes relevant.
>
> (Maynard Smith 1982, p. 1)

Although behaviour that is obviously social has been the main area of application, game theory is in principle also applicable to structure, colour, developmental pattern and so on. Having a beak of a certain shape could be as much a social and even frequency-dependent characteristic as forgoing reproduction and helping at the nest with sibs. Even wing-form might be subject to frequency-dependent selection: think of a racing-driver making for the slipstream, or think of the advantages of rare flying skills when predators have adjusted their tactics to the average. And then there's the growth of a plant:

> The optimal growth pattern for a plant depends on what nearby plants are doing. A plant growing by itself would not gain, in seed or pollen production, by having a massive woody trunk. Leaves may be selected as much for shading out competitors as for photosynthesis. In other words, functional analysis of plant growth is a problem in game theory, not in optimisation.   (Maynard Smith 1982, p. 177)

If the success of a beak or a wing or a plant is frequency-dependent, this raises once again the question of how classical Darwinism could be so successful without some such analysis.

Well, first, as one would expect given the continuity between classical and modern views, intraspecific, social and frequency-dependent forces were not ignored. A major example is the theory of sexual selection – although, as we shall see, it is quite atypical of classical Darwinism in several respects (and, by the way, is not readily amenable to game theoretical analysis). And there are other examples, such as the case of mimicry in butterflies – an early explanatory triumph for Darwinian theory (Bates 1862; [Darwin] 1863; Wallace 1889, pp. 232–67, 1891, pp. 34–90); it was recognised that the protective value to a palatable butterfly of mimicking an unpalatable species could be crucially affected by the relative proportions of its conspecifics and the mimicked species (e.g. Wallace 1891, pp. 58, 60).

Second, the assumption that the environment is unstrategic can be thought of as a limiting case. So Darwinian theory can get a long way without a frequency-dependent analysis, particularly if it is not attempting to explain complex social behaviour. Take again the contrast between wing form and dispersal behaviour. If wing form is not a frequency-dependent characteristic then game theory is in practice redundant. But it is still, strictly speaking, applicable. Selection for wing form could be regarded as a limiting case – a case in which the frequency-dependence reduces to zero, so that the optimum is not affected by allowing for the behaviour of others. Richard Dawkins illustrates this point with the case of optimal foraging theory, comparing analyses that can treat it as an individual activity with those that need to treat it as social (and frequency-dependent) behaviour:

> Our optimal foraging theorist assumes that it does not matter what the other predators are doing. This assumption might indeed be justified ... In this case it might seem superfluous to bother to speak of an ESS ... but it would not be strictly incorrect. If, on the other hand, it turned out that the presence of other individuals, all optimizing from their point of view, affected the optimum rule for any one individual, ESS analysis would become a positive necessity.
>
> (Dawkins 1980, p. 357)

Once again, then, classical Darwinism succeeded as a good approximation. On the one hand, it incorporated social behaviour to some extent; on the other hand, a wide range of characteristics can be treated as asocial.

Complexities and diversities

It is a long journey from the woodpecker's efficiently-engineered beak to the parasite's unscrupulous manipulation, from organisms and their offspring enjoying the benefits of auspicious adaptations to selfish genes cunningly outmanoeuvering one another with far-reaching phenotypic ramifications.

Throughout this journey, questions about adaptation have been a recurrent theme. In the previous chapter we saw that, when Darwin dealt with the evidence of adaptation, he could be thought of as playing off Paley and Owen against one another, perfectionist against imperfectionist approaches to adaptation. In this chapter, too, we have seen that there is a constructive tension between these two interpretations of the complex characteristics of living things. The two approaches reflect a divergence that has been a constant feature of the history of Darwinism: the contrast between adaptationist and non-adaptationist explanations.

Historically, although not logically, these alternative views have tended to go along with other differences of approach. Roughly, we can think of Darwin's and Wallace's two major problems, adaptation and likeness in diversity, as staking out two distinct areas of interest, two different and to some extent competing priorities, that have divided Darwinians from that time until now. Ernst Mayr, the eminent American evolutionist, has commented: 'There is a fundamental and rarely sufficiently emphasized difference among evolutionists whether diversity (speciation) or adaptation (phyletic evolution) holds first place in their interest' (Mayr 1982, p. 358; see also e.g. Simpson 1953, pp. 384–6). For Mayr, the diversity of species comes first. John Maynard Smith has said: 'For Darwin, the origin of new species was a central problem. Mayr would say that it was *the* central problem, but I am less sure. I think that for Darwin the most important problem was to provide a natural explanation for the adaptation of organisms to their ways of life' (Maynard Smith 1982a, p. 41). And so it is for Maynard Smith himself.

Understandably, non-adaptationism has been more congenial to those for whom speciation is central. These Darwinians have also tended to emphasise the conservative power that developmental constraints can exert on adaptation, the way in which embryology can clip the wings of adaptive initiatives. Darwinians whose paramount interest is in adaptation have had more confidence in the power of natural selection to shape evolutionary history. They see developmental constraints as themselves subject to selective forces – indeed, perhaps less as 'constraints' than as channels, as pioneering opportunities for new adaptive pathways. These divergent sets of sympathies are the Darwinian counterparts of the divergence between the two pre-Darwinian traditions, between idealists and utilitarian-creationists, between Owen and Paley.

These alignments have reverberated down the whole of Darwinian history. We shall find them cropping up often in the debates that we'll examine. They have been the cause of recurrent disagreement, sometimes intense and highly acrimonious disagreement, within the Darwinian camp. But we shall see that, once we understand its historical roots, much of this apparent dissension is revealed as nothing more than that – merely apparent. One debate in which these divisions have made themselves felt is a long-running discussion over the explanatory scope of natural selection and, in particular, the scope of adaptive explanations. And it is this that is the topic of the next chapter.

4

Demarcations of design

What characteristics should we expect natural selection to explain? The eye, the kangaroo's pouch, the human chin, the cheetah's sprint, the chameleon's camouflage? What about the peacock's tail, the bee's suicidal sting, the crimson of blood, the flash of colour on a bird's wing? Should we expect it to explain human altruism, our love of music, feelings of aggression, sexual jealousy? And divorce rates, wars, political oppression? What, in short, is the scope of natural selection? Can it explain all these things and, if not, what should the alternative explanations be?

Take human altruism. One answer is that it should not be explained biologically at all. Human social behaviour, it is commonly argued, is a candidate for, say, political, economic, social or cultural analysis but not for Darwinian explanation. Biology, it is urged, is too low on the explanatory hierarchy; if we take the question of why we are nice to one another down to that level, we'll lose sight of niceness and perhaps even of us. Others have argued that Darwinism can deal with the question but only if the Darwinism is of an unorthodox kind; it's been claimed that human altruism – and the bee's sting and many other cases of apparently self-sacrificial behaviour – involve selection at the level of the group, selection against some individuals but to the advantage of the group of which they are members. Similar claims have been made for the peacock's tail; ornamental characteristics, it is said, can be explained by a Darwinian force – sexual selection – but it is a force that is fundamentally different from natural selection. We shall look at these various views later. And we dealt earlier with the anti-Darwinian answers. Here I shall concentrate on the idea that natural selection is ruled out for some characteristics because they simply aren't candidates for any adaptive explanation at all.

We have seen that the question of adaptation was one of the two fundamental problems that Darwin attempted to explain. Indeed, it was the more fundamental of the two because it could not be explained merely on the assumption that living things had evolved. It is a major triumph of Darwinism that, of all theories yet proposed, it alone can explain how the organic world has come to bear a striking appearance of deliberate design without the intervention of a deliberate designer. So it may seem odd for a Darwinian to challenge apparent examples of adaptation. But, such dissenters say, they are merely trying to purge the Darwinian bestiary of red herrings and wild geese, to prevent a fruitless search for imaginary selective advantage. After all, they rightly point out, some characteristics are not the product of natural selection and, indeed, may have no selective value, either negative or positive. Darwinians should bear this in mind, they caution, and not be over-eager to assume that adaptive explanations will be appropriate. For a typical statement of this view one need go no further than Darwin himself. During the late 1860s he came to believe that he had formerly been too wedded to natural selection, seeing its hand in cases that probably weren't adaptations at all:

> I now admit ... that in the earlier editions of my *Origin of Species* I perhaps attributed too much to the action of natural selection ... I had not formerly sufficiently considered the existence of many structures, which appear to be, as far as we can judge, neither beneficial nor injurious; and this I believe to be one of the greatest oversights as yet detected in my work.   (Darwin 1871, i, p. 152)

I shall call this view 'non-adaptationism'. In one way, the name is unsatisfactory, for it suggests an anti-adaptationist approach – whereas Darwinians are, of course, committed to adaptive explanation. But in another way it is regrettably apt. For some Darwinians the term 'adaptationist' has become a term of abuse – a term associated with a stifling, narrow, strait-jacketed outlook. It's time to redress the balance and a start can be made by reclaiming respectability for the word 'adaptationist'. This makes 'non-adaptationist' the natural opposite. Yes, I know that terminology is unimportant in itself. But the names are useful reminders not to turn Darwinism too far on its head. And they shouldn't anyway be interpreted rigidly; they describe approaches, leanings, preferences, not hard-and-fast-claims to explanatory territory.

The Darwinian controversy over when-is-an-'adaptation'-not-an-adaptation is as old as the theory itself (Provine 1985). One reason for its long run is that some Darwinians have championed non-adaptationism as part of a broader crusade: an attempt to widen the explanatory options beyond a strict commitment to just natural selection and yet more natural selection. These pluralists, as they are sometimes called, offer us a vision of a more eclectic world, a world in which knee-jerk hyper-adaptationism gives way to a supposedly more subtle, complex analysis of the characteristics of living beings.

That was the position of the eminent nineteenth-century Darwinian George John Romanes, the most redoubtable opponent in his time of what he called ultra-Darwinism. He could not accept that natural selection alone, indeed any agent alone, could account for the whole of evolution: it is 'improbable that, in the enormously complex and endlessly varied processes of organic evolution, only a single principle should be everywhere and exclusively concerned' (Romanes 1892–7, ii, p. 2). So convinced a pluralist was he that his summary of the 'general conclusions' of Darwinism, in the form of twelve propositions (Romanes 1890), led off with Darwin's statement that natural selection has not been the exclusive means of modification. He reserved a particular disdain for what he took to be over-enthusiastic adaptationism (e.g. Romanes 1892–7, ii, pp. 20–2).

In spite of Darwin's recanting his 'great oversight', he didn't embrace non-adaptationism as wholeheartedly as Romanes was later to do. But this recantation did reflect an aspect of his thinking that he stressed from time to time (e.g. Peckham 1959, pp. 232–41). On this particular occasion his doubts were triggered largely by a paper of 1865 by the highly respected botanist Charles-Guillaume Nägeli, who worked in Germany. Nägeli, who was an idealist, had urged that many characteristics of plants – the arrangement of the leaves on the axis, for example – were of no adaptive value. For several of these supposedly useless features Darwin managed to come up with evidence of hitherto unrecognised functions, such as the astonishing variety of mechanisms for pollination that he had recently found in orchids. Nevertheless, Darwin was stumped by some of Nägeli's examples; he agreed that they really weren't adaptations and couldn't be explained as the direct result of natural selection. On the more general issue of pluralism, Darwin was at first a fairly straight-and-narrow Darwinian but, as successive editions of the *Origin* testify, he became more catholic as difficulties accumulated. Summarising his theory at the end of an early edition, he says that evolution has taken place 'by the preservation or the natural selection of many successive slight favourable variations' (Peckham 1959, p. 747). By the last edition this has been expanded: '... variations; aided in an important manner by the inherited effects of the use and disuse of parts; and in an unimportant manner ... by the direct action of external conditions, and by variations which seem to us in our ignorance to arise spontaneously' (Peckham 1959, p. 747). And he adds:

as my conclusions have lately been much misrepresented, and it has been stated that I attribute the modification of species exclusively to natural selection, I may be permitted to remark that in the first edition of this work, and subsequently, I placed in a most conspicuous position – namely, at the close of the Introduction – the following words: 'I am convinced that natural selection has been the main but not the exclusive means of modification'. (Peckham 1959, p. 747)

It was not long before many others were convinced of the same. The heyday of Darwinian non-adaptationism (and pluralism, too) followed close on the heels of Romanes. During the eclipse of Darwinism, most non-Darwinians believed that the prevalence and importance of adaptation had been exaggerated; from the mid-1890s, for about twenty years, 'the neo-Darwinian selectionist-adaptationist view ... suffer[ed] its deepest decline in the entire time between the first publication of the *Origin* and the present' (Provine 1985, p. 837). We have seen how orthogenetic and mutationist theories in particular tended to deny design; after the decline of Lamarckism they were the main alternatives to Darwinian theory. By the 1930s, non-Darwinians could recite a well-rehearsed catechism of purportedly neutral characteristics to demonstrate that the scope of Darwinian explanation was extremely limited (Bowler 1983, pp. 144–6, 202–3, 215–16). This view of nature so permeated evolutionary thinking that it was even absorbed to some extent within Darwinism (and remember that at this time some naturalists were so liberally pluralist that it's difficult to decide whether they could be called Darwinians at all).

In recent years, Romanes' position has undergone a revival. The Harvard biologists Stephen Gould and Richard Lewontin, in particular, have campaigned for more pluralism, in opposition to what they regard as today's unjustifiably complacent 'panselectionism' (Gould 1978, 1980, 1980a, 1983; Gould and Lewontin 1979; Lewontin 1978, 1979; see also e.g. Ho and Fox 1988). Unlike Romanes, they don't allow their pluralism to stray beyond Darwinian bounds. But like Romanes, they cannot accept that a single mechanism, natural selection, could be responsible for the prodigious complexity and variety of living things: 'At the basis of [the view that we should introduce a multiplicity of mechanisms] ... lies nature's irreducible complexity. Organisms are not billiard balls, propelled by simple and measurable external forces to predictable new positions on life's pool table' (Gould 1980, p. 16); to be 'pluralistic and accommodating [is] ... the only reasonable stance before such a complex world' (Gould 1978, p. 268). And heading this pluralist programme is a stand against doctrinaire adaptationism.

The parallel history of the adaptationist tradition takes us back to Wallace, for he was a leading figure – a firm, even proselytising, adaptationist and a strenuous anti-pluralist. Indeed, Wallace had the distinction of being singled out by Romanes as the arch-malefactor in that most serious of 'ultra-Darwinian' crimes, zealous adaptationism. Among Wallace's followers were some of the prominent Darwinians of their day, including E. B. Poulton, zoologist and Hope professor of entomology at Oxford, and E. Ray Lankester, zoologist, Linacre professor of comparative anatomy at Oxford and later the Director of Natural History at the British Museum (e.g. Poulton

1908, pp. xliv–xlv, 106–7). This school of thought was for a long time undermined by the eclipse of Darwinism. But with the consolidation of Darwinian theory – the grand synthesis – adaptationism gradually regained its nerve. Wallace's counterpart in this later generation was no less than R. A. Fisher: 'Fisher was more thoroughly a selectionist/adaptationist than any other evolutionist before him, and perhaps any after him' (Provine 1985, p. 856). Than any after him? Well, let's not prejudge just how successful adaptationism might prove to be.

History itself shows us that it's not merely a matter of historical interest to compare what adaptationism and non-adaptationism have to offer. But before we examine the more serious arguments, let's deal with the parody adaptationist who can be constructed from non-adaptationists' criticisms. It is he (for I shall let this one be a 'he') who has made 'adaptationist' something of a dirty word.

First, he is a Panglossian. He assumes that natural selection creates perfectly designed, optimally functioning organisms. In the words of William Bateson, a leading Mendelian of his time, writing at the turn of the century: 'Those who have lost themselves in contemplating the miracles of Adaptation ... [try] to discover the good in everything ... The doctrine "que tout est au mieux" ... [is] preached ... and examples of that illuminating principle ... discovered with a facility that Pangloss himself might have envied' (Bateson 1910, pp. 99–100; see also e.g. Gould 1980; Gould and Lewontin 1979). (If Bateson sounds unduly disapproving, even for a non-adaptationist, bear in mind that the early Mendelians were generally hostile to Darwinism.)

But we have seen that, on the contrary, it is natural to Darwinian theory to avoid perfectionist assumptions. Even the natural selection of classical Darwinism doesn't perform at all like a Panglossian optimising agent. And this is all the more true of Darwinism today. Adaptationists often complain that to associate adaptationism with perfectionism is to revert to the pre-Darwinian dark ages of utilitarian-creationists (e.g. Pittendrigh 1958). And certainly (for what it's worth – they could, after all, be inconsistent) adaptationists are not generally Panglossians. Ernst Mayr, for example, places himself in the adaptationist camp whilst strongly repudiating the Panglossian view (Mayr 1983); and Richard Dawkins, a self-confessed arch-adaptationist, devotes an entire chapter of his book *The Extended Phenotype* to discussing why Darwinians should not expect perfection (Dawkins 1982, ch. 3).

Second, our straw-adaptationist stands accused of explanatory imperialism, of making grossly inflated claims for the scope of adaptationist explanations, of assuming that *all* characteristics of organisms must be of adaptive advantage. Here is Wallace, for example, making the kind of claim that has provoked resounding cries of 'Imperialism!': It is 'a necessary deduction from the theory of Natural Selection ... that none of the definite facts of organic nature, no special organ, no characteristic form of marking, no peculiarities of instinct or of habit, no relations between species or between groups of species ... can exist, but which must now be or once have been *useful* to the individuals or the races which possess them' (Wallace 1891, p. 35). And again: 'the assertion of "inutility" in the case of any organ or peculiarity which is not a rudiment or a correlation, is not, and can never be, the statement of a fact, but merely an expression of our ignorance of its purpose or origin' (Wallace 1889, p. 137). Here is Darwin in the first edition of the *Origin* with a similar declaration of faith: 'every detail of structure in every living creature ... may be viewed, either as having been of special use to some ancestral form, or as being now of special use to the descendants of this form – either directly, or indirectly through the complex laws of growth' (Darwin 1859, p. 200).

But this is not explanatory imperialism. Critics are conflating the claim that natural selection is the only evolutionary force with the claim that all characteristics of organisms must be adaptive. Romanes, for example, reconstructs Wallace as holding the following view: 'Natural selection has been the *sole means* of modification ... Thus the principle of Utility must necessarily be of *universal* application' (Romanes 1892–7, ii, p. 6; my emphasis). A century later, Stephen Gould talks about 'what may be the most fundamental question in evolutionary theory' and then, significantly, spells out not one question but two: 'How *exclusive* is natural selection as an agent of evolutionary change? Must *all* features of organisms be viewed as adaptations?' (Gould 1980, p. 49; my emphasis). But natural selection could be the only true begetter of adaptations without having begot all characteristics; one can hold that all adaptive characteristics are the result of natural selection without holding that all characteristics are, indeed, adaptive. As we've already seen, side effects, 'unintended' phenotypic by-products of adaptations, are to be expected. So are time-lags; organisms inherit adaptations not to their own environments but to those of previous generations – and the two may be crucially different. And, of course, there's pathology to be allowed for; when Darwin criticised the Kantian view that the study of biology required teleology because adaptation was all-pervasive in organisms, he cited the inheritance of a harelip or a diseased liver as counterexamples (Manier 1978, p. 54). Then there's the fact that heredity can manifest itself atypically – and perhaps in a way that's selectively neutral or deleterious – outside the organism's standard environment; Darwin mentions how some parrots change plumage colour when they are fed on fat from certain fish or inoculated with poison from toads (Darwin 1871, p. 152). So it is obvious that, even if natural selection is taken to be the sole agent of evolution, the universal claim 'All characteristics are adaptive' cannot be inherent in adaptationism. A second glance at Wallace's and Darwin's supposedly imperialist statements supports this conclusion. Neither is making a sweeping claim about the ubiquity of adaptations. Both hedge their assertions with the reservations I have mentioned and more – with rudiments, correlations, past but not present utility, *definite* facts, *special* organs, *characteristic* markings, *peculiarities* of instinct and so on.

Finally, the adaptationist is purportedly a dogmatist. He shows an 'unwillingness to consider alternatives to adaptive stories' (Gould and Lewontin 1979, p. 581). Why, critics complain, won't he ever give up? Even when his claims are not ultra-imperialist, his practice is. He refuses to consider alternative explanations except in the most marginal or trivial areas. This is sheer dogmatism. It is sterile and it blinds Darwinians to the factors that are really at work.

Many adaptationists would not deny the charge of dogmatism – though they may prefer to call it, say, 'tenacity' or 'perseverance'. But they do deny, and most emphatically, the charge of sterility. On the contrary, they assert, their approach has proved highly fruitful. Their 'dogmatism' has been vindicated by history. This spirit is nicely captured by the following typically adaptationist declaration of faith: 'I am convinced, from the light gained during even the last few years, that very many structures which now appear to us useless, will hereafter be proved to be useful, and will therefore come within the range of natural selection'. That 'adaptationist' is Darwin, in the very passage from *Descent of Man* in which we earlier found him recanting his former commitment to natural selection. He added the comment to the second edition (p. 92), published only three years after the first. The light thrown by natural selection during that intervening period must have been remarkably bright.

With Darwin's experience in mind, let's leave our parody adaptationist and turn to more serious issues. Non-adaptationism certainly raises questions that Darwinians need to consider: When is a characteristic not an adaptation? And if it is not the result of natural selection, how else might it be explained? We'll concentrate on some of the answers that, historically, have been the typical favourites of non-adaptationists.

The scrapheap of chance

Chance can't explain adaptation. But if the problem is to explain characteristics that are supposedly without adaptive value, then it could come into its own.

Indeed, chance has a natural place in Darwinian theory. In each generation the genes in a population are only a sample of those of the previous one. Natural selection obviously constitutes non-random sampling. But there is also a possibility of some genes being eliminated and others taking over not by selection but merely through sampling error. And, as with sampling errors of any kind, this possibility is increased in small populations. This idea, known as genetic drift, is a standard part of modern Darwinian thinking. It can, of course, happily be incorporated into adaptive theories, with chance gene frequencies providing the initial material on which selection sets to work. An example is what Ernst Mayr called the founder principle (Mayr 1942, p. 237). This explains how a new group of organisms could evolve by the chance geographical isolation of particular genotypes. If the fragment that breaks off from the rest of the population is very small – perhaps even just one pregnant female – then the pioneering genes are highly unlikely to be representative of the parent population.

Incidentally, genetic drift should not be confused with the neutral theory of molecular evolution (Kimura 1983). This theory also assumes that chance is an evolutionary force but it is to do with changes at the molecular level that have no phenotypic effects, not evolution in the sense that we are concerned with – adaptive change. So it is not relevant to explaining the peacock's tail, the bee's sting or any other phenotypic characteristics.

In the light of the theory of genetic drift we can see that the question is not whether chance could play a role. It's agreed that it could. But how great a role has it in fact played? And how can its influence be detected in any particular case? These have been questions of heated, at times bitter, controversy among Darwinians – even to the extent of souring relations among the founding fathers of the modern synthesis (Provine 1985a). And matters are still far from settled. But in recent decades there's been a considerable shift in thinking. At one time the shape of a petal, the pattern on a shell or any other apparently unimportant or odd characteristics might well have found themselves written off to the indifferent hand of chance: 'a tendency developed in the 1940s and 50s to ascribe to genetic drift almost any puzzling evolutionary phenomenon' (Mayr 1982, p. 555); 'in North America particularly, genetic drift was very popular. If one could not think of an obviously adaptive function for a feature, then it was put down to drift' (Ruse 1982, p. 97). Since then, however, Darwinian adaptationism has been born again. And, time after time, phenomena that had been surrendered to genetic drift have been shown to be astonishingly intricate, finely-adjusted adaptations. This is certainly not to suggest that drift has played a negligible role in evolution; its role is still controversial. But Darwinian explanations have got further in the last thirty years by challenging chance than they got previously by acquiescing in it. I'll take just one example of natural selection rescuing phenomena from the explanatory clutches of genetic drift.

The snail *Cepaea nemoralis* is common in Britain and elsewhere in Europe. Its shell may be yellow, brown or pink and may be heavily striped with black bands, more sparsely banded or have no bands at all; the frequencies vary geographically. *C. nemoralis* is not unusual. In several genera of land snails, colour and banding vary within many of the species and from one species to another. In this variability lies a notorious, century-old Darwinian dispute (see e.g. Mayr 1963, pp. 309–10). It isn't the little molluscs themselves that have so excited Darwinian interest. It is the more general issues that they have gathered in their trail. Is polymorphism within populations adaptive? What about variation from one population to another within a species? And what about species-specific characteristics – differences between closely related species that are often as minor as a single spot of colour but so reliably distinct that taxonomists can use them as diagnostic criteria? In short, what's the point of all this variability and at all these levels? Is it adaptive? Or does it have no point at all – a matter of indifference to natural selection? Darwinian opinion has been so deeply divided on these questions that it has dignified the snail with a degree of notoriety – a notoriety that, in its time and in its own way, has rivalled the place of the eye or the peacock's tail in the litany of difficulties for Darwinism. The dispute first erupted in the nineteenth century but it has rumbled on intermittently until quite recently. Even now, although it's generally agreed that unaided chance isn't the answer and that natural selection is up to something, there's no consensus about what exactly that something is; theories are almost as polymorphic as the snails themselves.

From Darwinism's earliest decades, some Darwinians felt that many species-specific differences (particularly many that systematists could rely on to classify species) should not be explained adaptively. Small differences between species, they declared, were just that – mere differences, not adaptations (see e.g. Kellogg 1907, pp. 38–44, 136, 375). The differences arose, they claimed, because speciation begins with geographical isolation (or some other cause of abrupt reproductive isolation); and if a new species was formed by the chance isolation of a section of the population, then its differences from the parent species could result from what we would now call genetic drift (not to mention a number of unDarwinian non-adaptive forces, such as orthogenetic trends or marked variation without selection). From the 1870s to the 1890s, this view was forcibly argued by Romanes, to an increasingly receptive audience (e.g. Romanes 1886, 1886a, 1892–7, ii, pp. 223–6, iii, pp. 1–40). He promoted the work of the American naturalist the Reverend John Thomas Gulick, on snails of the genus *Achatinella*, from the Sandwich (now Hawaiian) Islands (Gulick 1872, 1873, 1890). Gulick had discovered an abundance of species and varieties within a very small, and what seemed to him uniform, geographical area. Unable to find an adaptive reason for such vast diversity, he attributed it to geographical isolation without the subsequent intervention of natural selection. Henry Crampton, Professor of Zoology at Columbia University, who, from 1906 devoted himself on and off for several decades to studying Polynesian snails from the genus *Partula*, found equally prodigious variation and concluded that it had been favoured (if not entirely caused) by geographical isolation and drift (e.g. Crampton 1916, p. 12, 1925, p. 2, 1932, p. 4). In England, Cyril Diver, a highly distinguished amateur naturalist (eventually Director General of the Nature Conservancy), who began his work in the 1920s, came to similar conclusions on discovering differences that he thought must be non-adaptive between local populations of *Cepaea* (Diver 1940, pp. 323–8).

Throughout this period, as the influence of Darwinism waned, non-adaptationists – not only non-Darwinians but Darwinians, too – enlisted the snails more and more on their side. 'Some of the best-known and most spectacular taxonomic work before the evolutionary synthesis was on land snails' (Provine 1985, p. 842); and this work became some of the best-known and most spectacular evidence in favour of non-adaptationism. By the 1920s and 30s, adaptive thinking was at such a low point that many of the characteristics used for classification, in both animals and plants, from the level of varieties, to species, up to the level of the genus, were widely thought to be largely non-adaptive. This view was reinforced by the most influential textbook on systematics at that time, *The Variation of Animals in Nature* by G. C. Robson and O. W. Richards, which asserted that many specific differences were useless (Robson and Richards 1936, e.g. pp. 314–15, 366) and that a great deal of specific divergence was the result of drift (e.g. pp. 371–2); snail polymorphism was cited as a case in which no signs of natural selection could be detected (pp. 99–100, 200–1, 203–4). Not until the coming of the synthesis of the 1940s did adaptationism gradually begin to be viewed more favourably. The dawn of this change could be seen in two of the textbooks that replaced Robson and Richards: *The New Systematics*, edited by Julian Huxley (1940), and Ernst Mayr's *Systematics and the Origin of Species* (1942) (see e.g. Huxley 1940, p. 2). But, even so, Mayr's book stated unequivocally: 'There is ... considerable indirect evidence that most of the characters that are involved in polymorphism are completely neutral, as far as survival value is concerned. There is, for example, no reason to believe that the presence or absence of a band on a snail shell would be a noticeable selective advantage or disadvantage'; 'The variation in color patterns, such as bands in snails ... are, by themselves, obviously of very insignificant selective value' (Mayr 1942, pp. 75, 32). And Huxley was still so inclined to invoke drift that, in his *Evolution: The modern synthesis*, which was published two years later (1942), his explanation of snail polymorphism relied heavily on Gulick and Crampton – so heavily that, as William Provine points out (Provine 1985, p. 858), he had to correct it in the second edition, twenty years further on, stressing instead 'the inadequacy of drift and the efficacy of natural selection in accounting for local differentiation, including that of snails like *Cepaea*' (Huxley 1942, pp. xxii–xxiii).

But, although the snails were slow to emerge from their non-adaptive cover, Darwinians did begin to take a fresh look at what they had to tell. Wallace had maintained from the first that they would teach us an adaptationist lesson (e.g. 1889, pp. 131–42, 144–50). He insisted that natural selection must have been responsible for the differences that Gulick had found, even though the snails' environments might appear to us to be much of a muchness. Naturalists, he pleaded poignantly, should think themselves into the snail's shell:

> It is an error to assume that what seem to us identical conditions are really identical to such small and delicate organisms as these land molluscs, of whose needs and difficulties ... we are so profoundly ignorant. The exact proportions of the various species of plants, the numbers of each kind of insect or of bird, the peculiarities of more or less exposure to sunshine ... at certain critical epochs, and other slight differences which to us are absolutely immaterial and unrecognisable, may be of the highest significance to these humble creatures, and be quite sufficient to require some slight adjustments of size, form, or colour, which natural selection will bring about. (Wallace 1889, p. 148)

And, in the case of *Cepaea nemoralis*, this empathetic snail's-eye-view of selection pressures, pressures unnoticed by humans, has proved prophetic, even down to the intuition about exposure to sunshine. These findings were pioneered in the 1950s by A. J. Cain and P. M. Sheppard (for summaries and subsequent findings see Jones *et al.* 1977; Maynard Smith 1958, pp. 156–9, 166–8; Sheppard 1958, pp. 87–91, 94–5).

One selective force is generated by the sharp eye of the song thrush – in particular, by the fact that the best way of evading detection constantly changes. There is some evidence that the camouflage provided by the different types of shell varies from season to season and place to place; pink and brown shells are favoured in spring, for example, whereas a background of summer foliage favours yellow; unbanded shells are less conspicuous where the background is comparatively uniform, such as short turf, whereas banded shells provide better camouflage in, say, hedgerows and rough herbage. But this isn't enough to account for the high levels of polymorphism; after all, the thrush's selective predation weeds out variation. The frequency-dependent advantage of rarity (so-called apostatic selection) might sometimes be the answer. If predators have to build up a 'search image' of their prey, they may find it hard to spot a form that they have not often encountered, even though it is apparently (to us) quite conspicuous. Another selective force, as Wallace guessed, is that the different forms of snail enjoy 'more or less exposure to sunshine', although their environments look the same to us. Dark-coloured (banded) shells absorb more solar energy than light (unbanded) ones; banded snails are at an advantage in cold, shady microclimates but liable to death from heat shock in warm, sunny places. Predictably, the different types can be found in the areas where the climatic conditions suit them. But why, then, are some populations mixed? This isn't clear, but it may be significant that the different types within the population create their own microclimates, spending different amounts of time in sunlight (Jones 1982a).

Remember, however, that no demonstration of selective forces entirely precludes the intervention of drift. In spite of exacting selection pressures, the snails certainly owe some of their polymorphism to chance. The founder effect appears to have played a role, for example, when *Cepaea* rapidly recolonised the low-lying fens in East Anglia in 1948 after extensive flooding had wiped out local populations; and the same happened in newly-drained Dutch polders (Jones *et al.* 1977, pp. 128–30; see also Cameron *et al.* 1980, Ochman *et al.* 1983, 1987).

Some commentators have objected to Wallace's insistence, in the face of no evidence, that there must be some point to different numbers of bands and varying colours. John Lesch describes it as 'a rather awkward interpretation of Gulick's data' (Lesch 1975, p. 497). Gould and Lewontin pillory it as a glaring example of the hyper-adaptationist rule, 'In the absence of a good adaptive argument in the first place, attribute failure to imperfect understanding of where an organism lives and what it does ... Consider Wallace on why all details of colour and form in land snails must be adaptive, even if different animals seem to inhabit the same environment' (Gould and Lewontin 1979, p. 586); and they go on to quote the passage that we have just noted.

I was rather dismayed to find that passage of Wallace's ridiculed by Gould and Lewontin. I had long admired the very same words for their sensitive understanding of how a Darwinian who is looking for adaptive explanations might think about other creatures whose worlds are so different from our own. (And, by the way, given that Wallace's suggestion about sunlight turned out to be right because the snails create their own microclimates, it's ironic to discover Lewontin preaching to would-be adaptationists that adaptive explanations can be problematic because 'Organisms do not experience environments passively; they ... themselves determine which external factors

will be part of their niche by their own activities' (Lewontin 1978, p. 159).)
I'm not drawing the moral that Wallace was right because he turned out to be
right in this particular case. He was right because he insisted that Darwinians
should make a serious, systematic and stringent attempt to apply adaptive
principles before consigning puzzling phenomena to what Lewontin himself
has called 'the scrap heap of chance' (Lewontin 1978, p. 169).

'Strange deviations tied together'

When we talk about a gene for, say, white fur, we are picking out just one of
the gene's phenotypic effects. But that gene may also happen to cause
changes in tail-length or shape of claws. These 'unintended' phenotypic
effects are regarded as side effects of selection, side effects in this case of an
adaptation for winter camouflage. According to non-adaptationists, all sorts
of characteristics that Darwinians strive valiantly to explain adaptively may
not be adaptations at all; they may be mere side effects (e.g. Lewontin 1978,
pp. 167–8, 1979, p. 13; see also Gould and Lewontin's idea of 'spandrels',
the automatic consequences of structural features of organisms (Gould and
Lewontin 1979, pp. 581–4, 595–7)).

It may seem that such explanations aren't much of a victory for non-
adaptationism. For they rely ultimately on natural selection being at work,
albeit indirectly. And so they allow adaptationists to stress the importance of
natural selection after all:

> We should bear in mind that modifications ... which are of no service to an
> organism ... cannot have been ... acquired [by natural selection]. We must not,
> however, ... forget the principle of correlation, by which ... many strange deviations
> of structure are tied together ... [so that] a change in one part often leads ... to other
> changes of a quite unexpected nature ... *Thus a very large and yet undefined
> extension may safely be given to the direct and indirect results of natural
> selection* ...   (Darwin 1871, i, pp. 151–2; my emphasis)

Indeed, some non-adaptationists have classed such 'extensions' as underhand
adaptationist stratagems – a means of clinging on to natural selection, even in
the face of what are agreed to be non-adaptive, unselected characteristics (e.g.
Romanes 1892–7, ii, pp. 171, 268–9n).

But if we look more closely at the assumptions behind such explanations,
we'll find that they're not really at all congenial to an adaptationist frame of
mind. Think of what it would mean to claim, for example – an extreme case,
but we'll see that it has been claimed – that the stag's massive, baroque
antlers are nothing but a side effect of natural selection, that they are not

adaptations but merely automatic consequences of natural selection's other activities. The antlers reach outsize proportions without the direct intervention of selection, so the argument goes, because they are tied in with the embryological development of some characteristic that natural selection is selecting for; they come along with some adaptation as part of its biological package. As Darwin put it, explaining what he meant by 'correlation of growth': 'the whole organisation is so tied together during its growth and development, that when slight variations in any one part occur, and are accumulated through natural selection, other parts become modified' (Darwin 1859, p. 143). Such a claim could be making one of two assumptions – one highly implausible, one more reasonable.

The unacceptable assumption is that the antlers are selectively neutral, neither advantageous nor disadvantageous. Although this is obviously hard to swallow in the case of a structure so ornamental, so conspicuous, so elaborately fashioned as antlers, it might at first sight seem more plausible in the case of less flamboyant characteristics. But we shouldn't be too ready to assume that it is likely even then. After all, we do know how keen-eyed natural selection can be, how it's able to elevate what seem to us to be minutiae into matters of life and death. And there's also a more weighty consideration than adaptationist intuitions about natural selection's vigilance. It is improbable that any one of a gene's 'unintended' side effects is neutral, so it's multiplying improbabilities alarmingly to assume that all of them are. If, then, a side effects explanation is making that assumption, or anything remotely approaching it, we can pretty well dismiss its chances of being right. It's safe to assume that non-adaptationists nowadays wouldn't make such a strong assumption about neutrality. But, as we'll see when we look at sexual selection and altruism, classical Darwinians probably at times had some such notion at the back of their minds when they talked of side effects (a consequence of their failure to appreciate costs).

The other, more plausible, assumption that could lie behind the claim that antlers are mere side effects is that these 'unintended consequences' of selection are not neutral – indeed, are deleterious – but are nevertheless unavoidable. They're unavoidable because they're so tightly and irrevocably tied up with the embryological development of some adaptation, that natural selection cannot sever the links; some phenotypes that embryology has joined together, selection cannot put asunder. On this view, side effects are costly but their costs are outweighed by the adaptive advantages with which they keep company. This involves a very strong assumption about the unavailability of the variation that natural selection would need in order to prise these phenotypic effects apart, and thus the inability of natural selection to favour phenotypes that are adaptive whilst dampening down unwanted side effects. Unlike the claim about neutrality, there's certainly nothing intrinsically unreasonable in this. It suggests that natural selection is a weaker force (and developmental constraints stronger) than most adaptationists would like to believe. But how much power selection does indeed have in any individual case is an empirical question.

As snails have been to genetic drift, so antlers have been to the question of side effects. We'll stay with these impressive structures, then, for an example of what adaptationist success can look like on this question.

It's not just the extraordinary size of some antlers that has posed an adaptive puzzle but, much more, the relationship between this size and that of the rest of the body. As deer get larger, antlers generally increase not in proportion to body size but faster; the antlers of large deer are not merely absolutely larger but also relatively larger (relative, that is, to body size) than those of small deer. This relationship holds across different species of the family (Cervidae); a large species such as the reindeer has disproportionately larger antlers than a small species like the muntjac. It also holds within species; large adults have exaggeratedly large antlers compared with small adults. In the early decades of the century, in the heyday of orthogenetic theories, these excesses were held up as prime examples of orthogenetic trends, the march of inexorable evolutionary forces. It was Julian Huxley (1931, 1932, pp. 42–9, 204–44) who wrested the phenomena from these unDarwinian clutches. Huxley explained the exuberant growth as the result of allometry. An allometric relationship is a regularity between different characteristics of an organism – traditionally concentrating on regularity in size between the whole body and some of its parts, but, more recently, on regularities between structure and behaviour, too. Huxley showed that behind the orthogeneticists' vague trends was a very precise, constant proportion: when antler size is plotted against body size, with both axes logarithmically scaled, 'the points ... fall nicely along a straight line' (Huxley 1931, p. 822); and the slope is greater than one – antler size shows a positive allometric relationship to body size. Huxley suggested that this allometry was a side effect of adaptation. In his view, body and antlers are so closely linked by a common developmental mechanism that relatively large antlers are an automatic consequence of selection for large body size. The massive antlers of large species and large individuals can be thought of as overgrown versions of their smaller counterparts. He admitted that the exact 'mechanism of this relation is at present obscure' (Huxley 1932, p. 49). But one can imagine, say, a growth hormone that influences the development of both bodies and antlers; selection for increased body size might result in increased production of growth hormone and larger antlers would follow as a side effect.

Although Huxley managed to oust the alien force of innate, straight-line tendencies, he didn't manage to install natural selection as a primary cause; his was still a non-adaptive explanation. Huxley rescued the antlers for Darwinism. But could they also be rescued for adaptationism? According to some non-adaptationists, most notably Richard Lewontin, one needn't try (Gould and Lewontin 1979, pp. 587, 591–2; Lewontin 1978, pp. 167–8, 1979, p. 13): 'Although allometric patterns are as subject to selection as static morphology itself, some regularities in relative growth are probably not under immediate adaptive control' (Gould and Lewontin 1979, p. 591); 'It is ... unnecessary to give a specifically adaptive reason for the extremely large antlers of large deer. All that is required is that the allometric relation not be specifically maladaptive at the extremes' (Lewontin 1979, p. 13). Unnecessary perhaps, but only insofar as any adaptive explanation may be unnecessary. If 'allometric patterns are as subject to selection as static morphology', why single them out for special non-adaptive treatment?

And it turns out that there is, indeed, an adaptive force behind the allometry of antlers. This force, as T. H. Clutton-Brock and P. H. Harvey have shown, is competition between males for females (Clutton-Brock 1982, pp. 108–13, 119–20; Clutton-Brock and Harvey 1979, pp. 559–60; Clutton-Brock *et al.* 1980, 1982, pp. 287–9, 291; Harvey and Clutton-Brock 1983). Huxley simply logarithmically plotted antler size against body size and found a straight line. But divide the deer species into three, depending on how fiercely males compete for females, and a different picture emerges. Antler size relates to body size only because both are independently related to intensity of competition between males; they are independent effects of a common cause. The greater the competition (the degree of polygyny), the more a male invests in size of body and, even more so, of antlers. So the species that form larger breeding groups tend to have larger body size; and, because larger antlers are better weapons than smaller ones, species that form large breeding groups also have relatively larger antlers than small-group species. The antler–body relationship is revealed not to be Huxley's nice single straight line at all, but three distinct straight lines. (Huxley's allometric relationship between antler size and body size is still there within each of the three mating categories. But that is not surprising. Larger species are more polygynous, and that will be true within any one of the categories.) Antler size, then, is not merely dragged upwards in the adaptive slipstream of body size; it is an adaptation in its own right.

This selective pressure might well, by the way, explain what the extraordinarily large antlers of the extinct Irish elk were all about. These were a favourite anti-adaptationist, anti-Darwinian example of a structure too gross, too disproportionate to have been the result of natural selection, a structure that must have been controlled by an orthogenetic trend, perhaps the force that eventually took the elk to extinction. But the colossal relative size of the antlers is to be expected (and so, although at first sight it seems counter-intuitive, is their palmate shape) if the elk was polygynous and used them as a weapon in battles between males (Clutton-Brock 1982, pp. 112–13; Clutton-Brock *et al.* 1982, p. 299).

Large as they are, antlers are only a small part of the allometric success story. Brain size in great apes (Clutton-Brock and Harvey 1980), tooth size in Old World monkeys (Harvey *et al.* 1978, 1978a), testis size in primates (Harcourt *et al.* 1981) and many another allometry has yielded to adaptationist scrutiny. What's more, these adaptive analyses can show up gaps and anomalies that would otherwise have remained hidden. Harvey and Clutton-Brock cite a telling instance:

> Roger Short ... had predicted that in [primate] species where females mate with more than one male during a reproductive cycle, the males would have larger testes for their body size than in species whose females had only a single mate per cycle. When the females were promiscuous, the sperm of each male would have to compete with those of other males, and the male producing the most sperm was most likely to generate offspring.
>
> Short's prediction fitted the data beautifully, except for the proboscis monkey. This species has small testes for its body size, even though the literature records that females associate with several males. But more recent, detailed field studies have shown that the proboscis monkey is not an exception after all. Females associate with only a single male during times when they are most likely to conceive.
>
> (Harvey and Clutton-Brock 1983, p. 315)

For Huxley, allometric constants were pretty well just that: constant. Or, at least, they lay beyond the reach of natural selection, immured in the constraints of developmental processes. But why assume, when it comes to growth mechanisms, that only embryology can get its hands on the controls? What, after all, tunes the controls of embryology itself? As Richard Dawkins has pointed out: 'constants on one time scale can be variables on another. The allometric constant is a parameter of embryonic development. Like any other such parameter it may be subject to genetic variation and therefore it may change over evolutionary time' (Dawkins 1982, p. 33) – and that change may be adaptive.

Up to now, we've gone along with the standard notion of pleiotropy ('unintended' phenotypic side effects). The time has come to challenge that notion, and to challenge it adaptively. We have already done this to some extent with the idea of extended phenotypes. The thickened shell of a parasitised snail, which would otherwise be viewed as merely an unfortunate side effect of its guest's activities, could turn out to be an extended

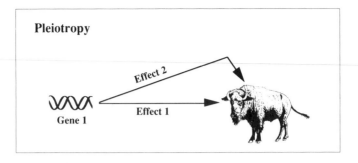

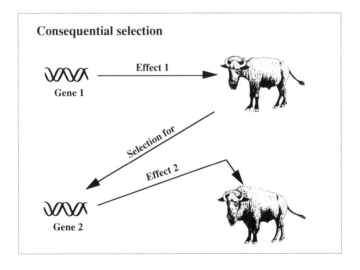

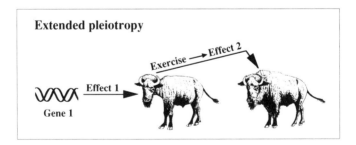

Extended pleiotropy: One of three ways for a large-headed bison to achieve a matching neck.

phenotypic effect of a gene in the parasite – not pleiotropy but adaptation. Following the style of argument that took us into the world of the extended phenotype, I should like to propose a line of reasoning that leads us into the realm of 'extended pleiotropy'. We shall find there that pleiotropic effects turn out to be more adaptive, and at the same time more common, than our usual idea of pleiotropy would lead us to suspect.

Consider, as we have been dealing with unusually large structures, the typically large head of a bison. Such a massive weight requires strong supporting muscles. By selecting for a large head, natural selection sets itself the problem of coming up with a functional match between head and muscles. How might this match be achieved?

It could be by what is standardly thought of as a pleiotropic effect. By a quirky stroke of luck, the gene for conferring larger heads could also confer larger muscles (where 'gene for' is, as always, a statement about genetic differences). But this would be an unlikely chance, no more likely than large heads going along with smaller muscles or no change in muscle size at all. It's likelier that natural selection would have to take a more active hand. As the gene pool filled up with genes for big heads, selection pressure would be set up in favour of genes for big muscles. This would be distinguishable from pleiotropy because in this case the match would be achieved only after a delay of several generations; natural selection could not make the match until the genes for big muscles just happened to crop up.

Those are two possibilities. But now let's think in terms of extended pleiotropy. A bison isn't simply endowed with neck muscles. It is also endowed with a tendency for those muscles to grow if they are exercised. So a large-headed bison will automatically tend to develop large neck muscles. Now, at first sight that doesn't sound like pleiotropy. But what is pleiotropy, after all, but the various phenotypic effects of a gene? The effect of the large-head gene on neck muscles is, strictly speaking, a phenotypic effect of that gene. In a normal environment, an environment in which a bison can exercise its head and neck normally, any individual possessing the large-head gene will tend to have large neck muscles. So that gene should be regarded as a gene for large neck muscles, just as much as a gene for a large head. If one wants to retain the category 'pleiotropic', then the enlarged neck muscles are pleiotropic. But what matters to us here is that, unlike our standard view of pleiotropy, they don't develop through a merely arbitrary, contingent, quirky embryological connection that just happens to turn out to be useful; they develop for adaptive reasons.

Now, it might be objected that the effect on head size is primary, whereas the effect on neck muscles is indirect and therefore secondary; pleiotropic effects, it might be argued, are normally direct, primary effects of a gene. But that's not so; any phenotypic effect, including that on head size, is indirect in the same kind of way (Dawkins 1982, pp. 195–7). 'Most gene effects seen by whole animal biologists, and all those seen by ethologists, are long and tortuous ... [What is] *any* genetic trait ... , morphological, physiological or behavioural, if not a "byproduct" of something more fundamental? If we think the matter through we find that all genetic effects are "byproducts" except protein molecules' (Dawkins 1982, p. 197). There is a long chain of causes and effects, hidden from us in our ignorance of embryological pathways, running from genes to proteins right up to the bison's massive head; it is only because of our ignorance of this chain that we call the head size a 'primary' effect of the gene. And it is only because we happen to know that the exercise effect is at work that we are tempted to call one particular effect of the gene, its effect on neck muscles, 'secondary'. The status of this 'secondary' link in the developmental chain is really no different from that of any other. If we knew that increased muscle size was in some way connected with the presence of the gene for increased head size, but we didn't know about the embryological details of the impact of exercise, we would simply designate the powerful muscles as a fortunate pleiotropic effect of the head-size gene, without even raising the question of whether this was only a 'secondary' effect; we would be unable to distinguish adaptive modification during an individual's lifetime from a standard pleiotropic effect. Indeed, for all we know, when the embryology of skull development is finally worked out we may discover that the effect that the gene has on head size is also a kind of developmental knock-on from some earlier, more 'primary' effect of the gene – which could perhaps, again, be 'exercise' in some sense. But even that earlier effect must be caused by something prior to it, and this, too, could be a kind of 'exercise effect'. We need, then, to extend our idea of what is pleiotropic and, at the same time, what is adaptive. Genes work in – to us, as yet – largely mysterious ways. When they seem to be throwing up mere chance connections at the phenotypic level, they may well be doing something far more adaptive; they may be using adaptive opportunities made available by natural selection.

Having subjected the idea of side effects to such relentlessly adaptive reinterpretation, I feel it is only fair to point out at least one way in which Darwinians may well be systematically underestimating the extent to which phenotypic effects are non-adaptive, underestimating the extent to which they are truly just side effects. When we talk about the side effects of a gene, we tend to pick on connections that we find intuitively plausible, passing over other possibilities because they seem less likely to be linked together. But our intuitions could well be too conservative a guide to what constitutes a side effect. Perhaps many of them are far from neat. We've noted that genes can exert their effects through all kinds of odd, unsuspected links, and it's likely that some of these connections would look bizarre to us at the phenotypic level. Are there, perhaps, side effects that we don't even think of as such because their connections with other phenotypic effects, with adaptive phenotypic effects, are hidden from us deep in embryological development? Are there perhaps genuine side effects that aren't recognised for what they are simply because they fall outside the intuitive categories of pleiotropy?

So far, our notion of a side effect has relied on pleiotropy. Pleiotropic effects arise through the intervention of embryology and development. But, a non-adaptationist might object, this idea focuses our attention too narrowly: the realm of side effects ranges far wider than 'pleiotropy' usually brings to mind. Take, for example, colour. Now, organisms must have some colour; it has to be admitted that even inorganic objects are coloured! So colour as such is not necessarily functional. It arises automatically through the workings of the laws of physics and chemistry. Perhaps, then, adaptationists search too readily for adaptive explanations of the colours of plants and animals. Given that the state of being coloured at all is merely a physico-chemical side effect, we need to be cautious about attaching significance to the particular colour that organisms happen to be. If we want to explain why, say, blood is red then we don't need to appeal to natural selection at all. The colour of blood is an incidental physico-chemical property of the haemoglobin molecule. It has no adaptive purpose. So physics and chemistry suffice. And perhaps many more characteristics of organisms are 'colour-of-blood'-like than Darwinians have suspected.

This non-adaptationist claim – that for some side effects a physico-chemical explanation is appropriate whereas an adaptive one is not – should not be confused with a claim that is often made about the purely practical difficulties of burrowing down to the appropriate reductive levels. Take, for example, the notorious problems involved in explaining us. It is hopeless, this practical argument goes, to attempt a biological explanation of a whole host of human characteristics, from altruism to divorce rates to wars; not only are we pitifully ignorant of how the relevant genes might be expressed in the unnatural environments of our modern world, but also the complexity of the phenomena may well put that detail indefinitely beyond our grasp. In principle, a biological explanation is appropriate. But in practice any attempt at so thorough a reduction would be far too ambitious. The physico-chemical side effects argument is quite different. It states that in the hierarchy of explanatory levels, natural selection is in principle at the wrong level of reduction to explain some characteristics, in principle (and this is the reverse of the practical argument) not reductive enough.

And that brings us to the difficult part. Which characteristics? Once again, we need to ask how we can tell mere side effects from The Real Thing. At first glance, common sense seems to be a good guide. But we shall see that a second glance tells a different story. Let's stay with the question of colour.

It's startling to realise that, in pre-Darwinian days, many aspects of animal and plant coloration that are now routinely regarded as adaptive were not viewed as functional at all. The responsibility for this lies, in part, with idealism. T. H. Huxley, for example, who, before he saw the Darwinian light, was strongly influenced by Continental idealism (Bartholomew 1975; Gregorio 1982; Hull 1983), denied that the colours of birds, butterflies and flowers were of any use to them:

> Regard a case of birds, or of butterflies ... Is it to be supposed for a moment that the beauty of outline and colour ... are any *good* to the animals? that they perform any of the actions of their lives more easily and better for being bright and graceful, rather than if they were dull and plain? ... Who has ever dreamed of finding an utilitarian purpose in the forms and colours of flowers ... ?   (Huxley 1856, p. 311)

Surely, one immediately wants to answer, utilitarian-creationists must have dreamed of it, must have tried to explain colour adaptively. So it's even more startling to find that on the whole they hadn't. It has been suggested that this was because, although adaptive explanations dealt admirably with drab cryptic coloration, they seemed inappropriate when it came to the gaudy and conspicuous (Kottler 1980, p. 205). There has been an attempt to show that, contrary to the subsequent priority claims of Darwinians, pre-Darwinian natural theology did incorporate a well-established tradition of explaining coloration adaptively (Blaisdell 1982). But the 'explanations' cited are so non-adaptive, so weak, so unconvincing that, albeit inadvertently, they bear out the Darwinian boast. I am not suggesting that pre-Darwinians saw coloration as the side effects of physics and chemistry. But they certainly weren't predisposed to view it as adaptive, in the way that Darwinians came to do.

Darwinism transformed naturalists' thinking on coloration. Wallace proudly singled this out as one of its greatest triumphs:

> Among the numerous applications of Darwinian theory ... none have been more successful ... than those which deal with the colours of animals and plants. To the older school of naturalists colour was a trivial character ... and it appeared to have, in most cases, no use or meaning to the objects which displayed it ... But the researches of Mr Darwin totally changed our point of view in this matter ... [H]is great general principle, that all the fixed characters of organic beings have been developed under the action of the law of utility, led to the inevitable conclusion that so remarkable and conspicuous a character as colour ... must ... in most cases have some relation to the wellbeing of its possessors. Continuous observation and research ... have shown this to be the case ...   (Wallace 1889, pp. 187–8)

Much of this success resulted from the efforts of Wallace himself, and against formidable opposition. His opponents were not confined to anti-Darwinians.

Many pluralist Darwinians thought that several distinctive aspects of coloration were non-adaptive; we've seen that species-specific differences became a major point of dispute. What's more, to Wallace's great dismay, Darwin attempted to shift much of the evidence into sexual selection. It's no wonder that, in his autobiography, Wallace picks out coloration as one of his two greatest victories in his battle to extend the scope of natural selection – indeed, this was an area in which he delighted in describing himself as more Darwinian than Darwin (Wallace 1905, ii, p. 22).

So far, so adaptive. But even Wallace was at pains to emphasise that the adaptationist should be wary of physico-chemical side effects:

Every visible object must be coloured, because to be visible it must send rays of light to our eye ... [I]n the inorganic world we find abundant and varied colours ... Here we can have no question of *use* to the coloured object, and almost as little perhaps in the vivid red of blood ... or even in the universal mantle of green which clothes so large a portion of the earth's surface. The presence of some colour, or even of many brilliant colours, in animals and plants would require no other explanation than does that of the sky or the ocean, of the ruby or the emerald – that is, would require a purely physical explanation only.   (Wallace 1889, pp. 188–9)

The green colours of foliage arise simply from the presence of chlorophyll; they are therefore 'unadaptive ... [They are] the direct results of chemical composition or molecular structure, and, being thus normal products of the vegetable organism, need no special explanation' (Wallace 1889, p. 302). In the case of blood, its colour could not have been subject to selective forces because the blood is concealed (Wallace 1889, p. 297). This, by the way, became a favourite argument of non-adaptationists; they commonly cited the coloration of microscopic organisms, of the inside of the snail's shell and of other recondite phenomena as undermining the view that colour was generally adaptive (see e.g. Bowler 1983, pp. 151, 203).

But Wallace, remember, was a committed adaptationist. So he was also interested in the question of when adaptive explanations should be employed. Pattern is one clue that colour is not merely an automatic consequence of physics and chemistry: 'It is the wonderful individuality of the colours of animals and plants that attracts our attention – the fact that the colours are localised in definite patterns, sometimes in accordance with structural characters, sometimes altogether independent of them; while often differing in the most striking and fantastic manner in allied species' (Wallace 1889, p. 189). Constancy also suggests that natural selection has been at work. Domestic selection provides independent evidence for this; colour is very constant in the wild but varies greatly under domestication, where selection pressures are lifted (Wallace 1889, pp. 189–90).

The criteria of pattern and constancy may sound so obvious and commonsensical as to be entirely uncontroversial. And they may seem to ensure clear-cut decisions in at least some cases. They certainly support our intuition that the colours of the peacock's tail require adaptive explanation whereas those of its internal organs do not. So here, surely, are some areas on which all Darwinians would agree.

Well, no. When it comes to pattern, it could be argued that distribution and intensity of colour are sometimes nothing more than the automatic outcome of physiological or structural features. In that case, one would expect the colour to be 'localised in definite patterns' and 'in accordance with structural characters'. Far from being a sign of adaptation, such coloration would be a diagnostic feature of a side effect. Wallace's criterion would be utterly misleading. We shall see that it was indeed urged by one leading nineteenth-century naturalist that the peacock's tail should be explained physico-chemically, not adaptively, on precisely these grounds. His argument admittedly now seems grossly inappropriate for the peacock's tail but it is not necessarily so in all cases. The leading naturalist, by the way, was Wallace.

The criterion of constancy has also come under attack. Take the dispute over species-specific characteristics, the characteristics used in classification, such as the distinctive flashes of colour in some species of bird. These characteristics are, of course, strikingly constant – hence their use in classification. Some non-adaptationists have reasoned that if an adaptive species-specific characteristic, say an ability to run fast, had an automatic physico-chemical side effect, say a red spot, then the red spot would be likely to remain constant as long as natural selection continued to act on running speed. And, these non-adaptationists have claimed, many a constant characteristic may well be a red spot, not an ability to run fast, and so have no adaptive value in spite of its constancy.

Pattern and constancy, then, are no guarantees against colours being side effects. Going the other way, the adaptive way, however, we shouldn't accept unquestioningly that 'not visible' suggests 'not adaptive', that just because blood is not seen, its colour can be put down to mere physics and chemistry. There are some obvious reasons for this. We typically think of an organism's colour as working on the sense organs of other organisms – camouflage, warning colours and so on. But inorganic agents also subject an organism's colours to selection – solar rays selecting for dark pigmentation, for example. And, even when we are thinking of organisms' senses as selective agents, our idea of what is visible should not stop at human vision. More generally, we should not give primacy to our human-centred idea of experience. After all, the ways in which organisms experience physical properties are highly species-specific, and the adaptive advantages of a characteristic may have nothing to do with how we humans experience them or indeed whether we experience them at all:

> Fluctuations in temperature do not reach the inner organs of a mammal as thermal signals, but as chemical signals ... Ants that forage in the shade detect temperature changes as such only momentarily, but over a longer term will experience sunshine as hunger ... [F]or bees, ultra violet light leads to a source of food, while for us it leads to skin cancer.
>
> (Lewontin 1983, p. 77; see Dawkins 1986, pp. 21–41 for a detailed example)

But, those more obvious reasons aside, there is another way in which biological function and what looks like 'side effect' coloration may be more closely connected than we generally appreciate. Consider again the redness of blood. Even so ardent an adaptationist as Wallace assumed that it is an entirely incidental property of the haemoglobin molecule, a property that can be given a physical explanation but not an adaptive one. And generations of Darwinians have trotted it out as a favourite example of a physico-chemical side effect. But perhaps it is less incidental than this exemplary status suggests. After all, adaptive function and colour are tightly linked:

> molecular resonance, visible as colour, [is] called forth by varying kinds and degrees of chemical unsaturation or unfulfilled valency. In many instances, unsaturated chromophoric groupings may impart both colour and increased reactivity or chemical instability to the same molecule. Such compounds may therefore assume more readily important biochemical roles ... or may constitute representative by-products of special metabolic processes ... *Colour and biochemical activity are, in such instances, two interlocked effects of the same fundamental molecular phenomenon.*
>
> (Fox 1953, pp. 4–5; see also p. 9; my emphasis)

Admittedly, the redness of blood as a colour visible to us is still best described as a side effect. But the properties that make it visible to us as red are intimately connected with its power of combining with oxygen and hence with its adaptive role. And such intimacy is belied by dismissing the colour as a mere side effect without further ado. The causal chain between an adaptation and the automatic workings of physics and chemistry can be shorter and less arbitrary than Wallace – and many a non-adaptationist – supposed.

Indeed, this example reinforces the argument that we have just noted, that to think of colour as being of interest to natural selection only because of its properties as a visible entity is a perception-centred prejudice. Once we think of the redness of blood not as a colour that we perceive but as molecular resonance of a particular frequency, we clearly have a property that natural selection could put to adaptive use regardless of whether it is seen. And we shouldn't automatically think of colour as something we perceive, we shouldn't automatically bring in the question of properties-as-experienced. The property that causes us to see colour could also perform other functions. The biological value of a 'colour' could, as we have seen, lie in its non-visual physical and chemical properties. The redness of blood as we experience it is certainly a side effect. But we shouldn't jump from there to a non-adaptive explanation. Natural selection may be indifferent to our experience. But it could be far from indifferent about whether blood is 'red' or some other 'colour'.

The point of all this is simply to note, once again, how subtle adaptive potential can be, how natural selection might scrutinise even 'hidden' colours, how we shouldn't let an apparently commonsensical notion like visibility of colours (particularly visibility to us) be our guide to adaptive purpose. The non-adaptationist should not feel too sanguine even about the colour of blood!

Artefacts of our minds

Up to now, the doubts about adaptive characteristics have all been about whether they are adaptive. But, when it comes to deciding what's an adaptation, there's also room for doubt about the characteristics themselves. It's all very well, a non-adaptationist might say, for a Darwinian to claim that some feature requires an adaptive explanation. But how does one decide what constitutes a feature in the first place? Nature doesn't come to us neatly marked out like a paint-by-numbers kit or a phrenologist's model skull. We have to divide up the organism before we can explain it. There's an analysis to be made before we get on to explanation. And if the descriptions resulting from that analysis are not right, what we're trying to explain may be no more than a mental construct, an artefact of our minds. This problem has been raised by Richard Lewontin:

> how [is] the organism ... to be cut up into parts in describing its evolution [?]. What are the 'natural' suture lines for evolutionary dynamics? What is the topology of phenotype in evolution? What are the phenotypic units of evolution?
> (Lewontin 1979, p. 7)

> The dissection of an organism into parts, each of which is regarded as a specific adaptation, requires ... [an] a priori [decision] ... [O]ne must decide on the appropriate way to divide the organism ... Is the leg a unit in evolution, so that the adaptive function of the leg can be inferred? If so, what about a part of the leg, say the foot, or a single toe, or one bone of a toe?   (Lewontin 1978, p. 161)

Or, one may add, what about some apparently even more arbitrary unit, such as the-shin-together-with-part-of-the-calf? Some attempts at adaptive explanation are misguided, Lewontin maintains, because the entity in question simply isn't an adaptive unit (Gould and Lewontin 1979, p. 585; Lewontin 1978, pp. 161–4, 1979, p. 7).

So when is an 'adaptive unit' really an adaptive unit? When is a category that's seen by us, seen by nature, too? The answer must be: When it's a unit that selection can work on. For classical Darwinism this would have been difficult to specify precisely. But for modern Darwinism, a unit is obviously a gene and the ramifying tree of all its phenotypic effects (in comparison with alternative forms of the gene, its alleles). If it should turn out that the bone of a toe and the shape of an eyebrow are pleiotropic effects of the same gene, then that bizarre combination is a respectable adaptive unit. Natural selection works on genetic differences in populations. If a genetic change that lengthens the bone also curves the eyebrow, then our adaptive explanation should recognise that; we should be interested in the genetic differences that give rise not merely to differences in toe-length but to differences in toe-length-plus-eyebrow-shape, even if eyebrow shape should turn out to be selectively neutral.

This is an answer that would not have been obvious to the organism-centred view of classical Darwinism but comes readily to a theory that is gene-centred. The question of adaptive units is a question about links between phenotypes. A gene-centred analysis tells us how to make these links. And, in so doing, it reminds us once more how arbitrary is our distinction between the adaptive effects of a gene and the pleiotropic side effects of that same gene. It is our distinction, and in many contexts a very useful one. But it is not one that is respected by natural selection, and we should not allow it to mislead us when the context is not what interests us but what interests natural selection.

That solution is all very well in principle. But unfortunately it is not, of course, of much help in any individual case (unless – which is highly unlikely – we can trace all the phenotypic effects of the relevant genes). And so we are as likely as ever to manufacture artefacts inadvertently and to set ourselves unsolvable puzzles. Indeed, the way is open for non-adaptationists to conjure up stubborn cases with alarming ease, cases that could put adaptationists permanently on the defensive. An adaptationist who could successfully explain why the leopard has spots and why they are their characteristic colour might run out of steam very rapidly if asked what advantage there was in eighty spots, as opposed to seventy-nine or ninety-one.

Might run out of steam. But I can't help thinking that if that adaptationist was the Israeli zoologist Amotz Zahavi, he'd have a ready reply and one that, whether right or wrong, would ensure that the leopard's spots, however he'd

The best way to introduce my theory is to give a simplified example; for this purpose I will use the advertisement of a disc. Imagine that you have a set of discs, all of which are more or less circular but some of which are more circular than others. Imagine also that you are the judge in a competition to evaluate the quality of the discs. High quality discs are perfectly circular, with less circular ones being of lower quality. Now, because of the limitations of your senses you may experience great difficulty in deciding just how perfect one particular disc is. But a dot in the centre of the disc may well help you to assess the disc's circularity, and make it easier for you to separate a perfect disc from one that is only nearly perfect. (This effect is shown Figure 1.)

If a dot in the centre helps the judges to assess the circularity of a disc, it will be to the advantage of a maker of perfect discs to put a dot in the centre of the discs. If the judges then decide to use the dot to discriminate in favour of perfect discs we have a coalition between the perfect discs (or their makers) and the judges as a result of which both benefit–the perfect discs because the judges

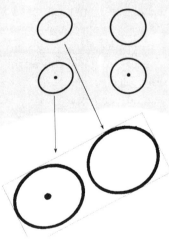

Figure 1 See how a dot helps you determine easily how perfect the circle is

Honest advertisement?

The handicap principle as illustrated in Zahavi's *Decorative patterns and the evolution of art.* But art can fake handicaps even if nature can't. It's not surprising that the circle with the dot looks less perfect: it *is* less perfect!

divided them up, never looked quite the same to us again. Actually, Zahavi has done what amounts to just that. Casting an adaptationist eye over striking markings like the leopard's spots and the zebra's stripes, he has, indeed, redrawn the suture lines of adaptive explanation (Zahavi 1978). Why, Zahavi asks, does an animal have the particular pattern it does, with its particular detail, and not another? Patterns are often explained as signals. But the connection between pattern and signal is generally thought to be arbitrary or, at most, based on some simple physiological effect, such as dazzling with a welter of lines. The zebra's stripes are generally thought to be for confusing predators or camouflage. But, as Zahavi points out, this can't explain why the stripes are placed precisely where they are. Suppose, however, that the zebra is using its stripes to advertise its quality to others. Suppose, say, that it is trying to let predators or potential mates know that it is large, muscular, sturdy or long-legged. In that case, the stripes will be strategically placed in such a way as to emphasise those very qualities; natural selection will be using '*particular* patterns to signal *particular* messages' (Zahavi 1978, p. 182). Zahavi pushes us into drawing new lines around adaptive characteristics.

Actually, he also invites us to do more. He applies a typically counter-intuitive idea of his, which has come to be known as 'the handicap principle', and which we'll be meeting again in both 'The Peacock' and 'The Ant'. Zahavi claims that, far from using stripes cosmetically, to hide and disguise deficiencies, to make its legs look longer or its muscles bigger than they really are, the zebra is potentially handicapping itself by using patterns that would show up inadequacies if it suffered from them, arrangements that would positively draw attention to them if they were there. What the zebra is doing is showing that it is big enough or muscular enough or long-legged enough to be able to afford to be honest about these qualities. 'An animal with a long neck may display the length of it by having a handicapping ring around the neck. Individuals with short necks will look even shorter-necked: "My neck is so long I can even afford to make it look short"' (Zahavi 1978, p. 183). So, not only Zahavi's idea of explaining *particular* patterns, but also his handicap principle invite us to redraw explanatory boundaries. All kinds of features that were previously overlooked or brushed aside as too odd or too costly to be the result of natural selection suddenly become at least plausible candidates for adaptive explanation.

That brings us to the end of this chapter. But I shouldn't like, by finishing here, to let it take its tone from such a startlingly unorthodox note (although, as we'll see, Zahavi's theory is gradually being thought to earn its stripes). The general point has been to illustrate how resourceful and subtle a tactician natural selection can be – even if not as perversely resourceful and subtle as

Zahavi supposes. Once this is appreciated, non-adaptive explanations cannot be treated as other than a last resort. And resolute adaptationists can be confident that 'The use of each trifling detail of structure is far from a barren search to those who believe in natural selection' (Darwin 1862, pp. 351–2).

PART TWO

The Peacock

5

The sting in the peacock's tail

Flying in the face of natural selection

At one time the eye, in its apparent perfection, gave Darwin cold shudders. The peacock's tail came to pose an even greater threat to his peace of mind: 'The sight of a feather in a peacock's tail, whenever I gaze at it, makes me sick!' (Darwin, F. 1887, ii, p. 296). For a Darwinian, that splendid tail has a sting in it. The eye is at least highly advantageous; nobody would question whether it is of benefit. But the peacock's tail is an extravaganza – flamboyant, bizarre, exaggerated, ornamental, apparently of no earthly use and actually damaging to its over-burdened bearer. And worse, 'peacocks' tails' abound throughout the animal kingdom. In species after species, particularly among birds and insects, the females are economically and sensibly dressed, obeying Darwinian dictates, whereas the males flagrantly flout the rules, flying in the face of natural selection and going in for gaudy colours, baroque ornamentation or elaborate song and dance routines. The peahen could have been designed by a hard-headed, cost-cutting engineer; her mate could have stepped off the set of a Hollywood musical.

The difficulty that such phenomena pose for Darwinism is obvious. What *good* is the peacock's tail? How can it possibly help him or his offspring in the Darwinian struggle? Indeed, how can it do other than hinder him? Darwin came to the conclusion that natural selection really was powerless to account for such apparently pointless splendour. His solution was his theory of sexual selection. He held that male ornamentation evolved simply because females prefer to mate with the best-ornamented males. This obviously gives these males a mating advantage and, ultimately, the likelihood of greater reproductive success. Thus, over evolutionary time, males develop ever-more exaggerated, immoderate flamboyance.

Darwin took sexual selection to cover any features that affected reproductive advantage over members of the same sex. This included direct rivalry among males for mates – threats, combat, and the weapons that accompany them. Unlike female choice, this form of sexual selection was thought to be easily assimilated by classical Darwinism; it appeared to call for characteristics – strength, sharp claws, a quick response – that natural selection would anyway favour. So this aspect of Darwin's theory was taken to be uncontentious (e.g. Groos 1898, pp. 229–30; [Mivart] 1871; Wallace 1905, ii, pp. 17–18) and played no part in the controversy over sexual selection. As Darwin said: 'Most ... naturalists ... admit that the weapons of male animals are the result of sexual selection – that is, of the best-armed males obtaining most females and transmitting their masculine superiority to their male offspring. But many naturalists doubt, or deny, that female animals ever exert any choice, so as to select certain males in preference to others' (Darwin 1882: Barrett 1977, ii, p. 278). This attitude – accepting direct male competition but rejecting female choice – predominated throughout most of the theory's history. We shall be looking at the controversy rather than the consensus. Female choice and male competition raise quite distinct theoretical issues. Notwithstanding the confident claims of Darwin's contemporaries, classical Darwinism was not able to explain why male rivalry often results in weapons so non-utilitarian that they appear not to be weapons at all. Why on earth should a peacock be threatened by another's tail? Claws and teeth, yes; feathers and song, no. But we shall examine this problem – conventional competition – under altruism. Here we shall concentrate on what most concerned Darwin and his critics: striking male ornaments and Darwin's claim that female choice was the selective force that shaped them.

Sexual selection is not, however, tantamount to female choice (or, more generally, mate choice; in some species the dimorphism is reversed – it is the female who has the 'peacock's tail'). Mate choice is certainly a crucial component of it. All sexual selection involves mate choice. (Remember that we're excluding direct male rivalry.) But not all mate choice gives rise to sexual selection. For sexual selection to occur, mate choice must, among other things, act as a selective force; it must bring about differential rates of reproduction favouring those individuals that bear the preferred characteristics (and that differ genetically in this respect from others of their sex). Assortative mating (the mating of likes or of unlikes), for example, depends on mate choice but does not necessarily give rise to a mating advantage and therefore selection.

Neither is sexual selection tantamount to the evolution of mating systems. Rather, the mating system affects, and is affected by, the action of sexual selection. Just think, for example, of how much more potential for sexual selection is offered by polygyny (several females mating with one male) than – all other things being equal – by monogamy.

Indeed, how does female choice manage to act as a selective force at all in monogamous species? How do the best ornamented males manage to achieve greater reproductive success than others if all males find a mate? Darwin knew of bird species, such as the British wild duck, bullfinch and common blackbird, in which the males looked typically sexually selected, and yet these species were monogamous. He rightly saw that they posed a problem for his theory (Darwin 1871, i, pp. 260–71, ii, p. 400). His answer was that attractiveness and reproductive success in males are connected through a link between early breeding and reproductive success in females. The females who are ready to breed the earliest become ready, he claims, because they are the best nourished and therefore the healthiest – and the healthiest obviously tend to have the greatest breeding success; so those males who mate earliest will also tend to have the greatest breeding success – and these, of course, will be the most attractive males. It seems that Darwin was right that sexual selection could operate under these conditions. R. A. Fisher (1930, pp. 153–4) pointed out that the female tendency to breed early would have to be non-hereditary (resulting from, say, variations in the food supply); otherwise there would be selection for ever-earlier breeding, rather than the stability in breeding dates that actually occurs. And he demonstrated quantitatively, albeit very briefly, how Darwin's theory might work. Recently, more detailed mathematical analysis has confirmed the Darwin-Fisher conjecture (Kirkpatrick *et al.* 1990).

Although sexual selection is about the evolutionary consequences of female choosiness, it is not about the ultimate evolutionary cause of that choosiness. Darwin provided no satisfactory solution to the question of why the females are generally the choosiest and why, indeed, there is choosiness at all. His rationale was a spurious argument (it's not often one can say that of Darwin!) – that nature's general rule is for the sperm to be carried to the egg but not vice versa, thus turning the males into indiscriminate searchers and the females into discriminating choosers (Darwin 1871, i, pp. 271–4; Darwin, F. and Seward 1903, ii, p. 76; see also Kottler 1980, p. 214, n60 for an unpublished letter from Wallace to Darwin).

Modern Darwinism recognises that choosiness results from a far more fundamental difference between the sexes (see e.g. Dawkins 1976, 2nd edn., pp. 300–1). Imagine a population in which there is sexual reproduction, but think away all the peacocks' tails, female choosiness, and everything else that makes the sexes asymmetrical. The only condition imposed by sexual reproduction is that matings have to be between the two different kinds of organism that make up the population – say, Blues and Pinks. Why should we expect choosiness to evolve? Think of any individual's reproductive effort as going into some trade-off between competing for mates and caring for

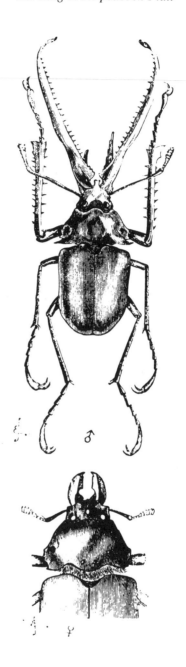

offspring. Now imagine that, among the Blues, mate competition happens to make a greater difference to reproductive success than caring for offspring; the gap between the most and the least reproductively successful Blue is established more by competing than by caring. And among the Pinks the opposite is true: being a good parent makes more difference to reproductive success than competing for mates. Blues, then, will get more return out of putting their effort into competing for Pinks than they will get out of parental care. And Pinks will benefit themselves more by investing their efforts in their offspring than by scrambling for mates. What is important is that these tendencies are self-reinforcing: once the Blues and the Pinks start to diverge, the divergence will escalate. The more that Blues divert their resources into mate competition rather than into being parents, the more it pays them to devote themselves even more singlemindedly to the task; a bit more effort spent in competing for mates could make a substantial difference to reproductive success, whereas however much effort a Blue puts into caring for offspring that effort will make a negligible difference between it and the next Blue. And vice versa for Pinks: the more that each generation lavishes reproductive resources on its offspring rather than on mating, the more, in successive generations, it becomes worthwhile to do so. Admittedly, we have built in an initial difference between the sexes. But, because the process is self-reinforcing, that initial difference can be very small and still the sexes will diverge into mate-competition-investors and parental-care-investors. So the whole thing could start from some minor chance fluctuation. Thus, even if Blues and Pinks started out alike, as soon as any difference arose in their reproductive investment strategies, it would be amplified into the kind of difference familiar to us as 'male' and 'female'. That, then, is why peacocks are more interested in impressing their rivals, in growing fine tails, and in competing fiercely for any female they can get, than in caring for their offspring. And why peahens don't bother much about rivalry but are very choosy about who fathers their offspring.

Elegant males, dowdy females

Chiasognathus grantii (upper figure male, lower figure female) (from Darwin's *The Descent of Man*)

"The great mandibles of the male Lucanidae ... are so conspicuous and so elegantly branched, the suspicion has sometimes crossed my mind that they may be serviceable to the males as an ornament ... The male Chiasognathus grantii of S. Chile – a splendid beetle ... has enormously-developed mandibles; he is bold and pugnacious; when threatened on any side he faces round opening his great jaws, and at the same time stridulating loudly; but the mandibles were not strong enough to pinch my finger so as to cause actual pain." (Darwin: The Descent of Man)

The career of a controversy

Darwin elaborated his theory in his *Descent of Man* in 1871. It immediately aroused considerable interest, not least disagreement. And it continued to do so until a few years after Darwin's death in 1882. Gradually, however, the theory came to be misunderstood and distorted – and, increasingly, neglected, underrated, ignored. Not until a century after the publication of *Descent of Man* did it start to be fully appreciated. Now, at last, it has been assimilated to mainstream Darwinian thinking. Indeed, it is undergoing a spectacular revival, having become a growing, lively, even fashionable area of research. A happy ending, then – so far, at least – to a chequered career.

What interest can these twists and turns of fate hold for us today? Well, for one thing, the earlier debates can help us to understand the modern science, for they anticipate present positions in unexpected ways. These historical continuities help us to see how the various currently competing theories of sexual selection relate to one another, and to view today's concerns (and what preceded them) in a new light.

The earlier discussions also touched on a number of questions that are only now being answered or are still being explored. In dealing with sexual selection modern Darwinism has in some ways been less successful than with the problem of altruism. We shall see that biologists can now explain, at least in principle and often in particular cases, why a bee forgoes reproduction and devotes her life to caring for her sisters or why a ground squirrel puts herself at risk to give warning cries. But how did the peacock acquire his flamboyant tail or the bower bird his predilection for decoration? Although the debate has advanced immensely and very excitingly since Darwin's and Wallace's time, many of their own questions, both theoretical and empirical, are no less pressing today.

What is more, sexual selection emerges as a telling case study for Darwinism as a whole. Its dramatic reversals of fortune reflect issues that have woven their way through Darwinian science for over a century – what an adaptive explanation should look like, for example, or where the limits to natural selection lie. And this history brings home to us how much has been gained from the revolution of recent decades, how ingenious are the solutions that have been found to some of the most acute problems of nineteenth-century Darwinism.

Finally, the history of sexual selection helps to remind us of the true magnitude of Darwin's achievement. In spite of biologists' renewed interest in sexual selection, historians and philosophers of science have paid it relatively little attention. It is barely mentioned in general histories of Darwinism; of the five listed by Michael Ruse (1979a) as being up to that time the standard general works, one (Eiseley 1958) does not mention sexual selection at all and the others (de Beer 1963; Greene 1959; Himmelfarb 1959; Irvine 1955) include only the most cursory of discussions, two of them being confined to sexual selection in humans and only one of them going beyond the debates in Darwin's lifetime. Ruse himself adds no more than a few brief comments to the history. Peter Bowler's *Evolution* (1984) – admittedly a general history of evolution rather than of Darwinism – grants the topic one paragraph. And a standard text on the history and general influence of Darwinism (Oldroyd 1980) ignores it entirely. The subject has been accorded a book of readings (Bajema 1984) but they stop at 1900 (although a volume of twentieth-century readings is promised). The history has also been treated to some extent in the more specialised literature but even there it is still a cottage craft compared with the vast outpourings of the rest of the Darwin industry.

Paradoxically, much of the debate about sexual selection, from the nineteenth century to the present day, hasn't been about sexual selection at all but about natural selection. Well, as we'll see, there's really no paradox. The issues raised by sexual selection, throughout the theory's history, have fallen into two categories. The first is the question of whether sexual selection is required to account for the phenomena at all or whether they can be explained instead by the standard forces of natural selection alone. For nearly a century, the majority of Darwinians saw this as the major issue. They sought almost any alternative to sexual selection, and it was on natural selection that they relied above all. The second category of questions concerns mate choice – in particular, the reasons for making the choice and how, or even whether, Darwinian forces could allow them to evolve. These questions were raised from the first, but it is only relatively recently that the role of mate choice has become the main focus of attention. It is now a flourishing line of research – and an enormously fruitful line it has proved to be.

The major nineteenth-century critic of sexual selection was Wallace. Indeed, according to Romanes: 'to consider the objections which have been brought against the theory of sexual selection ... is virtually the same thing as saying that we may now consider Mr Wallace's views upon the subject' (Romanes 1892–7, i, p. 391). Wallace pursued both lines of criticism, but he concentrated on the first – reducing sexual selection to the struggle for existence. He believed that sexual selection was not a 'proper' selective force and that by introducing it into Darwinian theory Darwin was fostering a grossly unDarwinian heresy. As Wallace said in the Preface to *Darwinism*:

my whole work tends forcibly to illustrate the overwhelming importance of Natural Selection over all other agencies ... I thus take up Darwin's earlier position, from which he somewhat receded in the later editions of his works ... Even in rejecting that phase of sexual selection depending on female choice, I insist on the greater efficacy of natural selection. This is pre-eminently the Darwinian doctrine, and I therefore claim for my book the position of being the advocate of pure Darwinism.   (Wallace 1889, pp. xi–xii)

Although Darwin and Wallace came to disagree strongly over sexual selection, they were not at first seriously divided over the central issue, female choice. Their divergence, although sharp, was largely confined to other questions about sex differences in coloration (see Kottler 1980). Their discussions, preserved in their correspondence, occurred mainly between 1867 and 1868, with a brief resumption in 1871. It was only from about 1871, after Darwin published the full-dress version of his theory, that Wallace began to marshall his major criticisms of the idea that female choice was an important evolutionary force; some of his strongest objections were not published until after Darwin's death. So, sadly, part of the Darwin–Wallace 'debate' over sexual selection wasn't really a debate at all.

Let's now turn to this debate. We'll begin with the attempt to assimilate sexual selection to the struggle for existence. (The principal sources for Darwin's and Wallace's own statements on sexual selection are as follows. Darwin set out his theory in *The Descent of Man* (1871, i, pp. 248–50, 253–423, ii, pp. 1–384, 396–402; the second edition (1874) is extensively revised throughout but there are no important changes to the theory; from 1877, reprints of this edition include (pp. 948–54) a paper from *Nature* (Darwin 1876a)). Incidentally, Darwin's ultimate concern in *Descent* is to apply his theory of sexual selection to the evolution of human races; the peacock's tail is partly just a means to this goal (see e.g. Darwin 1871, i, pp. 4–5, chs. 7, 19–21, 2nd edn., p. viii; Darwin, F. 1887, iii, pp. 90–1, 95–6; Darwin, F. and Seward 1903, ii, pp. 59, 62, 76). Darwin also published two brief papers on sexual selection after the second edition of *Descent* (1880, 1882). For references to sexual selection in the first edition of the *Origin* see pp. 87–90, 156–8, 468; for subsequent editions see Peckham 1959, pp. 173–6, 305–8, 367–72, 732. For the correspondence between Darwin and Wallace (and others) see Marchant 1916, i, pp. 157, 159, 177–87, 190–5, 199, 202–5, 212–17, 220–31, 256–61, 270, 292, 298–302; Darwin, F. 1887, iii, pp. 90–6, 111–12, 135, 137–8, 150–1, 156–7; Darwin, F. and Seward 1903, i, pp. 182–3, 283, 303–4, 316, 324–7, ii, pp. 35–6, 56–97. For Wallace's publications on sexual selection see his review of Darwin's *Descent* (1871); three essays written in the 1860s and 1870s, revised and reprinted in two collected works (1870, 1878) and finally in his *Natural Selection and Tropical Nature* (1891, pp. 34–90, 118–40, 338–94) (the first of these being on coloration only and the other two on both colour and sexual selection); his *Darwinism* (1889, pp. 187–300b, 333–7) (pp. 268–300b being on sexual selection and coloration, the rest being on coloration alone); Wallace 1890a; Wallace 1892; and his autobiography (1905, ii, pp. 17–20).)

6

Nothing but natural selection?

'The advocate of pure Darwinism'

Darwin strove hard to encompass a vast range of previously unrelated phenomena – colours, feathers, songs, dances – within the category of 'sexually selected'. In his wake, generations of Darwinians strove hard to dismantle that same category. For nearly a century, most of the work on sexual selection amounted to a concerted attempt to dispose of it entirely, and to rely on the more sober, utilitarian forces of ordinary natural selection to deal with Darwin's splendid array.

This demolition project began with Wallace. We shall see that, although he increasingly rejected the idea of female choice as a selective force, he didn't ever reject it entirely. Nevertheless, he tried to dispense with it as far as possible. He aimed to show that most 'ornaments' had been selected not because of female preference but because they were useful in other aspects of life. With his particular interest in colour, his main target was sexually dimorphic coloration. But he also touched on ornamental structures. On the sounds and scents that Darwin claimed were sexually selected he had little to say, even though Darwin thought that musical instruments in insects, for example, constituted striking evidence and Wallace himself had taken much the same view before he came to reject Darwin's position (Darwin, F. 1887, iii, pp. 94, 138; Wallace 1871).

As we have noted, Wallace's work on coloration made an outstanding contribution to Darwinism and he understandably took great pride in having pulled into Darwinian territory all sorts of phenomena that had previously been regarded as non-adaptive. Of all nature's coloration, the beautiful colours that Darwin explained by sexual selection had particularly been singled out as of no adaptive value. The standard account had come from natural theology, which maintained that this gorgeous assemblage had been created solely for the sake of its beauty in human eyes or those of its creator (see e.g. Wallace 1891, pp. 139, 153–6, 339–40). This allowed God's guiding hand to be slipped in even when no utilitarian purpose could be discovered. It is a mark of Wallace's great achievement that he could sweep up much of this evidence and much else about the colours of animals and plants into two Darwinian categories, protection and recognition (or attraction of pollinators in the case of plants), and thereby explain them adaptively. Wallace's views were for the most part developed independently of Darwin's work on sexual selection and some of them predated it. So Darwin's vast kaleidoscope of ornaments presented Wallace with a challenge to his explanatory scheme.

Under the category of protection Wallace could explain not only cryptic colours but also, less obviously, many instances of conspicuous coloration. These were broadly of two kinds. First, there were colours that might appear to be conspicuous but were actually cryptic in the animal's natural environment; he claimed that the zebra, tiger and giraffe, for example, merged into the background in their natural habitats (Wallace 1889, pp. 199, 202, 220, 1891, pp. 39, 368). Second, there were the conspicuous warning colours adopted by inedible creatures and their mimics. The other category, recognition, covered colours that enabled animals to recognise conspecifics; they helped members of social species to keep together and helped individuals to identify potential mates. This category, too, covered some conspicuous colours, such as the bright flash marks sported by many species of bird. These explanations of coloration may not always have been correct in detail but both at the time and subsequently they were broadly successful and they became standard Darwinian lines of thinking. This, then, was the main approach that Wallace took to 'sexually selected' coloration. Let's see how it fared.

Coloration for protection

A major problem in applying the principle of protection to 'sexually selected' colours is the need to explain why males and females look so unalike. Wallace's explanation was that they were subject to different selection pressures. We shall see that on the whole his approach works admirably for the females' muted colours but fails miserably when it comes to the males' bright hues – the very phenomenon that Darwin's theory is attempting to explain. Whereas Darwin asks 'What selective pressures cause the males to be brightly coloured?', Wallace turns this problem on its head, concentrating on the question 'What causes the females not to be brightly coloured?' We'll come back to his justification for this way of looking at things and his failure to answer Darwin's question. Let's look first at his success in dealing with the female side of the dimorphism.

Wallace argues that the female's need for protective coloration is greater than the male's because of her role in reproduction (1871, 1889, pp. 277–81,

The zebra's puzzling stripes: No black-and-white solution

Darwinians are divided over how the zebra got its stripes: individual recognition, grooming orientation, tsetse fly crypsis, thermo-regulation, handicap … ? Darwin and Wallace were also characteristically at odds:

"It may be thought that such extremely conspicuous markings as those of the zebra would be a great danger in a country abounding with lions, leopards and other beasts of prey; but it is not so. Zebras usually go in bands, and are so swift and wary that they are in little danger during the day. It is in the evening, or on moonlight nights, when they go to drink, that they are chiefly exposed to attack; and Mr Francis Galton, who has studied these animals in their native haunts, assures me, that in twilight they are not at all conspicuous, the stripes of white and black so merging together into a gray tint that it is very difficult to see them at a little distance." (Wallace; Darwinism)

"The zebra is conspicuously striped, and stripes on the open plains of South Africa cannot afford any protection. Burchell in describing a herd says, 'their sleek ribs glistened in the sun, and the brightness and regularity of their striped coats presented a picture of extraordinary beauty, in which probably they are not surpassed by any other quadruped'. Here we have no evidence of sexual selection, as throughout the whole group of the Equidae the sexes are identical in colour. Nevertheless he who attributes the white and dark vertical stripes on the flanks of various antelopes to sexual selection, will probably extend the same view to the … beautiful Zebra." (Darwin: The Descent of Man)

1891, pp. 78–82, 136–8). And, not surprisingly, he gathers some impressive evidence in support of his claim.

He focuses primarily on birds. The female's drabness, he says, can be explained by her need for protection whilst incubating the eggs: 'To secure this end all the bright colours and showy ornaments which decorate the male have not been acquired by the female, who often remains clothed in sober hues' (Wallace 1889, p. 277).

Often, but not always. Wallace cites two classes of apparent counterexamples. Sometimes both sexes are brightly coloured; and sometimes the females are bright and the males dull. But, Wallace is quick to point out, these 'very curious and anomalous facts ... fortunately serve as crucial tests' and 'can be shown to be really confirmations of the law' (Wallace 1891, pp. 131–2).

It's not rare for both male and female to be brightly coloured. But Wallace discovered that, in the cases he investigated, the nests were always concealed: 'When searching for some cause for this singular apparent exception to the rule of female protective colouring, I came upon a fact which beautifully explains it; for in all these cases, without exception, the species either nests in holes in the ground or in trees, or builds a domed or covered nest, so as completely to conceal the sitting bird' (Wallace 1889, p. 278; see also Wallace 1891, p. 124). As for the much rarer case of reverse dimorphism in coloration, there is an even more striking correlation, for the burden of incubation is also reversed: 'There are a few very curious cases in which the female bird is actually more brilliant than the male, and which yet have open nests ... [B]ut in every one of these cases the relation of the sexes in regard to nidification is reversed, the male performing the duties of incubation' (Wallace 1889, p. 281). (At one point Wallace was persuaded by Darwin's view that the difference in colour was too slight to afford greater protection (Wallace 1891, p. 379) but he eventually reverted to his original belief (Wallace 1889, p. 281).) And Wallace finds many additional telling correlations to support his view that protection is the selective force. In the *Megapodidae*, for example, an unusual family of birds that do not incubate their eggs, the sexes have the same coloration (some species being dull, some conspicuous) (Wallace 1891, p. 128). Wallace stresses that few of these correlations had been explained or even systematically noted until he investigated the evidence in the light of his theory of coloration for protection (Wallace 1891, pp. 81, 131–2).

Wallace goes too far in claiming that there were *no* exceptions to these rules; he does list some apparent counterexamples (Wallace 1891, pp. 133–5). But they don't greatly undermine his case. Only a few are what he calls 'positive' exceptions – bright female and open nest (as opposed to the 'negative' exceptions of dull female and concealed nest) – and on the whole he manages to deal with most of both the 'positive' and 'negative' cases. He shows, for example, that the bright female is protected in some other way or that what appears conspicuous may really be protective in the natural environment. So Wallace manages to establish a very plausible connection between coloration and type of nest.

Butterflies are another class of creatures that exhibit strikingly dimorphic coloration. Wallace again deals with it by stressing the female's need for protection, in this case whilst depositing her eggs: 'anyone who has watched these female insects flying slowly along in search of the plants on which to deposit their eggs, will understand how important it must be to them not to attract the attention of insect-eating birds by too conspicuous colours' (Wallace 1889, p. 272). And Wallace elaborates several lines of evidence. The females are as conspicuous as the males, for example, in species that gain protection by distastefulness and advertise it to predators by bright coloration (Wallace 1889, pp. 273, 278, 1891, p. 137). What's more, Wallace again manages to turn apparent counterexamples of reverse dimorphism into supporting evidence. First, showy colours can sometimes provide excellent camouflage. He cites one species, *Adolias dirtea*, in which the female has striking yellow spots; when she is seen in the collector's cabinet they make her as conspicuous as the male, but in the dappled forest sunlight of the natural habitat the creature's 'yellow spots so harmonise with the flickering gleams of sunlight on the dead leaves that it can only be detected with the greatest difficulty' (Wallace 1889, p. 271). Second, as Darwin himself admitted (Darwin 1871, i, pp. 394–5; Darwin, F. and Seward 1903, ii, p. 67), conspicuous patterns on the tip of the wing can be protective by attracting predators to that area instead of the body (Wallace 1891, p. 371). Third, in several reverse cases or cases of equally showy but sexually dimorphic coloration, the female is gaining protection by mimicking the bright warning colours of an inedible species (Wallace 1891, pp. 78–80, 136–8). In *Diadema missippus*, for example,

the male is black, ornamented with a large white spot on each wing margined with rich changeable blue, while the female is orange-brown with black spots and stripes – we find the explanation in the fact that the female mimics an uneatable Danais, and thus gains protection while laying its eggs on low plants in company with that insect. (Wallace 1889, p. 271)

What is more, claims Wallace, such cases show how markedly dimorphism is influenced by the female's greater need for protection. Even in some of the species that are so strong and fast-flying that the males have no need of mimicry, the females are nevertheless mimics; when both sexes are mimics it will always be found that the species is weaker and slower-flying, so that the males, too, have needed to adopt mimicry as protection; and there are no cases of the males alone being mimics.

It's unnecessary to proliferate examples. Even now, Wallace's work on protective coloration is recognised as an impressive contribution to Darwinian theory; it set a framework and a standard for a rich vein of research. This tribute from Darwin typifies many a Darwinian's appreciation of Wallace's achievement – and it reminds us of Wallace's success in the apparently counter-intuitive task of explaining the showy and the gaudy by protection:

> How ... are we to account for the beautiful or even gorgeous colours of many animals in the lowest classes? It appears doubtful whether such colours usually serve as a protection; but we are extremely liable to err in regard to characters of all kinds in relation to protection, as will be admitted by every one who has read Mr Wallace's excellent essay on this subject.   (Darwin 1871, i, p. 321)

But, as we have seen, Wallace's main achievement didn't lie in explaining the 'beautiful' and the 'gorgeous'. He excelled in understanding the dull, the drab and the dowdy. This is coloration that a Darwinian dazzled by the peacock's tail might take too much for granted – as indeed, we shall see, Darwin did. In principle no Darwinian would have denied that protection plays a major role in determining the colours of animals. But Wallace went further, emphasising the need to explain in precise detail not only extraordinary coloration but also the most ordinary – again, a task familiar nowadays, but at that time far from routine. Perhaps Wallace was alerted to this need more than most by his experiences in the Malay Archipelago, where he had found that the spectacular birds and insects that were so prominent in naturalists' collections comprised a relatively small proportion of species; the collector's taste for the large and exotic, and neglect of the small and obscure, grossly misrepresented nature's own interests (Brooks 1984, pp. 132–4, 176–7). By contrast, Wallace's own approach, as he himself rightly said, 'led to the discovery of so many interesting and unexpected harmonies among the most common (but hitherto most neglected and least understood) of the phenomena presented by organised beings' (Wallace 1891, p. 140).

However, impressive as Wallace's contribution is, it has so far dealt with only one half of his task. He aims to provide an adaptive explanation of sexually dimorphic coloration. And he has explained the female. But he has still to explain the nub of Darwin's 'sexually selected' phenomena: the conspicuous, flamboyant, ornamental colours of the males. Before examining how he tackled this problem we'll look at how he developed his second adaptive principle, recognition, and how he dealt with male display.

Coloration for recognition

Wallace claimed that certain kinds of dimorphic coloration (and some of the sounds, smells and structures peculiar to one sex – generally the male) had evolved as a means of recognition. Their major task was to keep social species together; sometimes they also promoted efficient mating by helping animals to recognise members of their own species of the opposite sex (but not by aiding mate choice within the species) (Wallace 1889, pp. 217–27, pp. 284–5, 298, 1891, pp. 367–8). Such characteristics would typically have a dual aspect – readily seen and easily recognised but at the same time as inconspicuous to predators as possible.

Wallace laid great stress on coloration for recognition, considering it to be very widespread and to play a crucial role: 'I am inclined to believe that its necessity has had a more widespread influence in determining the diversities of animal coloration than any other cause whatever' (Wallace 1889, p. 217). He was perhaps influenced by his own early attempts at species recognition; in collecting specimens in the Malay Archipelago, he had found that, for the taxonomist at least, structural coloration was highly reliable and generally of paramount importance in differentiating species (Brooks 1984, pp. 66–70, 84–93). Wallace used the idea of selection for recognition to mop up several Darwinian problems. First, recognition, along with protection, was central to his explanations of coloration. Second, he wielded it in his campaign for adaptive explanations, using it in particular to explain many of the distinctive species-specific markings that, as we have seen, were the subject of so much dispute between adaptationists and non-adaptationists. Third, as we shall see when we look at altruism, recognition was important in his solution to the problem of interspecific sterility. Wallace was casting around for ways of explaining the sterility adaptively. He was looking for reproductive (that is, non-geographical) barriers to mating. The ability to recognise conspecifics fitted the bill: recognition helped to prevent interspecific mating and the 'evils' of infertile crosses (Wallace 1889, pp. 217, 298, 1891, p. 154, n1). It is indicative of the importance that Wallace – rightly – attached to these explanations that he draws attention to all three of them as subjects of novelty or special interest in the Preface to his *Darwinism* (1889, p. xi). As with coloration for protection, the principle of recognition would have been important in Wallace's thinking even if he had not been searching for alternatives to sexual selection.

But although explaining bright coloration by recognition fits neatly within Wallace's total explanatory scheme, it predictably falls far short of explaining male coloration. First, much of the coloration that Wallace puts in this category is not dimorphic – not surprisingly when its function is to keep

Labels for species

Three species of African plovers (from Wallace's *Darwinism*)

"Some means of easy recognition must be of vital importance ... and I am inclined to believe that its necessity has had a more widespread influence in determining the diversities of animal coloration than any other cause whatever ... Among birds, these recognition marks are especially numerous and suggestive. Species which inhabit open districts are usually protectively coloured; but they generally possess some distinctive markings for the purpose of being easily recognised by their kind, both when at rest and during flight. Such are ... the head and neck markings in the form of white or black caps, collars, eye-marks or frontal patches, examples of which are seen in the[se] three species of African plovers." ('Wallace: Darwinism)

together all members of social species, male and female. He does allow that for insects, particularly butterflies and moths, the major function of recognition marks might be to facilitate mating with conspecifics; so in that case one could perhaps expect dimorphism. But he specifically denies that recognition for mating can get very far in explaining dimorphic coloration in birds (Wallace 1889, p. 224, n1; see also pp. 226–7, 1891, p. 354) (although he is not entirely consistent on this limitation (1889, p. 298, 1891, p. 154, n1)). The second difficulty is that selection for recognition may be able to explain the more modest cases of coloration but how can it explain those wilder excesses that so worried Darwin? Recognition would hardly be likely to produce the peacock's tail. As Wallace himself said: 'the resplendent train of the peacock ... exhibits to us the culmination of that marvel and mystery of animal colour' (Wallace, 1889, p. 299). Would natural selection have been so grossly inefficient as to have evolved elaborate, flamboyant adaptations merely for spotting potential mates, even if there is a pressing need to prevent confusion between species?

We shall see in a moment how Wallace answered these questions. First, we'll add just one more piece to his explanatory scheme.

Explaining away display

Protection and recognition may be able to absorb some conspicuous coloration but they cannot cope with one of its most salient aspects: male display. Many males are not merely gorgeously hued; they also have an element of show in their coloration and structure, and elaborate, stylised, ceremonial behaviour that seems designed to flaunt their glamour. As Wallace said of birds (at the period when he had not yet rejected sexual selection): 'It is a well-known fact that when male birds possess any unusual ornaments, they take such positions or perform such evolutions as to exhibit them to the best advantage while endeavouring to attract or charm the females' (Wallace 1891, p. 320).

Wallace couldn't ignore this. It demands explanation if he is to construct a comprehensive theory of coloration. And, equally troubling, it is prima facie convincing evidence, albeit indirect, of female choice. Indeed, in Darwin's view it was the best indication: 'the evidence is rendered as complete as ever it can be, only when the more ornamented individuals, almost always the males, voluntarily display their attractions before the other sex' (Darwin 1871, 2nd edn., p. 401).

Darwin made sure that this evidence was 'rendered complete'. He attempted to show that male display is not incidental or inadvertent but really is about exhibiting ornaments to the females. He argued, for example, that display is most common among the most sexually dimorphic groups, that the behaviour shows the characteristics at their best, that the males are apparently attempting to catch the females' attention or that they display only in the females' presence. This is his delightful description of the behaviour of a species of fish, a Chinese *Macropus*, during the breeding season: 'The males are most beautifully coloured ... and, in the act of courtship, expand their fins, which are spotted and ornamented with brightly coloured rays, in the same manner ... as the peacock. They then also bound about the females with much vivacity, and appear [to try to attract their attention]' (Darwin 1871, 2nd edn., pp. 522–3). And he notes how some male birds parade for the females: 'The rock-thrush of Guiana, birds of paradise, and some others, congregate; and successive males display with the most elaborate care, and show off in the best manner their gorgeous plumage; they likewise perform strange antics before the females, which, standing by as spectators, at last choose the most attractive partner' (Peckham 1959, p. 175). He cites the case of the butterfly *Leptalides*; both sexes have evolved protective mimetic coloration but the male has retained a patch of the original colour, which he displays only during courtship. Darwin quotes a striking comment made by the naturalist-explorer Thomas Belt, in his book *The Naturalist in Nicaragua*: 'I cannot imagine its being of any other use to them than as an attraction in courtship, when they exhibit it to the females, and thus gratify their deep-seated preference for the normal colour of the Order to which the Leptalides belong' (Darwin 1871, 2nd edn., p. 498; see Belt 1874, p. 385).

How does Wallace deal with all this? When he was still willing to allow a significant role to sexual selection he agreed with Darwin that the evidence from birds at least was compelling:

> birds ... [have] furnished Mr Darwin with the most powerful arguments ... Among birds is found the first direct proof that the female notices and admires increased brilliancy or beauty of colour, or any novel ornament; and, what is more important, that she exercises choice, rejecting one suitor and choosing another. There is abundant evidence too that the male fully displays all his charms before the females ...   (Wallace 1871, p. 179)

Subsequently he shifted ground, acknowledging that the evidence required explanation but denying that female choice was the answer. Here he is on birds again, now sounding less enthusiastic:

> There remains ... to be accounted for, the remarkable fact of the display by the male of each species of its peculiar beauties of plumage and colour – a display which Mr Darwin evidently considers to be the strongest argument in favour of conscious selection by the female. This display ... may, I believe, be satisfactorily explained ... without calling to our aid a purely hypothetical choice exerted by the female bird.   (Wallace 1891, pp. 376–7)

He fully admitted that the evidence certainly appeared to be in Darwin's favour: 'The extraordinary manner in which most birds display their plumage at the time of courtship, apparently with the full knowledge that it is beautiful, constitutes one of Mr Darwin's strongest arguments' (Wallace 1889, p. 287). Nevertheless, he claimed, these apparent displays may not really be displays at all. The male may merely be expending some of the surplus energy that he accumulates during the mating season – just like the gambolling of young animals:

> During excitement, and when the organism develops superabundant energy, many animals find it pleasurable to exercise their various muscles, often in fantastic ways, as seen in the gambols of kittens, lambs, and other young animals ... [A]t the time of pairing, male birds are in a state of the most perfect development, and possess an

enormous store of vitality; and under the excitement of the sexual passion they perform strange antics or rapid flights, as much probably from an internal impulse to motion and assertion as with any desire to please their mates.

(Wallace 1889, p. 287)

And why, he asks, if the males' activity is for display, do unornamented birds also behave in this way (Wallace 1889, p. 287, 1891, p. 377)? Far from supporting Darwin's theory, this connection between vigour on the one hand and structure and colour on the other seems to him to be evidence for his own theory (which we shall examine) that such connections are mere by-products of physiology: 'It ... indicates a connection between the exertion of particular muscles and the development of colour and ornament ... The display of these plumes will result from the same cause which led to their production' (Wallace 1889, pp. 287, 294). Similarly, he says, there is an inverse correlation between ornamental colours and structures on the one hand and the development of vocal power on the other. This, too, is what would be expected if song is merely an alternative outlet for superabundant energy (Wallace 1889, p. 284).

Wallace's arguments are utterly inadequate to their task. They do not account for the apparent purpose in display. And it is highly implausible to maintain that such elaborate and stereotyped behaviour is not shaped by selection. Wallace has adopted a non-adaptationist position and tried to push it too far. But there is worse to come.

Coloration without selection

So much for the females and the more soberly clad of their mates. But Wallace still has to account for the most conspicuously coloured males. This brings us back to how he turned Darwin's question on its head, claiming that it is the female's dull colours and not the male's bright ones that are most in need of explanation. It is here that his non-adaptationist arguments strain under their explanatory burden.

During the period up to about 1871, when Wallace accepted Darwin's theory of sexual selection (which at that time was mainly confined to birds and insects), he combined his theories about protection for the female with Darwin's explanation of the male's bright coloration by sexual selection (e.g. Wallace 1891, p. 89). Even when he began to have doubts about sexual selection, he allowed it some role, albeit subsidiary to natural selection: 'while sexual selection has ... been doing its work, the still more powerful agency of natural selection has not been in abeyance, but has also modified one or both sexes in accordance with their conditions of life' (Wallace 1871,

p. 180). So at this period Wallace had a selective theory – either natural or sexual selection – to cover both the males and the females. He held that the primordial colours were probably dull; over evolutionary time sexual selection had evolved showy males whereas natural selection had generally retained or enhanced the females' inconspicuous dress (Wallace 1891, p. 130). But when Wallace abandoned sexual selection, he needed an alternative explanation for the males. His solution was his physiological theory of conspicuous coloration (Wallace 1889, pp. 288–93, 297–8, 1891, pp. 359–60, 391–2).

We have already touched on Wallace's views about non-adaptive coloration. We saw that even this committed adaptationist took care to distinguish between colours that are purely physical or physiological and those that are biological, and to emphasise that the former do not require an adaptive explanation (Wallace 1889, pp. 188–9). It turns out, however, that his idea of what could be explained by physiology alone was very catholic indeed. He developed a theory that, over evolutionary time, if organisms were not impeded by natural selection, they would naturally tend to become multicoloured as a result of constant physicochemical changes: 'Colour may be looked upon as a necessary result of the highly complex chemical constitution of animal tissues and fluids' (Wallace 1889, p. 297); 'Many of the complex substances which exist in animals and plants are subject to changes of colour under the influence of light, heat, or chemical change, and ... chemical changes are continually occurring during ... development and growth ... [E]very external character is [also] ... undergoing constant minute changes; and these changes will very frequently produce changes of colour' (Wallace 1891, p. 359). So to be multicoloured is the 'normal' state: 'These considerations render it probable that colour is a normal and even necessary result of the complex structure of animals and plants' (Wallace 1891, p. 359). Indeed, were it not for the constraining hand of natural selection, animals would rejoice in splendid colours. After all, there are no such constraints on the insides of animals and they present a technicoloured array. Their outsides are subject to more change and so would naturally tend to even more gaudy hues:

> The blood, the bile, the bones, the fat, and other tissues have characteristic, and often brilliant colours, which we cannot suppose to have been determined for any special purpose, as colours, since they are usually concealed. The external organs, with their various appendages and integuments, would, by the same general laws, naturally give rise to a greater variety of colour.   (Wallace 1889, p. 297)

It is only the action of natural selection that prevents this polychromatic explosion. Domestication provides independent evidence of this. When selection pressures are lifted, colours appear that are unknown in nature. And in domestic fowl the patterns develop symmetrically, a 'crucial fact', according to Wallace, because it suggests the action of physiological laws of development rather than the selective forces that Darwin assumes (Wallace 1891, p. 375). (He holds that symmetries maintained by selective forces are typically inexact and often lost under domestication (Wallace 1889, pp. 217–18, n1).)

What is more, Wallace claims, the tendency to develop bright colours is generally stronger in the male for the same non-adaptive reason: coloration increases with physiological activity and the male, so he asserts, is usually the more vigorous (Wallace 1891, pp. 365–6). Wallace supports his claim with three points. First, bright, glossy colours are generally indicative of robust health. Second, the male's vitality is at its peak in the mating season and this is when his colours are brightest. Third, males tend to develop brighter colours than females even under domestication, in the absence of any selection for coloration. Wallace also came to attribute the brighter colouring of female birds in cases of 'role reversal' (the male incubating) to the female having more vital energy in these cases (Wallace 1891, p. 379).

Wallace accounts for the males' ornamental structures in the same way: they arise at points of heightened physiological activity. Many birds of paradise, for example, sport an immense tuft of feathers on the breast; this springs from the most powerful of the birds' muscles, the pectoral muscle, and at a point where it is most active. Wallace claims that Darwin's theory cannot explain why the ornaments occur at these particular parts of the body (Wallace 1889, pp. 291–3).

So this, according to Wallace, is why it is the female's sombre colouring but not the male's brilliant hue that requires adaptive explanation. Both sexes tend naturally to be brightly coloured (although males more than females) but the female is under selection pressures that damp down this physiological drive:

There seems to be a constant tendency in the male of most animals – but especially of birds and insects – to develop more and more intensity of colour, often culminating in brilliant metallic blues and greens or the most splendid iridescent hues; while, at the same time, natural selection is constantly at work, preventing the female from acquiring these same tints, or modifying her colours in various directions to secure protection by assimilating her to her surroundings, or by producing mimicry of some protected form.   (Wallace 1889, p. 273)

But surely, one wants to protest, the 'designed' appearance of the males' colours strongly indicates that they are adaptive. Wallace, however, comes to the opposite conclusion: the connections between colour and structure are further evidence on *his* side, evidence that colour is merely an inevitable, unselected side effect of physiology. After all, on his theory bright colours arise from physiological changes. And aren't patterns likely to emerge in the process? He calls attention to the fact that arrangements of colour generally coincide with structure: 'diversified coloration follows the chief lines of structure, and changes at points, such as the joints, where function changes' (Wallace 1889, p. 288). So the most flamboyant colours tend to be found on the most elaborate or altered structures: 'Brilliant colours usually appear just in proportion to the development of ... appendages ... Colour increases in variety and intensity as external structures and dermal appendages become more differentiated and developed' (Wallace 1889, pp. 290–1, 297). And changes in colour occur with a regularity that, according to Wallace, suggests not selection but the automatic side effects of developmental laws: 'There are indications of a progressive change of colour, perhaps in some definite order, accompanying the development of tissues or appendages ... [Such changes indicate a] law of development ... dependent on laws of growth' (Wallace 1889, p. 298).

Thus it is, argues Wallace, that the males are often both distinctively patterned and vividly coloured; their superior vitality favours the development of new structures and these will be accompanied by changes of colour (Wallace 1891, p. 366). Thus it is that butterflies and birds, whose surface structures have been subject to an extraordinary amount of change, so greatly exceed all other animals in the intensity and variety of their coloration (Wallace 1891, pp. 368–9). Thus it is that the most brilliantly coloured birds are those with the most enlarged and elaborate plumage (Wallace 1889, p. 291); the humming-birds, particularly the males, exhibit more vital energy and spectacular colours than most other groups, the most pugnacious of their species being the most showy (Wallace 1891, pp. 379–81). And thus it is, concludes Wallace triumphantly, that (at least in part) the peacock got his tail – and the Argus pheasant and the bird of paradise theirs (Wallace 1891, p. 375).

Wallace's arguments are certainly ingenious. He is wrong to turn Darwin's question entirely on its head – adaptive explanations shouldn't be reserved for females alone – but he is certainly right to call attention to the need to explain dull, mundane coloration. Darwinians should pay attention to the peahen's everyday hues as well as her mate's Sunday best. After all, she isn't merely dull but camouflaged. Conversely, as Wallace's argument about the insides of bodies suggests, brightness may be a 'natural' state – in which case Darwinians should not jump to the conclusion that it requires adaptive explanation.

The glory of humming-birds (from Belt's *The Naturalist in Nicaragua*)
For Wallace, the humming-bird's splendour was just the undirected outpouring of
surplus energy. Thomas Belt, a staunch adaptationist and sexual selectionist, took a
different view:

*"[The tail of] the beautiful blue, green, and white humming-bird (Florisuga mellivora, Linn.)
... can be expanded to a half circle, and each feather widening towards the end makes the
semicircle complete around the edge. [This show is] ... reserved for times of courtship. I have
seen the female sitting quietly on a branch, and two males displaying their charms in front of
her. One would shoot up like a rocket, then suddenly expanding the snow-white tail like an
inverted parachute, slowly descend in front of her, turning round gradually to show off both
back and front. The effect was heightened by the wings being invisible from a distance of a
few yards, both from their great velocity of movement and from not having the metallic lustre
of the rest of the body. The expanded white tail covered more space than all the rest of the bird,
and was evidently the grand feature in the performance. Whilst one was descending, the other
would shoot up and come slowly down expanded. The entertainment would end in a fight
between the two performers; but whether the most beautiful or the most pugnacious was the
accepted suitor, I know not." (Belt: The Naturalist in Nicaragua)*

Nevertheless, Wallace's arguments, however ingenious, fail quite
spectacularly. They are inherently implausible: is so remarkable, so
undeniable an appearance of design likely to arise without adaptation? They
fail in his declared aim of replacing Darwin's theory of female choice with

standard natural selectionist principles. And they are inconsistent with his own programme of holding out for adaptive explanations.

Just think what Wallace would have us believe: male coloration, with its fine detail, its striking patterns, its appearance of design, its constancy and its widespread occurrence throughout the animal kingdom, has arisen simply as a physiological side effect without the help of direct selection; and the end result of this physiological process is selectively neutral – neither advantageous nor deleterious – and is maintained by physiological forces alone.

Take first Wallace's assertion that because colour differences follow structural features the coloration is not the result of selection. Certainly, connections between colour and structure could originate in the way he suggests. But clearly this doesn't imply that whenever one finds colour and structure hand in hand this is the result of physiological laws alone, without the intervention of selective forces. One of Wallace's own criteria for natural selection having been at work was that 'the colours are localised in definite patterns, sometimes in accordance with structural characters' (Wallace 1889, p. 189). After all, one would expect natural selection to seize on and develop connections between structure and colour. A structure that is both differentiated and coloured to match is fitting raw material for, say, display or complex camouflage. E. Ray Lankester made a similar point in his review of Wallace's *Darwinism*: 'Mr Wallace seems scarcely to have succeeded in showing that Darwin's theory of sexual selection is inapplicable to the explanation of special developments of colour and ornament, although he has suggested additional causes which influence the primary distribution and development of colour' (Lankester 1889, p. 569). Indeed, Wallace himself later conceded (Wallace 1900, i, pp. 390–1) that it was more in keeping with his adaptationist goals to argue that colour and ornament had originated in the way he first suggested and had then been shaped by selection for recognition; but he didn't pursue this idea.

Next, Wallace himself ruled that 'the wonderful individuality of the colours of animals and plants' (Wallace 1889, p. 189) calls for an adaptive explanation. Surely his rule should apply to male coloration. Certainly none of the physiological reasons he gives goes far towards explaining its complexity and variety. Why, for example, if colour merely follows structure are, say, butterflies' wings so very similar structurally but so vastly different in their colour patterns? As the eminent comparative psychologist C. Lloyd Morgan said:

> It can hardly be maintained that the theory affords us any adequate explanation of ... *specific* colour-tints ... If, as Mr Wallace argues, the immense tufts of golden

plumage in the bird of paradise owe their origin to ... the arteries and nerves ... [why do] other birds in which similar arteries and nerves are found in a similar position ... have ... no similar tufts? (Quoted in Romanes 1892–7, i, p. 449)

Karl Groos, another comparative psychologist and professor of philosophy at the University of Basel, came to a similar conclusion: '[Wallace] sets out from the fact that the characteristic marks and appendages of animals are closely connected with their anatomical structure ... However, I for one can not quite conceive how such developments as, for instance, a peacock's tail, can be derived from beginnings so insignificant, simply by a superabundance of energy' (Groos 1898, pp. 235–6).

What is more, if there is a constant tendency to produce colour but selection is not at work, why do the males end up with brilliant colours rather than acquiring the monochromatic murkiness of mixed paints? The book on coloration that most influenced Wallace concluded that, as a result of this kaleidoscopic process, 'this colour would, if unrestrained and undirected, be indefinite, and could not produce definite tints, nor the more complicated phenomenon of patterns' (Tylor 1886, p. 29). Wallace could have maintained that the complexities of embryonic development might well produce such complicated phenomena (and this might have answered the point about 'wonderful individuality', too). But, on the contrary, Wallace himself held that a 'haphazard mixture' of pigments would produce 'neutral or dingy' colours (Wallace 1891, pp. 360–1). Indeed, he used a similar argument against Darwin's claim that female taste could be responsible for the males' well-defined colours: 'Successive generations of female birds choosing any little variety of colour that occurred among their suitors would necessarily lead to a speckled or piebald and unstable result, not to the beautiful definite colours and markings we see' (Wallace 1871, p. 182).

Even in the unlikely event that coloration could have developed in the way that Wallace suggests, its constancy over time and uniformity within species presents a problem. How is it maintained, unless by selection (what would now be termed stabilising selection – favouring the average type)? He gave no reason to suppose that the laws of physiology alone would ensure such constant effects. And Wallace himself had insisted that constancy was a sign of selection taking a hand (Wallace 1889, pp. 138–42, 189–90, 1891, p. 340): 'the minutest markings are often constant in thousands or millions of individuals ... [This] must serve some purpose in nature' (Wallace 1891, p. 340). Indeed, he cited the constancy of species-specific characteristics as his principal evidence against the view that they were non-adaptive. (Admittedly he mentions that secondary sexual characters tend to be variable (Wallace 1889, p. 138); but they are sufficiently constant to be candidates for adaptive explanation on his criterion and, as we shall see, he uses the fact of their relative constancy as evidence that they are not the result of female choice.) What is more, adaptationists were generally agreed that one would not expect selectively neutral characteristics to be highly stable, and this is also a fairly standard view among Darwinians today (e.g. Cain 1964; Maynard Smith 1978c; Williams 1966, pp. 10–11).

Consider, too, that Wallace himself rightly declared that the Darwinian principle of utility 'leads us to seek an adaptive ... purpose ... in minutiae which we should otherwise be almost sure to pass over as insignificant or unimportant' (Wallace 1891, p. 36). And when he is wearing his adaptationist hat he is reluctant to concede even that the colours of fruit could ever be a mere by-product rather than an adaptation for attracting animals (Wallace 1889, p. 308). Yet when he comes to male ornamentation, Wallace happily allows a host, not merely of minutiae but of 'peacocks' tails', to slip through the adaptive net.

Finally, Wallace himself insists that a theory of coloration should be judged on how comprehensively it covers the phenomena:

> to those who oppose the explanation now given of the various facts bearing upon this subject [coloration], I would ... urge that they must grapple with the whole of the facts, not one or two of them only. It will be admitted that on the theory of evolution and natural selection, a wide range of facts with regard to colour in nature have been co-ordinated and explained.   (Wallace 1891, pp. 139–40)

Yes, a wide range; but not wide enough. Wallace needs to explain the colours of both females and males as adaptations to (different) selective pressures. On the whole he carries out the first part of this programme extremely well. Indeed, he is a victim of his success: it shows up the need for an equivalent explanation for male coloration and his embarrassing failure to provide one. His explanations of coloration are weakest at precisely the points that for a Darwinian are the most puzzling. However impressive his explanation of female coloration may be, he cannot hope to replace sexual selection unless he explains both halves of the dimorphism. Romanes pointed out the gross discrepancy between Wallace's declarations on the universality of the principle of utility and his explanation of 'sexually selected' phenomena:

> Can it be held that all the 'fantastic colours', which Darwin attributes to sexual selection ... are to be ascribed to 'individual variability' without reference to utility, while at the same time it is held, 'as a necessary deduction from the theory of Natural Selection', that *all* specific characters must be '*useful*'? Or must we not conclude that we have here a contradiction as direct as a contradiction can well be?
>   (Romanes 1892–7, ii, p. 271)

If, says Romanes, Wallace appeals so readily to physiology rather than utility, then he is not as committed to adaptive explanation as he claims to be: 'it appears to me that the difference between Mr Wallace and myself, with respect to the principle of utility, is abolished' (Romanes 1892–7, ii, p. 222).

Most Darwinians made some concessions to non-adaptive coloration – sometimes even, in retrospect, unnecessarily generous concessions. But Wallace went much further. It was generally agreed that, as Darwin put it, 'the complex laboratory of living organisms' would probably give rise to splendid colours, just as the chemists' laboratories do (Darwin 1871, i, p. 323). We've seen that concealed colours, for example, such as the crimson of blood, were standardly explained in this way. Similarly, conspicuous colouring in the 'lowest' animals was commonly thought to be non-adaptive; Darwin himself was willing to put aside sexual selection in favour of physicochemical explanations in their case (e.g. Darwin 1871, i, pp. 321–3, 326–7; Romanes 1892–7, i, pp. 409–10). Even E. B. Poulton, whose work was largely devoted to discovering the adaptive significance of coloration, felt obliged to stress that colours may be 'incidental'; he made a prominent point of praising Darwin for recognising this and for warning against over-enthusiastic adaptationism (e.g. Poulton 1910, pp. 271–2). But all this was a far cry from claiming, with Wallace, that the peacock's tail was 'incidental'.

In short, for any Darwinian it is an appallingly weak tactic to relegate to non-adaptationism a phenomenon so widespread, so constant and so apparently designed as the striking coloration that Darwin claimed was sexually selected; for one who professes to favour adaptive explanations, particularly one who prides himself on his explanation of coloration as a major contribution, it amounts to unmitigated failure. And his failure is not surprising. As John Maynard Smith has aptly remarked: 'however much one may be in doubt about the function of the antlers of the Irish elk or the tail of the peacock, one can hardly suppose them to be selectively neutral' (Maynard Smith 1978c, p. 36).

Having talked of unmitigated failure, I shall nevertheless put in a small plea of mitigation for Wallace. His arguments appear in some ways so far-fetched, particularly for the more dramatic cases, that it is only fair to mention that he was not alone in holding any of the individual points (although he assembled and exploited them in his idiosyncratic way). Some of his more ingenious explanations of patterns were taken from a book on animal coloration by Alfred Tylor, an English geologist (Tylor 1886; Wallace 1889, p. 288). Tylor viewed the physiological effects as a base on which natural selection set to work rather than as nature's final offering (Tylor 1886, pp. 6–7, 17). But others were closer to Wallace's view. One critic of Darwin's theory of sexual selection, G. Norman Douglass, writing in the 1890s, thought that a theory like Wallace's was more scientific than Darwin's appeal to female taste:

> If the tendency of biology is to become a more exact science ... , the processes involved in the formation of animal pigments will soon ... show whether order cannot be brought into the 'fortuitous concourse of atoms of colouring matter' without external (female) sanction. I think it will be found that the harmonious distribution of tints on the feather of the argus pheasant merely continues a principle which the bilateral and radiate forms of all living organisms illustrate – the coincidence of symmetry with economy.   (Douglass 1895, pp. 404–5)

Wallace's theory that organisms tend naturally to bright coloration was also held by several others. A correspondent to *Nature* in the 1870s claimed that 'the colour-producing force which exists in the plant will break through all obstructions whenever the opportunity is presented ... This law holds good throughout the organic world, and accounts for colour wherever it is found' (Mott 1874, p. 28). Some thirty years later Jacob Reighard, Professor of Zoology at the University of Michigan, proposed a similar theory (Reighard 1908, pp. 310–11, 316–21). And thirty years after that one of his successors at Michigan, a geneticist, A. Franklin Shull (Shull 1936, pp. 179–80, 198), championed both Wallace's and Reighard's views. Incidentally, John Turner (Turner 1983, p. 152, n5) cites Reighard's theory as an example of a view that has gone unrecorded by historians because it has been discredited scientifically. It is just such filters of history that we should bear in mind before dismissing Wallace as an eccentric among his contemporaries.

As for Wallace's related claim that males were imbued with greater vitality than females and that this could give rise to elaborate structures and colours, it was standard thinking, both popular and scientific (see e.g. Farley 1982, pp. 110–28; Wallace 1889, pp. 296–7, n1). When *Descent* was published, one critic wrote to Darwin: 'Is it wrong ... to suppose that extra growth, complicated structure, and activity in one sex exist as escape-valves for surplus vigour, rather than to please or fight with ... ?' (Darwin, F. and Seward 1903, ii, p. 93). And Darwin acknowledged in his reply that he had been impressed with a similar suggestion that some extravagant male structures were 'produced by the excess of nutriment in the male, which in the female would go to form the generative organs and ova' (Darwin, F. and Seward 1903, ii, p. 94) (an argument that comes closer to the modern idea of males and females distributing their reproductive costs in different ways). Darwin also believed that gaudy colouring in males was correlated with pugnacity (Marchant 1916, i, p. 302). Later in the century Reginald Pocock, one of the British Museum's experts on spiders, suggested that some recent research claiming to demonstrate sexual selection in spiders could equally well be accounted for on Wallace's theory of male vitality:

> the cases that are quoted in this work ... are equally explicable by Mr Wallace's views. Thus ... it seems to be ... [the male] sex which excels in activity; and if activity be a criterion of high vitality we at once see the connection between high vitality and ornamentation ... Or again, if it be asked why it is that the males perform the strange antics in the presence of the females if it be not for display, it may be answered that the excitement of the males, always greater during the breeding season, attains to a maximum at that time in the society of the females, and shows itself in the performance of the strange antics ... (Pocock 1890, p. 406)

W. H. Hudson's popular book *The Naturalist in La Plata* (1892) dismissed Darwin's 'laborious' explanation of music and dancing in favour of Wallace's view that at 'the season of courtship, when the conditions of life are most favourable, vitality is at its maximum' (Hudson 1892, pp. 263, 285). Another widely-read book, *Animal Coloration* (Beddard 1892), published in the same year, also followed Wallace in attributing male colour to vitality. Douglass, the critic whom we met just now, declared:

> gestures and gambollings of all denominations throughout the various orders of saltatory nature – from the 'unusual antics and gyrations' of worms up to the contortions performed by the gilded youth in modern ball-rooms – will ultimately be found to be only the outcome of ... 'surplus vitality' ... Here lies, indeed, the root of the whole matter [of male ornament]. For surplus vitality is another name for the primary physiological processes that supply the material (be it colour, or structures, or exuberant activity, or song) whose subsequent elaboration, as incompatible with the principle of utility, is entrusted to female preferences. (Douglass 1895, p. 330)

(And, on his view, wrongly entrusted.) At the turn of the century, Vernon Kellogg declared the 'extra growth-vigour' theory to be the 'most appealing' alternative to the 'discredited' theory of sexual selection (Kellogg 1907, p. 352; see also p. 117). And it later became a standard view that 'the bright colours of the male ... are sometimes a sort of by-product of his vigour' (MacBride 1925, p. 218).

What is more, Darwin, like Wallace, occasionally resorted to the non-adaptive force of 'joie de vivre' or sheer pleasure in instinctive activity to deal with such awkward cases as peacocks 'displaying' when no females were present or robins singing at full force when the mating season had long passed (e.g. Darwin 1871, ii, pp. 54–5, 86). Indeed, animal activity has commonly been explained in this way, by naturalists as diverse as Paley (1802, pp. 454, 457, 458), Kropotkin (1902, pp. 58–9) and Julian Huxley (1923a, pp. 122–7, 1966). Perhaps most strikingly, a highly influential book, *The Evolution of Sex* (1889), by Patrick Geddes and J. Arthur Thomson, strongly supported many of Wallace's ideas. It urged that males are constitutionally predisposed to develop brighter colours, more elaborate structures and more vigorous behaviour than females because they are more active metabolically; natural and sexual selection played some role but a relatively minor one (pp. 11, 14, 16–31, 320, 324). Such examples could be multiplied. So Wallace's views, although they often seem far-fetched today, were not nearly as remote from mainstream thinking as one might imagine.

In spite of Wallace's failure, both he and others have, on occasion, been so carried away by his impressive explanations of female coloration that they have assumed that he achieved his wider aim (at least in principle) of elbowing out sexual selection entirely. The historian of science Peter Vorzimmer, for example, appears to think that the Darwin–Wallace debate ended in almost total victory for Wallace. He says in apparent agreement with Wallace:

Wallace's disinclination to accept the theory of sexual selection became complete disavowal as a result of his work on mimicry and coloration for protection. When he came to realize that the principle of natural selection would operate equally well to attain qualities of self-protection, and that most, if not all, secondary sexual characters could be thus explained, he saw no necessity whatever for calling upon such a process as sexual selection. The principle of natural selection, as originally postulated by both Darwin and himself, seemed perfectly adequate.

(Vorzimmer 1972, p. 197)

He dismisses Darwin's theory of sexual selection as 'not very important' (Vorzimmer 1972, p. 202), failing to notice that Wallace does not get to grips with the most salient phenomena that sexual selection is designed to explain. Here, in similar vein, is the botanist Verne Grant: 'As Wallace ... pointed out in a brilliant analysis of the problem [of secondary male characteristics], Darwin's theory of sexual selection [does not provide] ... a satisfactory explanation of the development of ornamentation and song in the male sex' (Grant 1963, p. 243). According to Grant this can be explained only by natural selection (mainly species recognition).

Wallace's own claims went even further. We have seen how he asserted that when it came to sexual selection he was more Darwinian than Darwin. He also claimed that his alternative theories of coloration widened the scope of natural selection: 'my view really extends the influence of natural selection, because I show in how many unsuspected ways colour and marking is of use to its possessor' (Wallace 1905, ii, p. 18; see also 1889, p. 268). He held that by replacing sexual selection with his theory, natural selection would be 'relieved from an abnormal excrescence and gain additional vitality' (Wallace 1871, pp. 392–3). And, unlike Darwin's theory, his has no need of a highly questionable assumption: it 'entirely dispenses with the very hypothetical and inadequate agency of female choice' (Wallace 1889, p. 334). Thus, he states, his theories can cover the whole range of Darwin's 'sexually selected' phenomena: 'I believe that I can explain (in a general way) *all* the phenomena of sexual ornaments and colours by laws of development aided by simple Natural Selection' (Marchant 1916, i, p. 298). He realises, he says, that this is a bold claim – but it is justified; and it will prove a 'relief' to naturalists (one can almost hear his sigh) to jettison sexual selection and stick to natural selection:

> It will perhaps be considered presumptuous to put forth this sketch of the subject of colour in animals as a substitute for one of Mr Darwin's most highly elaborated theories – that of ... sexual selection; yet I venture to think that it is more in accordance with the whole of the facts, and with the theory of natural selection itself ... The explanation of almost all the ornaments and colours of birds and insects as having been produced by the perceptions and choice of the females, has ... staggered many evolutionists, but it has been provisionally accepted because it was the only theory that even attempted to explain the facts. It may perhaps be a relief to some of them, as it has been to myself, to find that the phenomena can be conceived as dependent on the general laws of development, and on the action of 'natural selection' ...   (Wallace 1889, p. 392)

Wallace's stance reversed accustomed roles. The arch-pluralist Romanes, usually a stern critic of Wallace's enthusiasm for adaptive explanations, urged Wallace to concede that structures so elaborate and specialised as male ornament could not arise without selection (e.g. Romanes 1892–7, i, pp. 394–8). Meanwhile, the committed adaptationist Wallace exhorted Darwin to place less weight on his selectionist theory and more on 'unknown laws' of colour development and the like (Wallace 1871)!

It's odd, too, to realise that Wallace came full circle on ornamental colours. He originally developed his theory of coloration in part to overthrow natural theology's claims that beauty in nature lacked utility (e.g. Wallace 1891, pp. 153–6). Subsequently, he aimed to overthrow Darwin's claims that its utility lay in sexual selection. And that ended in his adamantly denying that 'sexually selected' colours had any utility at all.

Finally, there is a poignant irony in Wallace's triumphantly ultra-adaptationist, ultra-Darwinian protestations in the face of his manifest failure to deal with ornamental characteristics (see Kottler 1985, pp. 410–11). His fervour is perhaps that of the convert. For the younger Wallace, ornamental characteristics had been a providential light in a darkly utilitarian world. Here, for example, is the Wallace of 1856, before he had discovered natural selection. He has just argued that the orang utan's huge canines are of no use to it, and he continues thus: 'Do you mean to assert, then, some of my readers will indignantly ask, that this animal, or any animal, is provided with organs which are of no use to it? Yes, we reply, we do mean to assert that many animals are provided with organs and appendages which serve no material or physical purpose' (Wallace 1856, p. 30). Many beautiful colours and structures have been created simply for 'the love of beauty for its own sake' (Wallace 1856, p. 30). They are signs of design, the work of a supreme creator. This places severe limits on adaptive explanations: 'We conceive it to be a most erroneous, a most contracted view of the organic world, to believe that every part of an animal ... exists solely for some material and physical use to the individual' (Wallace 1856, p. 30). And he goes on to criticise over-zealous adaptationists: 'the constant practice of imputing ... some use to the individual, of every part of its structure, and even of inculcating the doctrine that every modification exists solely for some such use, is an error fatal to our complete appreciation of all the variety, the beauty, and the harmony of the organic world' (Wallace 1856, p. 31). The Wallace of later years had obviously travelled a long Darwinian road since then.

Males for Darwin, females for Wallace?

Now we come to a puzzle – one that I've not been able to resolve. The puzzle lies in Darwin's response to Wallace's strategy of explaining coloration by protection and recognition (particularly protection – Darwinian ideas on recognition were not as well developed in the 1870s as when Wallace applied them more fully a decade or so later). It's obvious that sexual selection and selection for protection could be complementary. Together they could furnish a complete explanation of both male and female coloration. Many cases of dimorphism could be dealt with in this way – the striking cases, nature's glorious peacocks and relatively drab peahens, above all. A contemporary of Darwin's and Wallace's captured this idea in a delightful image; remarking on a butterfly with a well-camouflaged underside to its wings and a gaudy top surface, he declared: 'We may give the under surface to Mr Wallace, but we must yield the upper surface to Mr Darwin' (Fraser 1871, p. 489). Darwin could have applied the same judgement of Solomon when it came to the two sexes. He could very happily have handed the dull females and some bright colours in both sexes to Wallace without seriously weakening his own claim to most of the flamboyant males.

But he didn't. Instead, he largely turned his back on selection and resorted to the laws of heredity and development alone (see Ghiselin 1969, pp. 225–9; Kottler 1980) – a non-adaptive explanation. At the beginning of their 1867–8 correspondence, Darwin largely accepted Wallace's view. But by the autumn of 1868, he had come to disagree. I don't want to exaggerate this. When Darwin wasn't dealing with sexual selection he of course made extensive use of selection for protection to explain coloration. And even in cases where he explained male coloration by sexual selection he most certainly made concessions to protective female coloration. He stated very clearly that both sexual selection and protection determine adaptive coloration: with 'animals of all kinds, whenever colour has been modified for some special purpose, this has been ... either for protection or as an attraction between the sexes' (Darwin 1871, i, pp. 391–2). But he was rather disinclined to view the non-sexually-selected half of the dimorphism as adaptive, as 'modified for some special purpose'. And so he relied less than one might expect of a Darwinian on the kind of explanation that Wallace favoured, and more than one might expect on his theories about inheritance without benefit of selection. There is a parallel here with Wallace's non-adaptive explanations of coloration. In Wallace's case the non-adaptive mechanism was physiology; in Darwin's it was heredity. In Wallace's case it was used to explain male coloration; in Darwin's it was to explain female coloration.

Darwin claimed that the female's coloration depended largely on how the typically male variations happened to be inherited when they first arose in the course of evolution. They could be carried and expressed (to any extent) by both sexes from the first; in this case the female would share the male's sexually selected colours. Alternatively, inheritance could be sex-limited (manifested in only one sex) from the first, in which case the sexes would be dimorphic if the males were sexually selected. Now to a crucial point: according to Darwin, natural selection would usually lack the power to shape variations in one sex alone if they were expressed in both – that is, natural selection could not convert equal into sex-limited inheritance. So if there was equal inheritance of the male's sexually selected colours, selection would have no power to dampen down the female's coloration. Thus, in many a case in which sexual selection was at work on the male, Wallace's selective forces of protection and recognition would play no role. Wallacean natural selection could account for the female's duller coloration only if the system of inheritance had happened to be such as to leave her free of the male's sexually selected colours. Admittedly, Darwin did think that, although inheritance by both sexes was the general rule, nevertheless the males' sexually selected characteristics tended to be sex-limited rather more often than most characteristics were. So that did leave a place for Wallacean forces. But, as we'll see, he made surprisingly little use of them. I must stress that Darwin of course agreed with Wallace both that the female's role in reproduction may subject her to stringent selection pressures and that the principles of protection and recognition were a legitimate part of Darwinian theory. But he disagreed that the female's coloration in cases of dimorphism could generally be explained along these lines. As Wallace said, Darwin 'recognises the necessity for protection [in some cases] ... but he does not seem to consider it so very important an agent in modifying colour as I am disposed to do' (Wallace 1891, p. 138).

The most telling indication that this was Darwin's position lies in the evidence of his gradual change of mind, a change from largely accepting Wallace's view to broadly disagreeing with it. This change is not hard to trace. In the fourth (1866) edition of the *Origin*, in which he expands his discussion of sexual selection, he allows that sexual dimorphism in birds can sometimes be explained by sexual selection on the male and natural selection on the female. He cites two cases – one of structure, the other of coloration. The peahen, he says, would be encumbered during incubation with a tail as long as the male's; and the female capercailzie would be dangerously conspicuous during incubation if she were as black as the male. But in the sixth (1872) edition this discussion is omitted (Peckham 1959, p. 372). And in *Descent* he recants the remarks he had made in the *Origin*, claiming that sexual selection on the male plus natural selection on the female is not generally the cause of sexual dimorphism in birds:

> In my 'Origin of Species' I briefly suggested that the long tail of the peacock would be inconvenient, and the conspicuous black colour of the male capercailzie dangerous, to the female during the period of incubation; and consequently that the transmission of these characters from the male to the female offspring had been checked through natural selection. I still think that this may have occurred in some few instances: but after mature reflection on all the facts which I have been able to collect, I am now inclined to believe that when the sexes differ, the successive variations have generally been from the first limited in their transmission to the same sex in which they first appeared.   (Darwin 1871, ii, p. 154)

Similarly, he says that formerly he 'was inclined to lay much stress on the principle of protection, as accounting for the less bright colours of female birds' (Darwin 1871, ii, p. 198) but he now takes the view that, although in some cases the females 'may possibly have been modified, independently of the males, for the sake of protection' (Darwin 1871, ii, p. 200), nevertheless 'Whether the females alone of many species have been thus specially modified, is at present very doubtful' (Darwin 1871, ii, p. 197). And he concludes that Wallace is wrong: 'I cannot follow Mr Wallace in the belief that dull colours when confined to the females have been in most cases specially gained for the sake of protection' (Darwin 1871, ii, p. 223). He deals with butterflies and moths in much the same way; he allows that the colours of some species, even bright ones, are protective but claims that, all the same, this is not generally so, even when the females are dull (e.g. Darwin 1871, i, pp. 392–3, 399, 409). And similarly he holds that although mammals are protectively coloured, 'yet with a host of species, the colours are far too conspicuous and too singularly arranged to allow us to suppose that they serve for this purpose' (Darwin 1871, ii, p. 299).

Darwin's preference for sexual selection plus the legacy of inheritance, rather than selection for protection, is also revealed in the way he deals with cases of dull coloration in both sexes. He concedes that in some species in which the sexes are alike and are dull there is no doubt that there has been selection for protection (Darwin 1871, ii, pp. 197, 223–6). But he nevertheless emphasises that even dull coloration in both sexes (whether the same or dimorphic) does not necessarily indicate that selection for protection, rather than sexual selection, is the cause – tempting as it might be to slip into that assumption. After all, as he rightly points out, the females of those species might prefer dull males (a reasonable assumption, but not one that he usually makes):

> I wish I could follow Mr Wallace to the full extent; for the admission would remove some difficulties ... It would ... be a relief if we could admit that the obscure tints of both sexes of many birds had been acquired and preserved for the sake of protection ... [that is, in those birds] with respect to which we have no sufficient evidence of the action of sexual selection. We ought, however, to be cautious in concluding that colours which appear to us dull, are not attractive to the females of certain species ...   (Darwin 1871, ii, pp. 197–8)

He adds that even if sexual selection has not been at work in these cases, it is better to plead ignorance than to conjecture about selection for protection in the absence of further evidence: 'When both sexes are so obscurely coloured, that it would be rash to assume the agency of sexual selection, and when no direct evidence can be advanced shewing that such colours serve as a protection, it is best to own complete ignorance of the cause' (Darwin 1871, ii, p. 226). Such caution is surely out of place (and, by the way, uncharacteristic of Darwin); it is plausibility, not proof, that is at issue.

Significantly, Darwin is far more ready to invoke protection when sexual selection could not be involved. He notes that the colour of the eggs of two Australian cuckoos is more closely matched to that of the cuckoos' hosts when the nest they parasitise is open than when it is covered (Peckham 1959, p. 393). And he seizes on Wallace's solution to the puzzle of strikingly conspicuous caterpillars – that they are advertising their distastefulness to predators (Darwin, F. 1887, iii, pp. 93–4; Darwin, F. and Seward 1903, ii, pp. 60, 71, 91–2; Marchant 1916, i, pp. 235–6).

So much for the evidence of Darwin's position. Now to the difficulty that it raises: Why did he adopt it? As I have said, I can't come up with a clear-cut answer. But there are some considerations that seem relevant.

Darwin felt driven to his conclusions about the importance of heredity by his own empirical investigations (published in *The Variation of Animals and Plants under Domestication* (1868)), particularly his findings on equal inheritance and sex-limitation. These were results that he considered to be experimentally very well founded. It seemed to him that to have conceded entirely to Wallace's position would have been inconsistent with his discoveries. But this can't be the entire answer. Surely, even given these findings, Darwin was unnecessarily non-adaptationist when it came to the workings of heredity. There is, after all, every reason to assume that the processes he examined (what, in particular, we would now recognise as hormonal influences (Ghiselin 1969, p. 226)) would themselves have been subject to natural selection over evolutionary time. What is more, in adopting the view he did, Darwin was taking up a position that was somewhat unusual for him. We have noted that, within Darwinian thinking, there has been a long-standing division between those who stress the strong influence of heredity and the developmental constraints it imposes, and those who stress the immense power of selection (these two positions being the Darwinian descendants of idealist and utilitarian ways of thinking). Darwin can generally be found in the power-of-selection camp. On this issue, he leaned unaccountably far in the other direction.

Contrast Darwin's position with that of Wallace. Wallace came down firmly in favour of the importance of selection: 'I think selection more powerful than the laws of inheritance, of which it makes use' (Darwin, F. and Seward 1903, ii, p. 86). Wallace complained that, on Darwin's explanation, the female's coloration bore no relation to the selective forces acting on her; it became a matter of mere 'chance':

> If this is explained solely by the laws of inheritance, then the colours of one or other sex will be always (in relation to the environment) a matter of chance ... It is contrary to the principles of *Origin of Species*, that colour should have been produced in both sexes by sexual selection and never have been modified to bring the female into harmony with the environment.
>
> (Darwin, F. and Seward 1903, ii, pp. 86–8)

Once again, it seemed to him that Darwin was unjustifiably diminishing the scope of natural selection – this time because he was overlooking how useful dull coloration could be:

> Your view appears to me to be opposed to your own laws of Natural Selection and to deny its power and wide range of action. Unless you deny that the general

dull hues of female birds and insects are of *any use to them*, I do not see how you can deny that Natural Selection must tend to increase such hues, and to eliminate brighter ones. I could almost as soon believe that the *structural adaptations* of animals and plants were produced by 'laws of variation and inheritance' alone, as that what seem to me equally beautiful and varied adaptations of *colour* should be so produced.

(Unpublished letter from Wallace to Darwin, quoted in Kottler 1980, p. 217)

As Wallace said in his review of *Descent*, Darwin is 'unnecessarily depreciating the efficacy of his own first principle when he places limited sexual transmission beyond the range of its power' (Wallace 1871, p. 181).

Darwin's empirical findings on heredity were his stated reason for not following Wallace. But his position also reflects a lack of interest in protective coloration, particularly dull coloration, that permeates all his work. The contrast with Wallace poignantly parallels their own places in the history of Darwinism. Whereas Wallace worried away at the selective forces that gave rise to the dowdy and cryptically coloured, Darwin was captivated by the gaudy and conspicuously decorated. Their different interests emerge even in the first public statement of their theory, the joint communication of 1858. Darwin brings in sexual selection but does not mention cryptic coloration; Wallace does the opposite (Darwin and Wallace 1858, pp. 94–5, 102, 106). And a decade later Wallace gradually retreated from sexual selection after initially accepting it, whereas Darwin increasingly withdrew from explaining the duller half of sexual dimorphism by selection for protection. At one point in their correspondence Darwin wrote to Wallace: 'I have formerly paid far too little attention to protection' (Darwin, F. and Seward 1903, ii, p. 73). But further comments from this correspondence more accurately sum up his final position and show the contrast with Wallace: 'I am fearfully puzzled how far to extend your protective views with respect to the females in the various classes. The more I work, the more important sexual selection apparently comes out' (Darwin, F. 1887, iii, p. 93) and 'the farther I get on [with sexual selection] the more I differ from you about the females being dull-coloured for protection' (Darwin, F. and Seward 1903, ii, p. 84). It is also typical that Wallace takes the cause of coloration in birds to be the type of nest, whereas for Darwin the colour is the cause and the nest is the effect; in Wallace's view, colour is eminently modifiable by natural selection whereas for Darwin it is so fixed by laws of inheritance that, if the bird is to obtain protection, natural selection must modify behaviour (the type of nest built) (Darwin 1871, ii, pp. 171–2; Wallace 1891, pp. 135–6; see also Wallace 1889, pp. 278–9).

One can't criticise Darwin for pursuing what most interested him. But this pursuit did blinker him. As Wallace rightly objected, Darwin tended to treat dowdiness in females as mere dowdiness, rather than as camouflage. Darwin tended to overlook the fact that in many cases of dimorphism the female is not merely dowdier than the male, she is cryptically coloured; her coloration is apparently adaptive. In those cases in which the dimorphism is slight it may be plausible to explain her colours by the incomplete hereditary transference of the male's sexually selected characteristics (although even that assumes greater indifference on the part of natural selection to slight variations than Darwin generally assumed). But what of those cases in which the female is not merely dull but camouflaged, and in fine detail? Wallace pointed out that this was true of the majority of birds (Darwin, F. and Seward 1903, ii, p. 87). Why, he asked, 'should the colour of so many female birds seem to be protective, if it has not been made protective by selection [?]' (Darwin, F. and Seward 1903, ii, p. 87; emphasis omitted). Similarly, he complained that Darwin did not satisfactorily explain why some females were the same as their bright males and others were dull and entirely different:

> this theory does not throw any light on the causes which have made the female toucan, bee-eater, parroquet, macaw, and tit in almost every case as gay and brilliant as the male, while the gorgeous chatterers, manakins, tanagers, and birds of paradise, as well as our own blackbird, have mates so dull and inconspicuous that they can hardly be recognised as belonging to the same species.
>
> (Wallace 1891, p. 124)

And it was the same with dimorphism in butterflies. Selection for protection must have been at work for it is an 'otherwise inexplicable fact, that in the groups which have a protection of any kind independent of concealment, sexual differences of colour are either quite wanting or slightly developed' (Wallace 1891, p. 80).

Wallace had good grounds for this complaint. Darwin of course accepted in principle that when colours appear to be protective an adaptive explanation is called for. But he had little interest in pursuing the principle when it came to dull coloration. Sometimes he even seemed to slip into the assumption that sexual selection and heredity alone were sufficient to explain 'many' cases of coloration. He remarked to Wallace, for example, that 'variations leading to beauty must often have occurred in the males alone, and been transmitted to that sex alone. Thus I should account in many cases for the greater beauty of the male over the female, *without the need of the protective principle*' (Darwin, F. and Seward 1903, ii, p. 74; my emphasis). But the fact that the female's colour is less flamboyant than the male's doesn't eliminate the need to explain it adaptively. Darwin's theories of sexual selection and inheritance may jointly explain both male coloration and the occurrence of dimorphism. But they cannot explain the female's coloration in the many cases in which she appears to have been protected; these appear to demand an adaptive explanation.

Or was Darwin, after all, meeting that demand? Was he simply assuming, without stating, that both sexes started off protectively coloured before sexual selection came on the scene? In that case the protective principle would indeed not need to be reinvoked to explain the females' continuing dull colours. Malcolm Kottler, the commentator who has written most extensively on this aspect of Darwinian history, suggests that Darwin neglected the question of whether natural selection *kept* the females protectively coloured, whilst concentrating on showing that natural selection could not *make* them so (Kottler 1980, p. 204). So perhaps some of Darwin's statements need to be interpreted more generously; perhaps they are premised on the assumption that the females are already protectively coloured. After all, it turns out that Darwin was not alone in making such statements.

Take, for example, Kottler himself. Although he makes the distinction explicit, even he sounds all too often just like Darwin on this point. He tells us, for instance, that Darwin was right, in cases of sex-limited inheritance, not to invoke selection to explain the dull colours of female birds because sexual selection and the laws of inheritance alone could account for them: 'female choice alone, in conjunction with sex-limited inheritance from the first of the variations sexually selected in the male, would produce a conspicuous male and inconspicuous female; in such cases, *natural selection for the sake of protection of the sex in greater danger was unnecessary*' (Kottler 1980, p. 204; my emphasis). He seems to be forgetting (but surely he's not) that the female is likely to be not merely 'inconspicuous' but camouflaged. Kottler also says approvingly that 'Darwin attributed the coloration of the less conspicuous sex to the sex-limited inheritance from the first of the color variations sexually selected in the more conspicuous sex' (Kottler 1980, p. 204). Again, he appears to be overlooking the many cases in which it is not enough to 'attribute' her coloration merely to the fact that she does not inherit his. Similarly, discussing correlations between coloration and type of nest, he concludes: 'The results were just as Wallace described, but they had been produced *without the action of natural selection for protection*' (Kottler 1980, p. 219; my emphasis). That sounds as if he's ignoring the fact that without natural selection one can explain some dull colours, but not protective ones. Is it merely that Kottler begins with the assumption that the females are protectively coloured, so that no further explanation is required for their side of the dimorphism? He certainly refers to their coloration as a 'clearly adaptive trait' and even agrees with Wallace that it is 'manifestly adaptive' (Kottler 1980, pp. 204, 217). And yet it's not so manifest, for, in the same breath, he castigates Wallace for over-zealous adaptationism in explaining female coloration (e.g. Kottler 1980, pp. 204, 219). Perhaps he really does feel that Darwin was 'without the need of the protective principle'?

Michael Ghiselin (1969, pp. 225–9, 1974, pp. 131, 178), too, if I understand him aright, feels little need to explain the female's dull colours adaptively. He notes that Wallace's explanation was adaptive whereas Darwin's was not, but sees this as a strength of Darwin's position as compared with Wallace's dogmatic adaptationism. He commends Darwin for relying on the laws of heredity to explain characteristics that appeared 'adaptively neutral' or 'maladaptive' (e.g. Ghiselin 1974, p. 178). Darwin does use heredity to explain characteristics of this kind, such as the horns of the female reindeer, which could be viewed as such a case (Marchant 1916, i, p. 217); but dimorphism also covers cases that call for an adaptive explanation.

Incidentally, talking of horns on female deer, it might appear that Darwin could have explained conspicuous coloration in females by sexual selection. His theory did not exclude the possibility of male choice of mates. Indeed, he even assumed that male choice was routine in humans and occurred occasionally in other animals. But he stressed that in general mate choice was almost exclusively female choice (his rationale, as we have noted, being based on the idea that sperm is generally carried to the egg – a very poor rationale indeed).

An obvious consideration in explaining Darwin's position is that Wallace's attempts to apply the principle of protection simply weren't very compelling. The principle certainly is more convincing for explaining the gross dimorphism of a gaudy, long-tailed male bird and his modestly-dressed mate than for explaining the many cases (which occur particularly among reptiles and mammals) in which the female's colours are only slightly duller. Darwin and Wallace discussed such points in some detail (see Kottler 1980). They raised the question, for example, of why in some cases when the male bird incubates, the differences between the relatively bright female and dull male are so very slight; how can this afford the female significantly greater protection? Conversely, if even these slight differences are for protection, why are female reptiles a little less conspicuous than males even though they are not apparently in need of extra protection because they do not incubate the eggs? And why in some species of fish is the non-incubating female less conspicuous than the incubating male? (On that one, Wallace suggested that the males are protected in some other way (Marchant 1916, i, pp. 177, 225).) Such questions remind us that Wallace concentrated on the two groups – birds and butterflies – in which sexual dimorphism in colour is generally most marked. He considered that they provided the best test for deciding between his views and Darwin's (Wallace 1889, pp. 275–6, 1891, p. 353). But from the point of view of selection for protection it is the apparently marginal differences that pose the greater problem. (Although one notorious case of very marked dimorphism – mimicry in butterflies being so often limited to the females – is still unexplained today (see e.g. Turner 1978); Darwin claimed for at least one species that the conservatism of female taste kept the males in their original colours, whereas Wallace explained the dimorphism as resulting from the females and males inhabiting different environments (Wallace 1891, p. 373).) Nevertheless, even allowing for all that, Darwin could have accepted selection for protection more wholeheartedly. For surely he, of all people, appreciated how significant in the eyes of natural selection even very slight differences can be.

And so we end up with no very satisfactory explanation of Darwin's preference for heredity over selection for protection. Its consequences parallel the consequences of Wallace's intransigence over sexual selection. The obvious solution to the disagreement between the two of them was to give the males to Darwin and the females to Wallace. But neither took this simple course. Wallace, in his eagerness to dispense with sexual selection, tended to overlook the fact that the males were not merely bright but 'designed'. Darwin, in his eagerness to explain colour by sexual selection, tended to overlook the fact that the females were not merely dull but 'designed'. Wallace explained female coloration adaptively but could not adequately account for male coloration. Darwin explained male coloration adaptively but was half-hearted over accounting for female coloration. One can appreciate how Wallace backed himself into his position. But Darwin's is a minor mystery.

Wallace's legacy: A century of natural selection

I'd be the first to recognise Wallace as a Darwinian to whom we are indebted; he was an imaginative thinker and had a deeply empathetic grasp of natural selectionist principles. But when it comes to sexual selection, he has a lot to answer for. Or, rather, he and his successors do. For their legacy to Darwinism was – to exaggerate only slightly – one hundred years without sexual selection, one hundred years in which natural selection was made to account for all the lavish beauty, all the ornamental flourishes that Darwin attributed to mate choice. Sexual selection wasn't ruled out entirely. Most Darwinians happily conceded that it played a role in evolution. But that role was seen as a very minor one indeed. Natural selection was viewed as the real driving force; sexual selection was just an uninfluential frill, a marginal extra, not something that made a real difference to the trill of a bird's song or the colour of its feathers. So dismissive, so dogged an attitude seems astonishing now, since our eyes have been opened to sexual selection in the last decade or so. But this was indeed the attitude that dominated almost a century of Darwinian thinking.

Darwin had admittedly been unashamedly imperialistic in staking claims for sexual selection; at the end of his life even he declared: 'It is ... probable that I may have extended it too far' (Darwin 1882: Barrett 1977, ii, p. 278). Whether he was right or not, he certainly extended it far too far for the majority of Darwinians from then until very recently. They turned instead to Wallace's natural selectionist alternatives. Wallace had led the way with the principles of protection and recognition. Gradually, those who followed him refined these principles and extended the list to include other selective forces. Eventually not only coloration but sounds, scents, structures and other characteristics that Wallace had largely neglected were swallowed up by natural selection. I should stress that such explanations were not necessarily wrong. Indeed, they were very likely right in many cases. What was wrong was to view them as replacing sexual selection, to view them as precluding sexual selection from making any significant contribution to evolution. Generations of Darwinians were brought up on the single-minded view that natural selection was the only force that really mattered, and that it could eventually win back all or most of the cases that Darwin had given to sexual selection.

Let's get a picture of how this natural selectionist programme looked. What might an undergraduate have been taught about the importance (or, rather, unimportance) of sexual selection even just a couple of decades ago?

Consider the question that proved so tricky for Wallace: male display. Darwin asserted firmly that the only alternative to explaining display by sexual selection was to assume that it had no purpose. He said of birds: 'To suppose that the females do not appreciate the beauty of the males is to admit that their splendid decorations, all their pomp and display, are useless; and this is incredible' (Darwin 1871, ii, p. 233; see also i, pp. 63–4, ii, p. 93). Discussing the display of fish, he asked: 'Can it be believed that they would thus act to no purpose during their courtship? And this would be the case, unless the females exert some choice and select those males which please or excite them most' (Darwin 1871, 2nd edn., p. 524). Similarly, with butterflies: 'on any other supposition the males would be ornamented, as far as we can see, for no purpose' (Darwin 1871, i, p. 399; see also 2nd edn., pp. 505–6).

Rather than concede that display was the result of sexual selection, Wallace did indeed opt for no selection at all. But his successors were stauncher

Wallaceans. Rejecting Darwin's dichotomy of sexual display or no purpose, they looked for alternative ways of explaining such features adaptively.

The category of threat, for example, was thought to cover many of the more flamboyant characteristics (an idea that Wallace entertained but eventually rejected (Wallace 1889, p. 294, 1891, p. 377)). Here is how Julian Huxley, one of the founders of the modern Darwinian synthesis, who came to be regarded as an authority on sexual selection, dealt with the problem half a century after Wallace: 'Many conspicuous characters (bright colors, songs, special structures or modes of behavior), to which Darwin assigned display function, have now been shown to have other functions ... Of these, that of threat characters includes a large number, probably the majority, of the cases adduced by Darwin as subserving display and therefore evidence of the existence of sexual selection' (Huxley 1938, p. 418). R. W. G. Hingston even went so far as to claim that all conspicuous coloration and male ornament were conventional signals for threatening conspecifics and members of other species; they were 'intimidating machinery' of which a 'rival will know the meaning' (Hingston 1933, pp. 11–12).

Another solution lay in the category of 'epigamic' characteristics. This came to refer to characteristics that are to do with mating and are typically associated with display but do not generally involve female choice. The idea was that the male's finery is needed to interest the female in mating – it was usually felt that females are 'coy' and difficult to arouse – but that her response is too automatic and passive to be deemed 'choice'. This approach was anticipated by the notoriously anti-Darwinian critic St George Mivart: 'the female does not select; yet the display of the male may be useful in supplying the necessary degree of stimulation to her nervous system' ([Mivart] 1871, p. 62). The notion was developed particularly by Poulton, the most important of Wallace's immediate successors in the natural selectionist tradition. Although he originally stressed the role of female choice in explaining male ornament (Poulton 1890), he later came to place less emphasis on it. It was he who coined the term 'epigamic' (Poulton 1890, pp. 284–313, 1908, 1909, pp. 92–143, 1910).

This solution became a mainstay of natural selectionists. At the turn of the century Karl Groos, in his widely-read book *The Play of Animals*, stated:

As sexual impulse must have tremendous power, it is for the interest of the preservation of the species that its discharge should be rendered difficult ... [T]he hindrance to the sexual function that is most efficacious ... is the instinctive coyness of the female. This it is that necessitates all the arts of courtship, and the probability is that seldom or never does the female exert any choice. She is not the awarder of the prize, but rather a hunted creature ... there is choice only in the sense that the

hare finally succumbs to the best hound, which is as much as to say that the
phenomena of courtship are referred at once to natural selection.

(Groos 1898, p. xxiii)

This view, he said, 'does away with all choice, and relegates the whole
subject to the sphere of natural selection'; sexual selection simply becomes 'a
special case of natural selection' (Groos 1898, pp. xxii, 244, 271). A few
years later, we find Kellogg, in his summary of the objections to sexual
selection, stating that the kind of evidence Darwin relied on often turns out to
be

more illustrative of sexual excitation of females resulting from the perception of
odour or actions, than any degree of choice by females ... [Colours, serenades and so
on] probably do exercise an exciting effect on the females, and are probably actually
displayed for this purpose. But does this in any way prove, or even give basis for a
reasonable presumption for belief in a discriminating and definitive choice among
the males on the part of the female?   (Kellogg 1907, pp. 115, 117–18)

A well known entomologist, O. W. Richards (he later became Professor of
Entomology at Imperial College, London), claimed that in many insects the
function of the male's behaviour and structure was to overcome female
'coyness'; the advantage was the saving of time spent on mating (Richards
1927). 'It has become obvious since Darwin wrote', he concluded, 'that
display-characters are probably acquired most often as a result of Natural
rather than Sexual Selection' (Richards 1927, p. 300). Shull dismissed sexual
selection, claiming that Darwin's phenomena were merely to do with
arousing sexual excitement (Shull 1936, pp. 194–8). And here is Julian
Huxley again: 'display may induce a psycho-physiological state of readiness
to mate, irrespective of any possibility of choice. In birds, display may
synchronize male and female rhythms of sexual behaviour ... and initiate
physiological changes leading to ... ovulation ... These effects directly
promote effective reproduction and need no special category of "sexual
selection" to explain their origin' (Huxley 1938, pp. 422–3; see also 1914,
1921, 1923). (Incidentally, in favour of this view he stressed first, that in
some bird species most of the mating ceremonies take place only after the
birds have paired up for the season – a point that Darwin did not deal with –
and, second, that both partners perform the display – which was the kind of
monomorphism that Darwin attributed to the laws of inheritance; so, in such
cases, Huxley argued (though not in these words), 'female choice' was
neither female nor choice (e.g. Huxley 1914, 1921, 1923).) For several
decades it was widely felt that the idea of epigamic characteristics was 'the
best solution of the riddle' (MacBride 1925, pp. 218–19) and that sexual
selection could largely be dispensed with. It was conceded that it might
sometimes be at work, but only work of the most marginal kind. The category

'epigamic', at least to the satisfaction of these critics, managed to preclude choice and thereby preclude sexual selection.

Incidentally, these arguments must have a familiar look to anyone who has come across the current debates about 'active' and 'passive' choice, or 'preference' and 'choice' (e.g. Arak 1983, pp. 192–201, 1988; Halliday 1983a, pp. 19–28; Maynard Smith 1987, pp. 11–12; Parker 1983, pp. 141–5). It was all very well for earlier Darwinians to rule out sexual selection on the grounds that females weren't really choosing. But what, their successors are now asking, is real choice? Consider, say, a hind who is corralled into the harem of a red deer. If she makes no attempt to move out of his harem is she choosing him for her mate? Does her apparent passivity rule out sexual selection? If a female natterjack toad, surrounded by a chorus of males, moves towards the call that she hears as loudest and mates with that caller, has she made a choice and, again, are we seeing sexual selection? Suppose that she is simply trying to cut the costs of delays in mating by going for the nearest male, using loudness as her cue. In that case, does this 'passive attraction ... [merely provide] a simple natural selection benefit' (Arak 1988, p. 318), the benefit of a quick choice? Should sexual selection be ruled out if there is no 'relationship between call characteristics and immediate or long-term benefits provided by males' (Arak 1988, p. 318) – if a loud call is no indication of, say, size or vigour? We might want to say that there is choice: 'Mate choice may be operationally defined as any pattern of behaviour, shown by members of one sex, that leads to their being more likely to mate with certain members of the opposite sex than with others' (Halliday 1983a, p. 4). In the case of the toads, 'females have a behavior (moving up a sound gradient) which makes them more likely to mate with loud-calling males' (Maynard Smith 1987, p. 11). Nevertheless, perhaps 'if we wished to model this situation, we would treat it as a simple case of male–male competition' rather than female choice (Maynard Smith 1987, p. 11). Well, how can we resolve such questions? That is an issue that we'll return to later.

The category that eventually came to be most favoured for dealing with display was that of ethological isolating mechanisms (e.g. Dobzhansky 1937; Grant 1963; Lack 1968, pp. 159–60; Mayr 1963). These are species-specific behavioural and structural characteristics that enable members of a species to mate only with their own kind. Like epigamic characteristics, they were seen as allowing potential mates to indulge in all kinds of showy display without being tainted by sexual selection. Ethological isolating mechanisms were confined to choosing a mate of the right species; they had nothing to do with choice of mate within a species. So they could be seen as conceding no more than Wallace's familiar category of recognition but applied to courtship. Any such 'choice' of mate, it was felt, was so much to do with speciation – and, what's more, so involuntary – that it fell comfortably within natural selection. An early (somewhat different) version of this idea was boosted by an influential book, *The History of Human Marriage* (1891), by the renowned Finnish anthropologist and sociologist Edward Westermarck. He argued that the purpose of ornamental structures was twofold: to facilitate finding mates and to prevent inbreeding (by attracting individuals from a distance) (Westermarck 1891, pp. 481–91). Ethological isolating mechanisms were congenial to Darwinians whose prime interest was speciation and it was through this influence that, from about the time of the grand synthesis, they became the most popular explanation of display. As late as the 1960s, Ernst Mayr, for example, was claiming that courtship patterns 'all ... ultimately serve, directly or indirectly, as isolating mechanisms' (Mayr 1963, p. 96; see also pp. 95–103, 126–7). And much the same plea for species recognition rather than sexual selection can still be heard today:

> I draw the sharpest possible distinction between mate selection and specific-mate recognition: I wish to accept no greater a commitment to conscious judgement in sexual organisms than is involved in the recognition of an antigen by an antibody, unless compelled by evidence to do so ... [M]y credulity is tested to the full when asked to consider mate selection in very many plants, fungi, protistans, and even animals like oysters.   (Paterson 1982, p. 53)

The natural selectionist view also came up with several extensions of Wallace's suggestions as to how conspicuousness, which is apparently so dangerous, could actually be protective against predation. James Mottram (a medical expert with an interest in camouflage), for example, appears to have had some such mechanism in mind when he claimed to have found a 'correlation between extra-sexual dimorphism among birds and their vulnerability to enemies' (Mottram 1915, p. 663); as a general rule, he said, bird species that are most liable to predation are more sexually dimorphic than fiercer, larger or more sociable species. Although he made no attempt to explain how such dimorphism could be protective (elsewhere he proposed a curious group-selectionist alternative to sexual selection (Mottram 1914)), he concluded that 'Darwin's theory ... in no way can account for the correlation' and that sexual differences were probably less to do with sexual selection than with 'escape from enemies' (Mottram 1915, pp. 674, 678). Hugh B. Cott, a Cambridge zoologist, later developed a version of Mottram's idea (Cott 1946). He noticed one day, when preparing bird skins in Egypt, that hornets were feasting on the discarded skins of Palm Doves but avoiding those of the Pied Kingfisher. The more conspicuous species was apparently distasteful. He conjectured that vulnerable birds – species that are small, ground-living, lacking in defensive weapons and so on – had been 'forced ...

along one of two lines of specialization: those which are relatively palatable seeking safety in concealment; those which are relatively distasteful, in advertisement' (Cott 1946, p. 506). Careful tests on the preferences of hornets, backed with information on those of cats and humans, seemed to support this conjecture. Cott appreciated that the members of his tasting panel were not the birds' natural predators; but 'this concurrence of taste appears all the more remarkable when found ... in three creatures so utterly different in organization and habits' (Cott 1946, p. 465) – remarkable enough to suggest that natural predators would also be likely to concur. Cott allowed that sexual selection could have been at work when a relatively non-vulnerable species sported showy colours. But relatively vulnerable birds that were conspicuous, he predicted, would generally turn out to be unpalatable; their florid hues were to do with the 'interspecific struggle for safety, as opposed to the intraspecific struggle for reproduction' (Cott 1946, p. 501).

More recently, Robin Baker and Geoffrey Parker (1979), going further along this trail, also concluded that predation has been far more important than sexual selection in the evolution of showy plumage. According to their 'unprofitable prey' theory, birds evolve bright coloration in order to advertise themselves to predators. This is not as counter-intuitive as it might at first sight seem. The claim is that it is those birds that are the most difficult to catch – the swiftest, say, or most sharp-sighted – that are also the most brightly coloured; they are informing potential predators that attempted predation will yield a low return compared with prey that are striving to look inconspicuous: 'You can't catch me. Go for the ones that are trying to hide'. Baker and others have subsequently attempted to interpret a variety of data as supporting evidence (Baker 1985; Baker and Bibby 1987; Baker and Hounsome 1983). This has generated a vigorous discussion and not much agreement (Andersson 1983a; Krebs 1979; Lyon and Montgomerie 1985; Reid 1984). To take just two examples out of the range of difficulties: What exactly would constitute evidence – is, say, predation by domestic cats relevant given that it would not have been a selection pressure over evolutionary time? And what about data that the theory apparently cannot handle, such as the timing of moults in birds with seasonal changes to their conspicuous plumage?

Incidentally, one nineteenth-century naturalist claimed that natural selection was trying to achieve the exact opposite of an 'unprofitable prey' result. Jean Stolzmann (Stolzmann 1885) maintained – in all seriousness – that male flamboyance among birds was natural selection's way of getting rid of excess males. Eggs developed more readily into males than into females because male embryos required less nourishment. But these spare males used up resources and bothered the females to no evolutionary advantage. Natural selection had hit on several clever solutions, which Darwin had mistaken for sexual selection. Conspicuousness helped predators to spot their prey and females to spot their persecutors; song and dance rituals kept males busy and out of the females' way; cumbersome feathers impeded flight and left more insects for the females. Stolzmann appreciated that all this would not be good for the males themselves; but, he insisted, it would undoubtedly be good for the species. And at least, he said (rather smugly), his explanation stuck to natural selection and didn't resort to 'un agent aussi artificiel que la sélection sexuelle' (Stolzmann 1885, p. 429).

Finally, sexual dimorphism has also been explained by an idea that Wallace appealed to extensively – that of different selective forces acting on males and females (for reasons other than mating). This now has the name 'ecological differentiation' – the two sexes being adapted to different ecological niches. Wallace's application of the theory was very limited. He was rather unwilling to attribute the differences to anything other than the female's greater need for protection (e.g. Wallace 1889, p. 271, 1891, p. 80). And generally, as in the case of the relatively dull colour of nesting female birds, he failed to account for both sexes. Nowadays, ideas of ecological differentiation range much wider. It has been suggested, for example, that some bird species follow a Jack Sprat principle of males and females exploiting different food resources because both benefit from the reduced competition for limited supplies (Selander 1972; cf. Darwin 1871, ii, pp. 39–40).

This, then, was how the natural selectionist programme developed. For nearly a century it was the Darwinian orthodoxy on extravagant male characteristics. Mate choice as a selective force was not ruled out but it was generally agreed that Darwin had vastly overestimated its scope. Most of Darwin's evidence, however dazzling, however extravagant, was put down to protection, threat, isolating mechanisms, or some other utilitarian pressure. Looking back now, since the spectacular revival of sexual selection in the last decade or two, it seems hardly credible that so many Darwinians for so long could believe that, as Mayr put it in the 1960s, 'The song of the nightingale belongs here [with natural selection] and so does the strutting of the peacock' (Mayr 1963, p. 96). Nevertheless, this was the tradition that held sway over both theoretical and empirical research during that century-long period. 'To such an extent does the enticing idea of the all-puissance of natural selection dominate the minds of scientific men that but few of them have paid any attention to the question of sexual selection ... "Natural selection explains everything, why then investigate further?" seems to be the general attitude of our present-day naturalists' (Dewar and Finn 1909, p. 308). This was the comment of Douglas Dewar (later a notorious anti-Darwinian) and Frank

Finn, two commentators who were critical of the majority view, writing at the turn of the century. Their description turned out to characterise not only their day but several decades to come.

The position adopted by E. B. Poulton was typical; it was also highly influential, for he was the foremost exponent of Darwinian theories of coloration in the decades around the turn of the century (Poulton 1890, 1908, 1909, pp. 92–143, 1910). He did not reject sexual selection. Indeed, in his *Colours of Animals* (1890) he defended the theory and stressed the role of female choice. It is perhaps for this reason that he is often mistakenly seen as a staunch supporter of Darwin's position and a theorist of sexual selection (e.g. George 1982, p. 77; Kottler 1980). But Poulton lost his initial enthusiasm for the theory. Whilst still allowing that sexual selection occurred, he came to relegate it to a very minor position, maintaining that it was 'relatively unimportant' in evolution (Poulton 1896, p. 79) and making only grudging concessions to its role: 'Probably the majority of naturalists are convinced by Darwin's arguments and his great array of facts that the principle of sexual selection is real, and accounts for certain relatively unimportant features in the higher animals, and they further accept Darwin's opinion that its action has always been entirely subordinate to natural selection' (Poulton 1896, p. 188). Poulton instead devoted most of his energies to subsuming Darwin's ornaments under Wallacean selective pressures. And subsequent Darwinian experts on coloration followed him in the view that sexual selection was relatively unimportant (e.g. Beddard 1892, pp. 253–82).

By the 1930s, Julian Huxley had come to be regarded as one of the major experts on sexual selection. And yet he was writing Darwinian position papers on the current standing of the theory that were almost entirely devoted to repackaging it as natural selection (Huxley 1938, 1938a). According to him, much of Darwin's evidence had nothing to do with mating, let alone sexual selection; Darwin, he declared *ex cathedra*, 'persistently attached too much weight to the view that bright colours and other conspicuous characters must have a sexual function ... it has now become clear that the hypothesis ... is inapplicable to the great majority of display characters ... Darwin's original contention will not hold' (Huxley 1938a, pp. 11, 20–1, 33). Indeed, much of the neglect of R. A. Fisher – a crucial figure, who, we shall see, brilliantly vindicated Darwin – has been attributed to Huxley's influence (O'Donald 1980, pp. ix, 2, 10–15; Parker 1979). Huxley's position is epitomised by his assessment of Cott's book, *Adaptive Coloration in Animals* (Cott 1940). He praised it as 'a worthy successor to Sir Edward Poulton's *The Colours of Animals* ... The one was a pioneer study, the other is in many respects the last word on the subject' (Cott 1940, p. ix); yet Cott specifically confined his study to prey–predator relations and excluded any discussion of selective forces within species.

Even as late as the centenary celebrations for *Descent of Man*, the natural selectionist alternative still emerged as the majority view. Darwin's theory of mate choice had become so comprehensively eclipsed that, astonishing as it now seems, in the preface to one of the few celebratory volumes (Campbell 1972), Julian Huxley could still be cited as the standard authority, with R. A. Fisher not even given an honourable mention. Mayr's contribution to this volume is typical. Citing Huxley and Richards as his two authorities, he claims: 'It is now evident ... that there are three major ... selection pressures which favor the development or enhancement of sexual dimorphism, without requiring sexual selection' (Mayr 1972, p. 96) and he lists epigamic selection, isolating mechanisms and the utilisation of different niches by male and female. The only paper in the volume that stands out against this trend is that of Robert Trivers, one of the leading figures in the recent Darwinian revolution. As he remarks there: 'most writers since [Darwin] ... have relegated [female choice] ... to a trivial role ... With notable exceptions the study of female choice has limited itself to showing that females are selected to decide whether a potential partner is of the right species, of the right sex and sexually mature' (Trivers 1972, p. 165). Even in a collection of papers published a few years later, on sexual selection in insects (Blum and Blum 1979), the historical survey describes Huxley's views with enthusiasm (Otte 1979), although the major contribution of that particular expert was to promote the natural selectionist movement.

It is clear, then, that Darwin's theory of mate choice was widely dismissed. But why? And how did it eventually get revived? It is these questions that we shall now answer.

7

Can females shape males?

Alternatives to sexual selection were not enough. Wallace would strengthen his position if he could also undermine the very idea that female choice could be a selective force, let alone a force powerful enough to create the peacock's tail. And this was, indeed, his second line of attack: an onslaught on the central mechanism of sexual selection.

He proposed three arguments: that female choice requires an aesthetic sense that few animals, perhaps none, possess; that even if females do prefer the ornaments of some males to those of others, nevertheless this doesn't influence their choice of mates; and that even if females did choose their mates on aesthetic grounds their taste would be too undiscriminating and fickle to give rise to the males' intricate adornments.

Only humans can choose

Wallace maintained that female choice called for aesthetic powers that only humans were likely to possess. Such refined choice was probably beyond the capacities even of those animals closest to us, and definitely went far beyond the capacities of, say, fish, insects and other 'lower' animals – certainly beyond those of lowly butterflies, which Darwin relied on as an important source of evidence. Wallace had begun to have doubts about insects even at the period when Darwin confined sexual selection largely to birds and insects and Wallace accepted it for birds: 'Passing ... to the lower animals ... the evidence for sexual selection becomes comparatively very weak; and it seems doubtful if we are justified in applying the laws which prevail among the highly organized and emotional birds, to interpret somewhat analogous results in their case' (Wallace 1871, p. 181). According to his autobiography, this was why he originally came to reject sexual selection (Wallace 1905, ii, p. 18) (though he finds other reasons elsewhere (e.g. Wallace 1891, p. 374)).

Many other critics felt the same way. For Stolzmann, the idea of aesthetic taste in birds was a prime reason for rejecting sexual selection (and replacing it, as we have seen, with his idiosyncratic alternative): 'Au premier abord, il nous est difficile d'admettre chez les femelles des oiseaux la présence d'un goût esthétique si fortement développé comme le signale Darwin' (Stolzmann 1885, p. 423). Groos felt that the whole Darwinian enterprise would be the better for rejecting the idea:

> It would ... be absurd to affirm that all bird-songs originate in a conscious aesthetic and critical act of judgement on the part of the female. A conscious choice either of the most beautiful or the loudest songster is certainly not the rule, and probably never occurs at all ... The Darwinian principle ... is materially strengthened by [eliminating the idea of] ... conscious aesthetic choice on the part of the female.
>
> (Groos 1898, pp. 240, 242)

Lloyd Morgan, whilst not wanting to submerge sexual selection entirely within natural selection, nevertheless also objected to human-like choice in general and aesthetic choice in particular: 'Both upholders of sexual selection and critics of that hypothesis, have been too apt to regard the choice of a mate in animals from too anthropomorphic a point of view – to look upon it as the outcome of rational deliberation, of weighing in the aesthetic balance the relative attractiveness of this suitor and of that' (Morgan 1900, p. 266). 'Aesthetics involve ideals; and to ideals ... no brute can aspire' (Morgan (1890–1, p. 413). He used the analogy of a chick 'choosing' a juicy worm; it is, he said, an

> unnecessary supposition that the hen bird must possess a standard or ideal of aesthetic value, and that she selects that singer which comes nearest to her conception of what a songster should be. One might as well suppose that a chick selected those worms which most nearly approached the ideal of succulence that it had conceived. The chick selects the worm that excites the strongest impulse to pick it up and eat it. So, too, the hen selects that mate which by his song or otherwise excites in greatest degree the mating impulse; and there is no more need to suppose the existence of an aesthetic standard in this case than there is to hypothecate a gustatory ideal in the case of the chick that eats a juicy worm.
>
> (Morgan 1896, pp. 217–18)

Kellogg, a decade later, objected equally to aesthetics in insects. Such choice 'implies a high degree of aesthetic development on the part of the females of animals for whose development in this line we have no (other) proof. Indeed this choice demands aesthetic recognition among animals to which we distinctly deny such a development, as the butterflies and other insects ... Similarly with practically all invertebrate animals' (Kellogg 1907, p. 114). In the 1920s, Nordenskiöld's *History of Biology* claimed that one of the reasons why sexual selection had been rejected was Darwin's 'tendency to attribute without criticism purely human ideas to the animal kingdom, to believe in "beauty competitions" among butterflies, beetles, fishes and newts, or that grasshoppers and crickets have a musical ear' (Nordenskiöld 1929, p. 474).

Darwin was aware that to talk of an 'aesthetic' sense was to invite such criticism. But he insisted that something of that kind was needed for mate choice, and that humans were by no means the only animals to have evolved it; although the particular tastes of other animals might differ from ours, most of them did possess a sense of beauty (e.g. Darwin, F. and Seward 1903, i, p. 325).

Darwin maintained that, implausible as his position might seem, there was evidence to back it up: 'No doubt this implies powers of discrimination and taste on the part of the female which will at first appear extremely improbable; but I hope ... to shew that the females actually have these powers' (Darwin 1871, 2nd edn., p. 326). A jelly-fish, say, or a sea-slug wouldn't have them; but, on the whole, they would be increasingly likely to crop up as one went from insects, to birds, to mammals (e.g. Darwin 1871, i, p. 321). It might seem that the extraordinary degree of sexual dimorphism and ornamentation in birds and, even more so, butterflies might not fit comfortably into this scheme. Darwin's answer was twofold. First, 'strong affections, acute perception, and a taste for the beautiful' (Darwin 1871, ii, p. 108) do not depend on intellectual development; conversely, intelligent animals such as snakes may lack these qualities (Darwin 1871, ii, p. 31). Second, it seemed that even ants and beetles were more generously endowed with such sensibilities than might be supposed; so butterflies could not be excluded merely on grounds of 'lowliness':

> We know that ants and certain lamellicorn beetles are capable of feeling an attachment for each other, and that ants recognise their fellows after an interval of several months. Hence there is no abstract improbability in the Lepidoptera, which probably stand nearly or quite as high in the scale as these insects, having sufficient mental capacity to admire bright colours. (Darwin 1871, i, p. 399)

Darwin's position, then, was not to deny that sexual selection demanded aesthetic appreciation on the part of females but to argue that they could indeed show such appreciation. We'll come back to this point.

Not choosing, just looking

But even if females do admire showy feathers, oversized crests or bursts of song, nevertheless, Wallace argued, they do not choose their mates on that basis. To enjoy and appreciate such features is one thing. To let them influence choice of mate is altogether another. He drew an analogy with female taste in humans:

A young man, when courting, brushes or curls his hair, and has his moustache, beard, or whiskers in perfect order, and no doubt his sweetheart admires them; but this does not prove that she marries him on account of these ornaments, still less that hair, beard, whiskers, and moustache were developed by the continued preferences of the female sex. So, a girl likes to see her lover well and fashionably dressed, and he always dresses as well as he can when he visits her; but we cannot conclude from this that the whole series of male costumes, from the brilliantly coloured, puffed, and slashed doublet and hose of the Elizabethan period, through the gorgeous coats, long waistcoats, and pigtails of the early Georgian era, down to the funereal dress-suit of the present day, are the direct result of female preference.

(Wallace 1889, p. 286)

And so it is, he argued, with birds: 'In like manner, female birds may be charmed or excited by the fine display of plumage by the males; but there is no proof whatever that ... [this has] any effect in determining their choice of a partner' (Wallace 1889, pp. 286–7). So what are the females up to? Nothing at all, according to Wallace: they're just looking.

Wallace's argument assumes an implausible conjunction of non-adaptive forces. The females pay close attention to the males – to no adaptive purpose. They exercise discrimination in judgement – without selective effects. They become charmed and excited – with no evolutionary implications for their choice of mate. Certainly each of these circumstances could arise for other reasons. But for all of them to arise together is highly improbable.

'The instability of a vicious feminine caprice'

Finally, Wallace argued that even if the females did exercise choice it would not have the power to create 'sexually selected' characteristics. The eagle's sharp eyesight could account for a nesting bird merging inconspicuously into her background. But could mere aesthetic choice be so exacting or constant as to account for the intricate markings of a butterfly's wing or the complex melody of a bird's song?

'Well, why not?' you might be thinking. 'Why presume that aesthetic judgements will be less discriminating and stable than a choice of what to eat or where to nest?' We'll come back to that, and to other issues about taste. For now, let's hold back from criticising Wallace, and see instead what Darwin thought.

Darwin felt that, at first sight, aesthetic preferences may indeed not seem to be a sufficiently powerful evolutionary force for his purposes: 'It may appear childish to attribute any effect to such apparently weak means' (Darwin 1859, p. 89). Nevertheless, he claimed, the idea was not implausible. After all, look at domestic selection. There we find the systematic application of aesthetic criteria achieving the desired results: 'if man can in a short time give elegant carriage and beauty to his bantams, according to his standard of beauty, I can see no good reason to doubt that female birds, by selecting, during thousands of generations, the most melodious or beautiful males, according to their standard of beauty, might produce a marked effect' (Darwin 1859, p. 89; see also Darwin 1871, i, p. 259, ii, p. 78).

Wallace raised two objections. First, it seemed to him unlikely that females could discriminate between differences that were only very slight. How, he asked, could natural selection have evolved such exacting powers? Wallace willingly conceded that creatures such as birds and insects could distinguish different colours. How could the expert on protective coloration and the evolution of colour in flowers (e.g. Wallace 1889, pp. 304, 306–8, 316–19) think otherwise? But, according to him, animals require no more than 'a perception of *distinctness* or *contrast* of colours' whereas sexual selection demands an 'appreciation of ... infinite variety and beauty, of ... delicate contrasts and subtle harmonies of colour' (Wallace 1891, p. 409). Female powers of discrimination, he said, are too weak to distinguish such slight variations: 'I do not see how the constant *minute* variations, which are sufficient for Natural Selection to work with, could be *sexually* selected. We seem to require a series of bold and abrupt variations. How can we imagine that an inch in the tail of the peacock, or 1/4–inch in that of the Bird of Paradise, would be noticed and preferred by the female' (Darwin, F. and Seward 1903, ii, pp. 62–3). And neither does Darwin's evidence show that the females are in fact employing such fine discrimination:

> Such cases do not support the idea that males with the tail-feathers a trifle longer, or the colours a trifle brighter, are generally preferred, and those which are only a little inferior are as generally rejected, – and this is what is absolutely needed to establish the theory of the development of these plumes by means of the choice of the female. (Wallace 1889, p. 286)

His paradigm is the wonderful precision of the Argus pheasant's curious ornamentation (which Darwin remarked was 'more like a work of art than of nature' (Darwin 1871, ii, p. 92)): 'The long series of gradations by which the beautifully shaded ocelli on the secondary wing-feathers of this bird have been produced, are clearly traced out, the result being a set of markings so exquisitely shaded as to represent "balls lying loose within sockets"' (Wallace 1891, p. 374). Could so fine a pattern be appreciated by a mere bird?: 'it was [this] ... case ... which first shook my belief in "sexual" ... selection' (Wallace 1891, p. 374). In short, Wallace balks at the possibility of any creatures other than humans discriminating so minutely as to shape the

'More like a work of art than of nature'
The male Argus pheasant

detail and intricacy of male ornament. Other critics argued along similar lines. Even if we concede that birds and mammals have aesthetic feeling, they said, 'is this feeling to be so keen as to lead the female to make choice among only slightly differing patterns of song?' (Kellogg 1907, p. 114). Pointing to the fact that butterflies are attracted to such crude aesthetic stimuli as gaudy paper or monochromatic flowers, they asked whether females have a double standard – one for the refinements of male ornamentation and one for other objects (e.g. Geddes and Thomson 1889, pp. 29–30).

Darwin replied that the females could bring about exquisite ornamentation without discriminating finely, just by choosing for some general impression:

> I presume that no supporter of the principle of sexual selection believes that the females select particular points of beauty in the males; they are merely excited or attracted in a greater degree by one male than by another, and this seems often to depend, especially with birds, on brilliant colouring. Even man, excepting perhaps an artist, does not analyse the slight differences in the features of the woman whom he may admire, on which her beauty depends.
>
> (Darwin 1876a: Barrett 1977, ii, p. 210)

Similarly, he tells Wallace: 'In regard to sexual selection. A girl sees a handsome man, and without observing whether his nose or whiskers are the tenth of an inch longer or shorter than in some other man, admires his appearance and says she will marry him. So, I suppose, with the pea-hen; and the tail has been increased in length merely by, on the whole, presenting a more gorgeous appearance' (Darwin, F. and Seward 1903, ii, p. 63). Thus one need not suppose 'that the female studies each stripe or spot of colour; that the peahen, for instance, admires each detail in the gorgeous train of the peacock – she is probably struck only by the general effect' (Darwin 1871, ii, p. 123). And the same can be said for Wallace's example of the Argus pheasant: 'Many will declare that it is utterly incredible that a female bird' (his emphasis is on bird, not on female!) 'should be able to appreciate fine shading and exquisite patterns ... [But] perhaps she admires the general effect rather than each separate detail' (Darwin 1871, ii, p. 93). He draws an analogy with what he calls 'unconscious' domestic selection; people could develop a population of swift-running dogs, even though they used very diffuse criteria of selection and never systematically bred for swiftness (Darwin 1876a: Barrett 1977, ii, p. 210).

Surprisingly, Wallace subsequently argued explicitly for a similar view of the workings of selection (Wallace 1893). In artificial selection, he said, breeders do not pick out a particular bone, muscle or limb – they select for overall 'capacities' or 'qualities', such as speed, strength or agility; natural selection acts in the same way. So, at least at this later period, Wallace should not have excluded the possibility of females making the kind of non-specific judgement that Darwin proposed. (It has been claimed, incidentally (e.g. Ghiselin 1974, pp. 131, 178–9; Gould 1983, pp. 13, 369; Gray 1988, pp. 213–14; Lewontin 1978, pp. 160–1, 1979a, p. 7), for reasons I find obscure, that committed adaptationists such as Wallace assume that natural selection somehow works in a piecemeal way, selecting separately on each characteristic; insofar as I understand this claim, Wallace's later position and that of many modern adaptationists (e.g. Dobzhansky 1956; Mayr 1983) surely undermine it.) Wallace also subsequently (e.g. Wallace 1900, pp. 379–81) changed his mind about the size of variations within a species, coming to the view that they were large enough to be 'easily seen and measured by any one who looks for them' (Wallace 1900, p. 381). But he did not reexamine the question of whether the peahen could easily see and measure variations in the peacock's tail.

Wallace's second objection to female choice as a selective force is that preferences are unlikely to remain sufficiently constant within and between populations or over time to produce the results that Darwin attributes to them. Female choice as a selective force 'has none of that character of constancy and of inevitable result that attaches to natural selection ... [It] is unlikely that all the females of a species, or the great majority of them, over a wide area of country, and for many successive generations, prefer exactly the same modification of ... colour or ornament' (Wallace 1889, pp. 283, 285). And if female choice is not uniform, how can it produce uniform results (Wallace 1871, p. 182)? Wallace found it 'absolutely incredible' (Wallace 1891, p. 374) that the feathers of the Argus pheasant, for example, could have resulted from such choice. Several authorities went further, emphasising the notorious fickleness of females. According to Mivart, 'such is the instability of a vicious feminine caprice, that no constancy of coloration could be produced by its selective action' ([Mivart] 1871, p. 59). Geddes and Thomson were of the gloomily misogynistic opinion that permanence of female taste was 'scarcely verifiable in human experience' (Geddes and Thomson 1889, p. 29).

Wallace's criticism spurred Darwin into action:

> Your argument ... that taste on the part of one sex would have to remain nearly the same during many generations, in order that sexual selection should produce any effect, I agree to ... I have recognised for some short time that I have made a great omission in not having discussed ... its permanence within pretty close limits for long periods.   (Darwin, F. 1887, iii, p. 138; see also i, pp. 325–6)

Darwin replied with two points. (He incorporated some of his arguments into the second edition of *Descent* (pp. 755–6)).

First, there was the issue of constancy among individuals (Darwin 1876a: Barrett 1977, ii, pp. 209–11). Darwin suggested that this would be maintained by lack of consumer choice. The females 'cannot have an unlimited scope for their taste [because] ... although the range of variation of a species may be very large, it is by no means indefinite' (Darwin 1876a: Barrett 1977, ii, p. 210). Moreover, even if female taste does vary, intercrossing between offspring of mates picked for slightly different qualities will bring about uniformity in the males – just as, conversely, there is often considerable divergence between the sexually selected characteristics of males in two closely related but non-intercrossing populations. Darwin also tentatively suggested that female taste might be shaped by the environment, in which case one would expect what indeed occurs – constancy within geographically divided populations and divergence between them. Darwin was not alone, by the way, in entertaining the view that female taste is influenced by surroundings; among others, the naturalist and author Grant Allen developed a theory of aesthetics along similar lines (Allen 1879, particularly pp. vi, 4, 280; see also e.g. Darwin, F. 1887, iii, pp. 151, 157; Wallace 1889, p. 335). Finally, Darwin claimed that females could anyway select for a variety of characteristics as long as their preferences did not conflict (Darwin 1882: Barrett 1977, ii, p. 279).

Second, there was the matter of constancy over time. Darwin admitted that the relatively unrefined taste of animals, like that of 'savages', is somewhat changeable and even that novelty is loved for its own sake. He denied, however, that taste is volatile. 'We may admit that taste is fluctuating but it is not quite arbitrary' (Darwin 1871, 2nd edn., p. 755). As we can see in our own more sophisticated fashions, there is a liking for small changes but a dislike of very great ones. Thus, although tastes may change, the shifts will always be gradual:

> Even in our own dress, the general character lasts long, and the changes are to a certain extent graduated ... [Like humans, animals dislike sudden change, but this] would not preclude their appreciating slight changes ... Hence ... there seems no improbability in animals admiring for a very long period the same general style of ornamentation or other attractions, and yet appreciating slight changes in colours, form, or sound.   (Darwin 1871, 2nd edn., pp. 755–6)

Incidentally, Darwin's characterisation of how aesthetic preference operates is similar to an idea that was originally developed for understanding the psychology of what we find pleasant and unpleasant (McClelland *et al.* 1953, pp. 42–67), but has now been fruitfully applied to studies of mate choice (e.g. Bateson 1983, 1983b); it is the 'optimal discrepancy hypothesis' – the idea that the most attractive object is one that differs just a little from a familiar standard.

Darwin's answers do go some way towards meeting Wallace's criticisms. If female choice is based on aesthetic preference as he characterises it, then it is at least rescued from the sheer chaos that would undoubtedly ensue if it were merely the product of personal whim, of something entirely arbitrary. Nevertheless, is this enough? After all, on Darwin's own admission, it is in the nature of aesthetic preferences that, in spite of some continuity, they are ultimately capricious. He says that, among humans, one expects 'the most capricious changes of customs and fashions' (Darwin 1871, i, p. 64); and similarly other 'animals are ... capricious in their affections, aversions, and sense of beauty' (Darwin 1871, i, p. 65). Why, then, is female choice not more fickle than it is? Again on his own admission (Darwin 1871, ii, pp. 229–32), sexual selection is typically 'capricious' in creating differences between species because it 'depends on so fluctuating an element as taste' (Darwin 1871, ii, p. 230). Why, then, does it apparently act so very steadily within species?

Wallace has hit on a weak spot in Darwin's position. There is undoubtedly something missing in Darwin's theory of aesthetic female choice. But the problem stems from something deeper than any of the criticisms that Wallace touched on; these were merely the consequences of Darwin's omission. In Darwin's theory of sexual selection there is a major premiss that cries out to be explained. What is missing is any explanation of how female choice itself evolved.

The trouble with taste

Darwin's theory of sexual selection stops short at this crucial point. It simply assumes female choice as a 'given'. It doesn't explain what adaptive advantages there are in such choice – what selective pressures have given rise to these preferences and how they are maintained.

With the 'sensible' selective pressures of natural selection such problems don't arise. It is obvious why the need for efficient foraging exerts a selection pressure – and a demanding, precise pressure – on the woodpecker's beak. It is not at all obvious, however, what selective forces are involved in aesthetic choice. After all, the females are choosing characteristics that are of no use and indeed very likely downright disadvantageous, and they are doing so consistently and exactly. But if there are no adaptive advantages in choice, then there is no reason why it occurs at all. And if there is no more 'rational' reason for any particular choice than mere taste what is there to keep it so very constant and precise? Unless there are selection pressures directing it, what is there to prevent it from changing pretty well arbitrarily?

Wallace – and, indeed, most other critics until well into the twentieth century – concentrated on the problem of how aesthetic choices could be made. They asked how mere birds and butterflies could show such fine appreciation, and how the vagaries of taste could operate as a selective force. In reply, Darwin pointed to aesthetic taste in humans as a model of widespread, longlasting and discriminating appreciation of apparently useless ornament. And that was all very well as far as it went. But it raised even more acutely the question of how aesthetic taste – whether in humans or other animals – had evolved at all. Darwin claimed it had evolved from female choice of mates. But what evolutionary forces had brought about a choice based on mere aesthetics? To say that the females exercise choice for the love of beauty is still to leave that choice as an unexplained 'given' as far as selection – natural or sexual – is concerned. Darwin's aim was to provide an adaptive explanation for 'peacocks' tails'. And, if we grant him female choice for beauty, he succeeded. But ask how that choice evolved and one finds that at the very core of his theory there is no explanation at all.

Why, then, did Darwin insist so strongly on the aesthetic nature of female choice? If he had assumed, for example (as we'll see later Wallace and many other Darwinians did), that the females were choosing the strongest, or the healthiest, or the most vigorous males, their choice would have been easy to explain (although, again as we'll see, it would have raised other problems). So why did he tie down mate choice so specifically to appreciation of beauty?

Well, let's first establish that this is what he did indeed do. There is a wealth of evidence in *Descent* that the passages we've seen were typical of his thinking and not occasional lapses. He emphasises repeatedly that mate choice involves emotions sufficiently strong to allow of preferences and an aesthetic faculty sufficiently developed to guide them: 'Sexual selection ... implies the possession of considerable perceptive powers and of strong passions' (Darwin 1871, i, p. 377); females could select a male, he says, 'supposing that their mental capacity sufficed for the exertion of a choice' (Darwin 1871, i, p. 259). Birds, for example, have 'strong affections, acute perception, and a taste for the beautiful' (Darwin 1871, ii, p. 108; see also ii, pp. 400–1). And, he stresses, this is true of most animals, however unlikely it seems: 'I fully admit that it is an astonishing fact that the females of many birds and some mammals should be endowed with sufficient taste for what has apparently been effected through sexual selection; and this is even more astonishing in the case of reptiles, fish, and insects' (Darwin 1871, ii, p. 400; see also ii, p. 401). He didn't need to assume that taste came into it, even allowing that females were discriminating between potential mates. He would, after all, credit animals with remarkable abilities to discriminate between poisonous and nutritious foods, or between the shadow of a hawk and that of a seagull, without feeling the need to argue that they were exercising 'good judgement'. And, as his critics pointed out, he specifically denied any need to ascribe mathematical reasoning to bees to account for their wonderfully constructed honeycombs. But he did feel the need to attribute aesthetic sense to birds and insects to account for the beautiful colours of their mates (e.g. Darwin, F. and Seward 1903, i, pp. 324–5, n3; Wallace 1889, pp. 336–7). It is clear, then, that in Darwin's eyes the idea of aesthetic choice was an important aspect of his theory.

This brings us back to the question of why it was. Let's begin by clearing away what he wasn't doing. He wasn't assuming that peacocks' tails must appear glamorous to peahens just because they are so strikingly beautiful to us. He was well aware that, although the paradigm cases of sexual selection impress us with their sheer beauty, the males' characteristics are sometimes far from alluring to humans, and may even look grotesque:

> no case interested me and perplexed me so much [in my study of sexual selection] as the brightly-coloured hinder ends and adjoining parts of certain monkeys ... I concluded that the colours had been gained as a sexual attraction. I was well aware that I thus laid myself open to ridicule; though in fact it is not more surprising that a monkey should display his bright-red hinder end than that a peacock should display his magnificent tail.
>
> (Darwin 1876a: Barrett 1977, ii, p. 207; see also Darwin 1871, ii, p. 296)

The ornaments of birds are 'not always ornamental in our eyes' (Darwin 1871, ii, p. 72). Think of the macaw's blue and yellow plumage and harsh screams; they appeal to his mate's aesthetic sense but to us they are in deplorably bad taste (Darwin 1871, ii, p. 61; Darwin, F. and Seward 1903, i, p. 325). (The poor old macaw particularly offended Darwin's sensibilities; in

Beauty is in the eye of the species

The male proboscis monkey (*Nasalis larvatus*) has a nose like a gigantic, pendulous cucumber – in marked contrast to the sharply turned-up noses of the females and young. The monstrous growth starts at about seven years old and continues with age, eventually reaching about seven inches. In older males, this swollen, trunk-like appendage droops down over its owner's mouth, sometimes almost touching his chin, so that he has to push it aside with his hands in order to eat. The proboscis seems to have evolved at least in part as an amplifier. In the dense mangrove forests of Borneo, where these monkeys live, calling is the best way of communicating at a distance; the sound that resonates through the male's nose is reminiscent of a double bass. But, ludicrous as he may look to us, perhaps he goes to these lengths to satisfy female taste.

his first London house he used to 'laugh over the ugliness of ... the furniture in the drawing-room which he said combined all the colours of the Macaw in hideous discord' (Clark 1985, p. 64).) Even different species of bird find very different sounds attractive, only some of which are beautiful to us. Thus, he cautions, 'we must not judge of the tastes of distinct species by a uniform standard; nor must we judge by the standard of man's taste' (Darwin 1871, ii, p. 67). (Not that human taste is uniform or even always mutually comprehensible!)

So it wasn't the beauty of 'peacocks' tails' in our perception that led Darwin to characterise female taste as aesthetic. Indeed, sexual ornaments provided him with a splendid means of challenging the claim of natural theology that many structures had been created for the sake of their beauty in human eyes (Darwin 1859, p. 199; Peckham 1959, pp. 369–72). Later Darwinians, incidentally, did sometimes stress the idea that sexually selected characteristics were distinguished by their beauty to humans. This seems surprising until one realises that the aim was to narrow the scope of sexual selection. Julian Huxley, for example, argued that most conspicuous male coloration, being merely 'striking', could be attributed to recognition, threat and so on; it was only 'beautiful' coloration – intricate, delicate and most effective at close range – that could be attributed to sexual selection (Huxley 1938, 1938a).

I suggest that there were two reasons why Darwin took sexually selected characteristics as evidence that females exercised aesthetic sense. The first is not hard to find. It was part of his case about human evolution (including the evolution of human races) (Darwin 1871, i, pp. 63–5; see also de Beer *et al.* 1960–67, 2 (3), [C] 178). It had been claimed that a faculty for aesthetic appreciation was uniquely human, and Darwin wanted to counter this claim by establishing a continuity between humans and other animals: 'Sense of Beauty. – This sense has been declared to be peculiar to man ... but assuredly the same colours and the same sounds are admired by us and by many of the lower animals' (Darwin 1871, i, pp. 63–4). 'Beauty ... cuts the Knot', as he put it in one of his notebooks (Gruber 1974, p. 272, [M] 32). Mate choice was the only evidence he could find of such appreciation: 'With the great majority of animals ... the taste for the beautiful is confined, as far as we can judge, to the attractions of the opposite sex' (Darwin 1871, 2nd edn., p. 141). So our love of paintings, music, scenery doesn't, after all, distance us from other animals. On the contrary, it brings us closer to them. It springs from common behaviour, behaviour that, with us as with them, has had important selective effects in the evolutionary past.

Incidentally, Darwin's continuity argument perhaps gave Wallace extrascientific reasons for rejecting sexual selection. Wallace was one of those Darwinians who wanted to confine aesthetic sense to humans. In his early days, when he accepted sexual selection, he acknowledged that any such continuity would be 'a fact of high philosophical importance in the study of our own nature and our true relations to the lower animals' (Wallace 1891, p. 89). It was just such implications that he later wanted to deny. We shall see, when we examine human altruism, that where Darwin hoped to establish links between humans and other animals, Wallace was anxious to carve a gulf. He came to maintain that several faculties were exclusively human; aesthetic appreciation was one of them. Some commentators (e.g. Fisher 1930, p. 150; Selander 1972) have suggested that this was the reason why Wallace could not accept sexual selection – that he thought of our aesthetic faculties as being part of our 'spiritual nature' (although as Kottler points out against this claim, Wallace adopted these extrascientific views in the 1860s but he continued to accept sexual selection, at least in birds, until about 1871 (Kottler 1980, p. 225)). According to Kottler, Wallace's spiritual beliefs would anyway not have prevented him from accepting sexual selection at least in humans; after all, Wallace readily acknowledged that we exercise aesthetic taste – indeed, that we do so uniquely (Kottler 1980, p. 225). But surely Wallace couldn't have accepted that we practice sexual selection if this involved accepting Darwin's view that females choose on aesthetic grounds. For this would provide an evolutionary role for our aesthetic sense and, as we shall see from his views on altruism, Wallace took it as a hallmark of our 'specially endowed' faculties that they could not have been evolved because they were surplus to evolutionary requirements. Indeed, he emphasised that, unlike the primitive ability merely to distinguish colours, which we share with other animals, our enjoyment and appreciation of colour cannot be explained on 'purely utilitarian principles' (Wallace 1891, p. 415). (In typical contrast, Darwin characterises aesthetic judgement in other animals as involving pleasure, thus making them *more* like us (Darwin, F. and Seward 1903, i, p. 325).) In later years (Wallace 1890; see also Fichman 1981, pp. 141, 148–53), Wallace did come to believe that mate choice by intelligent women could select for social qualities in human societies; but such choice was not an aesthetic judgement, it was a sensible choice. We shall see that Wallace was prepared to allow non-aesthetic mate choice even in 'lower' animals; he stressed the role of assortative mating in interspecific sterility and (as we've noted) of recognition of mates in locating conspecifics, and he was even prepared to allow mate choice on a variety of other grounds as long as those grounds were 'sensible' rather than aesthetic.

My second suggestion, complementary to the first, as to why Darwin held so tenaciously to the idea that female choice is nothing but aesthetic is more conjectural. Darwin's view perhaps reflects his Darwinian intuition that there really is something absurd, something indulgent, something blaze-of-glory-ish, about nature's peacocks' tails. This is certainly captured by the idea of sheer aesthetic preference for its own sake, a preference apparently free of utilitarian strings. And how right Darwin was, as Fisher eventually showed. But we'll come back later to that intriguing issue.

Oddly to our modern eyes, it was many decades before Darwin's failure to explain how female choice had evolved was seen as a major objection to his theory. The problem was raised, but in a piecemeal way and by critics of various persuasions, rather than as an accepted item on the Darwinian agenda. One theistically-minded reviewer of *Descent*, for example, complained that female choice was 'a cause which will seem to most men more needful of explanation and more worthy of it, than the effect itself' (Anon 1871a, p. 319). Where, he asked, had animals acquired their aesthetic sense if not from God? In his view, Darwin's continuity argument made more, not less, work for divine hands; not only humans but many a beetle, butterfly or bird had to be endowed with a taste for the beautiful. A similar article went further, doggedly maintaining, in spite of Darwin's denials, that 'the splendidly coloured snakes and birds of tropical forests ... are never what our taste would call vulgarly coloured, never coarsely patched with frightful patterns, such as you constantly see on gaudy gowns, showy wall-papers and glaring carpets'; this was in marked contrast with the 'preferences of the least cultivated classes of civilized human beings ... English sailors or ... maidservants' – think, after all, of the 'hideous but showy whirligigs of yellow, such as a British cook would select for the pattern of her Sunday dress'; a taste so impeccable must surely spring full-blown from a divine source (Anon 1871, p. 281). Two decades later, Edward Westermarck was led from much the same premises not to God but to natural selection (Westermarck 1891, pp. 477–91). He complained that, according to Darwin, male secondary sexual characteristics, 'depend upon an aesthetic sense, or taste, in the females, the origin of which we do not know'; female preference is 'an inexplicable tendency' (Westermarck 1891, pp. 478, 490). He concluded that ornament should 'be explained by the principle of the survival of the fittest' (Westermarck 1891, p. 479). By contrast, at the turn of the century, Thomas Hunt Morgan used female choice as evidence against any selection, sexual or natural. Morgan, as we have noted, believed at one time that mutation alone could do much of the work of evolution without the help of selection (see e.g. Bowler 1983, pp. 202–5). Like other mutationists at that period, he tried to show that all kinds of adaptive explanations were really inadequate. So Darwin's explanation of male ornament was, to him, not a special case but just one of the many characteristics that selection could not explain:

The development, or the presence, of the aesthetic feeling in the selecting sex is not accounted for on the theory. There is just as much need to explain why the females are gifted with an appreciation of the beautiful, as that the beautiful colors develop in the males ... Darwin assumes that the appreciation on the part of the female is always present, and he thus simplifies, in appearance, the problem, but he leaves half of it unexplained. (Morgan 1903, p. 216)

Not until 1915 was the question of how female taste evolved both raised explicitly and satisfactorily answered. The question was put by R. A. Fisher, who was not only one of the major architects of the Darwinian synthesis and a pioneer of statistics and population genetics, but also a central figure in the history of sexual selection. This is how Fisher spelt out the problem (unfortunately in species-level language but fortunately not in species-level thinking):

'Whence', it may be asked, 'has this extremely uniform and definite taste for a particular detailed design of form and colour arisen?' Granted that while this taste and preference prevails among the females of the species, the males will grow more and more elaborate and beautiful tail feathers, the question must be answered 'Why have the females this taste? *Of what use is it to the species that they should select this seemingly useless ornament?*' (Fisher 1915, pp. 184–5; my emphasis)

It is to Fisher that we shall return for the answer.

So, for almost half a century, there loomed over Darwin's theory of sexual selection the unanswered question of why it is adaptive for females to choose the best-ornamented males. Could mere aesthetic choice be selectively advantageous? Or was the choice perhaps not aesthetic? And if not, how could it be explained? This question, 'Why do females choose as they do?', brings us to the last lap of the nineteenth-century debate and carries us into the present, into the theory's most exciting and fruitful stage.

8

Do sensible females prefer sexy males?

Good taste or good sense?

So why do females choose as they do? Throughout the history of sexual selection, Darwinians have offered two very different answers to that question. The first we can think of as a 'good taste' solution. On this view, which was Darwin's, females choose solely for beauty; so their choice is maladaptive by natural selection's standards. The other answer can be thought of as a 'good sense' solution. According to this view, females choose along the same utilitarian lines as natural selection; so their choice is adaptive and unproblematic. This was the position that Wallace adopted. Admittedly, up to now I have depicted Wallace as being implacably opposed to the very idea of female choice once he had rejected Darwin's theory of sexual selection. But the time has come to modify that impression. In spite of all his protestations, Wallace didn't ever rule out female choice entirely. What he did do was to argue strenuously against Darwin's view of it and to propose an alternative, 'good sense', view. Once again, then, we find Darwin and Wallace on opposite sides. Let's begin with Darwin.

Darwin's solution: Beauty for beauty's sake

We have seen that, according to Darwin, females are interested only in what pleases them aesthetically; the characteristics they favour are purely ornamental and serve no other function: 'a great number of male animals ... have been rendered beautiful *for beauty's sake*'; 'the most refined beauty may serve as a charm for the female, and *for no other purpose*'; 'that ornament and variety is the *sole object*, I have myself but little doubt' (Peckham 1959, p. 371; Darwin 1871, ii, pp. 92, 152–3; my emphasis).

We have seen, too, that Darwin recognised some of the difficulties that arise from the 'irrationality' of such choice and that he tried to deal with them by appealing to the model of aesthetic sense in humans. He did not, however, face up to the most serious aspect of this irrationality: the fact that the choice is for costly and often grossly extravagant characteristics. There seems to be

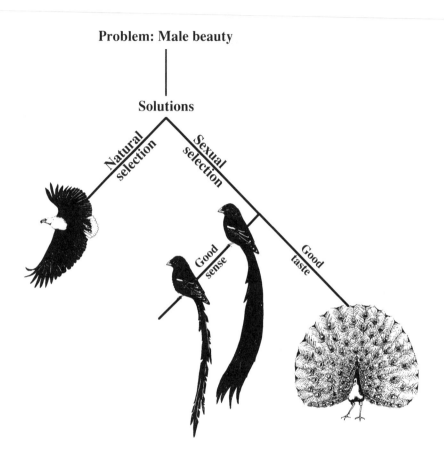

Problem: Male beauty

Solutions

Natural selection

Sexual selection

Good sense

Good taste

Three ways to beauty. The osprey's wings: are they just aerodynamically elegant? The widowbird's tail: does it reveal his quality? The peacock's fan: is it a whim of runaway female fashion?

no good reason for the female to choose as she does – and, worse, there seem to be many good reasons for her not to do so. To require the male to deck himself out in gaudy colours, or sport a long tail, or sing and dance for hours on end is to impose a heavy burden on him. It puts him at a disadvantage in his own struggle for existence. And if he is to help his mate then she, too, suffers. Not only a wife and children to support, but wife, children and tail! What's more, if his sons inherit his ornaments, they and their mates will undergo the same fate. Surely, then, the female has good reason not to choose 'beauty for beauty's sake'.

It may seem unjust to accuse Darwin of not facing up to the costliness of the males' characteristics. Surely sexual selection is the one area in which classical Darwinism systematically acknowledges that adaptations can be costly. After all, the theory was constructed expressly to deal with apparently non-utilitarian characteristics – characteristics that appear not to be adaptive by the standards of natural selection. In the *Origin*, the very heading under which Darwin discusses male ornamentation queries its usefulness: 'Utilitarian Doctrine how far true: Beauty, how acquired' (Peckham 1959, p. 367). And Darwin says explicitly that it 'can be called useful only in rather a forced sense' (Darwin 1859, p. 199). He also states that sexually selected characteristics 'have been acquired in some instances at the cost not only of inconvenience, but of exposure to actual danger' (Darwin 1871, ii, p. 399). Some birds, for example, with their conspicuous colours or ornamentation are easier prey or hampered in fights (Darwin 1871, ii, pp. 96–7, 233, 234); similarly, the structures built by male bower birds 'must cost the birds much labour' and toucans must be 'encumbered' by their immense beaks (Darwin 1871, ii, pp. 71, 227).

What is more, and again unusually within classical Darwinism, Darwin saw sexually selected characteristics as the product of a trade-off – survival chances being lowered in exchange for mating advantages:

> The development ... of certain structures ... has been carried ... in some instances to an extreme which, as far as the general conditions of life are concerned, must be slightly injurious to the male. From this fact we learn that the advantages which favoured males derived from conquering other males in battle or courtship ... are in the long run greater than those derived from rather more perfect adaptation to the external conditions of life. (Darwin 1871, i, p. 279)

Similarly he says that characteristics that would harm juveniles could, in older males, be outweighed by their reproductive advantages (Darwin 1871, i, p. 299).

Nevertheless, Darwin is cavalier about the extent of the burden. He assumes that male ornamentation never seriously threatens survival because natural selection always intervenes to curb its wilder excesses: 'Sexual selection will ... be dominated by natural selection for the general welfare of the species' (Darwin 1871, i, p. 296; see also pp. 278–9). According to him, even plumage as overdeveloped as the Argus pheasant's would not impede the birds in their search for food (Darwin 1871, ii, p. 97). It is clear that in Darwin's view 'ornaments' are above all useless rather than disadvantageous. When he describes sexually selected characteristics as non-utilitarian he has in mind merely that they are of no particular advantage. They are 'extraordinary', 'beautiful', 'curious', 'elegant', 'singular' and 'diversified' (e.g. Darwin 1871, ii, pp. 307, 312). But they are not necessarily costly. Well, Darwin may be right that sexually selected characteristics are less burdensome than may at first appear. But this can't just be taken for granted, without demonstration. And even if the costs turn out to be low, there's still a need to indicate how the benefits manage to outweigh them. So, although the theory of sexual selection does incorporate an idea of costs, it underestimates their magnitude and significance.

Once the possible costs of ornamentation have been taken into account, the lack of an adaptive explanation for female choice becomes even more pressing. Disadvantages to the male may be compensated by female choice. But this only throws the problem even more squarely back into the females' court. Why do they insist on making such costly choices? Unless female choice can itself be shown to be the product of selective forces there is, at the heart of Darwin's theory, not merely a mechanism that is not adaptively explained but a mechanism that appears to be grossly maladaptive.

Wallace's solution: Not just a pretty tail

According to Wallace, female choice is not to do with good taste, but with good sense. Insofar as females choose their mates at all, Wallace argues, they go for useful qualities like vigour, health or stamina. They choose along the same 'sensible' lines as natural selection. And they do this because it obviously pays them: they are getting a high quality mate. So their taste is just a straightforward product of natural selection.

Wallace admits that females often appear to be opting for taste rather than sense, to be making an aesthetic choice rather than a down-to-earth one. But this, he says, is because beauty and quality tend to coincide, the most energetic, healthy males also generally being the most decorative: the 'most vigorous, defiant, and mettlesome male' is 'as a rule the most brightly coloured and adorned with the finest developments of plumage' (Wallace 1891, p. 375; see also p. 369). So a peacock isn't just a pretty face – or tail, or song, or whatever. Females who use quality as their guide will be picking the most splendidly ornamented males as an automatic side effect. They're not judging the male by his ornaments but by the sensible qualities that accompany them. Put a colourblind peahen in a group of potential mates and even she will go for the most gaudy, rainbow-hued of the lot – not because she appreciates his beauty (sadly, she is necessarily indifferent to that) but because she is going for quality, and beauty comes along as an incidental part of the package. Of course, for Wallace, although this connection is not the result of selection, it is no mere chance. Remember his physiological theory that vigour and health give rise to bright colours and elaborate structures.

As for Darwin's evidence of female choice, Wallace rightly points out that the females may not be interested in the characteristics that Darwin focuses on. In the absence of more detailed knowledge, it is an open question whether a female is choosing a mate for his glamour or for more useful endowments. Take the example of butterflies. His own interpretation, he argues, is as plausible as Darwin's: 'Among butterflies, several males often pursue one female, and Mr Darwin says, that, unless the female exerts a choice the pairing must be left to chance. But, surely, it may be the most vigorous or most persevering male that is chosen, not necessarily one more brightly or differently coloured' (Wallace 1889, p. 275; see also Wallace 1871). Similarly, with birds, even when the female chooses we do not know the grounds for her choice and 'it by no means follows that ... differences in the shape, pattern, or colours of the ornamental plumes are what lead a female to give the preference to one male over another' (Wallace 1889, p. 285; see also 1891, pp. 369, 376). Thus 'the choice of ... more ornamental male birds by the females ... is an *inference* from the observed facts of ... display; ... the statement that ornaments have been developed by the female's choice of the most beautiful male *because he is the most beautiful*, is an inference supported by singularly little evidence' (Wallace 1905, ii, pp. 17–18). Darwin tells of female birds having strong likes or dislikes for particular males but he fails to show 'that superiority or inferiority of plumage has anything to do with these fancies' (Wallace 1889, p. 286). And Wallace quotes from Darwin himself citing experienced observers who don't believe that beauty of plumage affects female choice (Wallace 1889, pp. 285–6). One such expert, for example, is of the firm opinion 'that a gamecock, though disfigured by being dubbed, and with his hackles trimmed, would be accepted as readily as a male retaining all his natural ornaments' (Wallace 1889, p. 286). Again, this opens the way for Wallace to step in with his alternative view. He quotes the conviction of one of these authorities 'that the female almost invariably prefers the most vigorous, defiant, and mettlesome male' (Wallace 1889, p. 286). Indeed, says Wallace, the female generally pays so little attention to the male's display of finery that 'there is reason to believe that it is his persistency and energy rather than his beauty which wins the day' (Wallace 1889, p. 370).

According to Wallace it follows from his view that female choice has little or no importance in evolution. If the female's choice is sensible then it largely coincides with the choice of natural selection, in which case it will not be a significant selective force. And if her choice does not coincide, it will be selected against. So, if she were to choose the most ornamented male, either her choice would be redundant or it would be eliminated. On the one hand, the 'extremely rigid action of natural selection must render any attempt to select mere ornament utterly nugatory, unless the most ornamented always coincided with "the fittest" in every other respect ... [and] if they do so coincide, then any selection of ornament is altogether superfluous' (Wallace 1889, p. 295). On the other hand, 'If the most brightly coloured and fullest plumaged males are *not* the most healthy and vigorous ... they are certainly not the fittest, and will not survive' (Wallace 1889, p. 295). Thus female choice has no significant evolutionary effect: 'The action of natural selection does not indeed disprove the existence of female selection of ornament as ornament, but it renders it entirely ineffective' (Wallace 1889, pp. 294–5). It can be no more than a marginal force in evolution, forever subordinate to utilitarian forces.

At most, Wallace says, female choice can reinforce natural selection. With birds, for example, natural selection will favour the most vigorous males, and elaborate plumage will develop as an automatic side effect; if the females also choose the most vigorous males – the 'sensible' choice – then 'sexual selection will act in the same direction, and help to carry on the process of plume development to its culmination' (Wallace 1889, p. 293). Wallace does not explain exactly how female choice would help. Perhaps he imagined that it would narrow down the range allowed by natural selection or that it would coincide with natural selection's choice but increase the cost of deviating from it.

Although he dismisses the idea that the females would ever choose beauty just for beauty's sake, Wallace does touch on the possibility of beauty nevertheless being used as a criterion for choice. He recognises that if there is a close, reliable connection between ornament and 'sensible' qualities, as there is on his physiological theory of ornamentation, then the female could use ornamental display as a marker for the qualities that she is really after: 'The display of the plumes, like the existence of the plumes themselves, would be the chief external indication of the maturity and vigour of the male, and would, therefore, be necessarily attractive to the female' (Wallace 1889, p. 294). Sadly, Wallace didn't develop this concept. We'll see that modern Darwinism has done so, to great advantage. Nevertheless, to suggest that Wallace could have exploited the notion of markers is not to foist a twentieth-century outlook on him. It is obvious, even without venturing into the labyrinths of epistemology, that many of the experiences of organisms are to some extent vicarious. And explanations of adaptations allow for this as a matter of course. Fruit tastes sweet, not nutritious. Wallace himself used the idea of warning colours. What are they if not markers?

Wallace's position is clearly the antithesis of Darwin's. According to Wallace, females may seem to go for splendour but they're really going for quality. According to Darwin, females are going for splendour and nothing else. So it is odd to find some passages in *Descent* in which Darwin sounds just like Wallace. He occasionally falls back on the claim that a female may choose both for beauty and for quality. And he goes further than Wallace. On Wallace's view her 'choice' of beauty is a mere side effect; on Darwin's view she is making a genuine dual choice.

> The females are most excited by, or prefer pairing with, the more ornamented males, or those which are the best songsters, or play the best antics; but it is obviously probable ... that they would at the same time prefer the more vigorous and lively males ... [T]hey will select those which are vigorous ... and in other respects the most attractive.   (Darwin 1871, i, p. 262; see also i, pp. 263, 271, ii, p. 400)

Why does Darwin at times adopt a position that is so out of keeping with his theory? The reason is to do with his explanation of how sexual selection can operate in monogamous species. As we saw earlier, he rightly argues that female choice can be an effective selective force if the most resplendent males mate with the healthiest and therefore earliest-breeding females. This solution is adequate as it stands. But Darwin nevertheless seems to feel the need to give sexual selection an extra boost by postulating that the males who are the most attractive, and therefore breed the earliest, are also the healthiest (like the earliest breeding females). He even summarises his theory of sexual selection this way: 'I have shewn that this [greater breeding success of the more attractive males] would probably follow from the females ... preferring not only the more attractive but at the same time the more vigorous ... males' (Darwin 1871, ii, p. 400). This assumption is unnecessary for solving Darwin's problem. It is also unfounded, given the evidence available to him (for, unlike Wallace, Darwin offers no reason for assuming that beauty and quality will go hand in hand). And it is alien to his theory, his central idea being that females choose nothing but beauty for beauty's sake.

I stress this point because, before the recent revival of interest in sexual selection, Darwin's theory was often mistaken for the view that female choice combines 'good taste' and 'good sense'. Misunderstandings like the following were common; this is from a purportedly authoritative review of the state of Darwinian theory in the 1920s: 'the struggle to ... [find a mate] leads to the success of the most vigorous and attractive male; a result which Darwin called sexual selection' (MacBride 1925, p. 217; emphasis omitted). Fifty years later, a leading Darwinian was still taking this to be a central tenet of Darwin's theory. Mayr accused Darwin of assuming 'rather naively', 'with no tangible evidence' that attraction and vigour generally go together; he even bracketed Darwin with Wallace in this respect (Mayr 1972, pp. 97, 100). His evidence was the atypical passages from *Descent* that we have just noted. Such passages do seem to support this misinterpretation. So it is crucial to bear in mind that the assumptions that Darwin makes in them are neither necessary to, nor typical of, his theory of sexual selection. Darwin's theory bore no resemblance to Wallace's; when it came to the question of why females choose the mates they do, the two of them were poles apart.

And now back to Wallace. His theory that females choose sensibly certainly avoids Darwin's major problem of leaving female taste unexplained. Unfortunately, it also lands him in an obvious difficulty: accounting for the costliness of male ornamentation, the sheer extravagance of maintaining an oversized tail, baroque horns or hours of elaborate song. On the face of it, it's highly implausible to claim that a female who prefers so profligate a male is making a sensible choice of mate. Wallace's answer is that she isn't choosing the costly characteristics; they are merely an unavoidable accompaniment to her sensible choice. But this just shows how far Wallace underestimates what the extent of the costs might be – a much more serious underestimation than Darwin's. Wallace seems blithely unaware that male extravagance, at least on first analysis, undermines his view that female choice is sensible. When he is dealing with female coloration, Wallace implicitly acknowledges that the males' conspicuous colours could be disadvantageous; after all, he assumes that the females have suppressed them for the sake of protection. But when he deals with the males' ornaments explicitly, he airily dismisses the idea that they could threaten the survival of their bearers. Take his discussion of the peacock's tail (Wallace 1889, pp. 292–3). He says that in some cases accessory plumage is useful, having been developed by natural selection for protection in combat. He admits that, nevertheless, this cannot account for all cases of apparent costs: 'The enormously lengthened plumes of the bird of paradise and of the peacock can, however, have no such use, but must be rather injurious than beneficial in the bird's ordinary life' (Wallace 1889, pp. 292–3). But according to Wallace the injury can't amount to much because the birds seem to manage in spite of it. Indeed, the peacock's extravagance merely supports his claim (remember his physiological theory of ornament) that the males have so great a reserve of energy that they can afford to bear what would otherwise be a burden. The fact that the plumes

> have been developed to so great an extent in a few species is an indication of such adaptation to the conditions of existence, such complete success in the battle for life, that there is, in the adult male at all events, a surplus of strength, vitality, and growth-power which is able to expend itself in this way without injury.
>
> (Wallace 1889, p. 293)

And he points to the fact that these species are highly successful – abundant and wide-ranging – as evidence that the males' flamboyance does not impede their struggle for existence. So, although there are costs, Wallace concludes, they must be negligible.

Without some means of accounting for the males' seemingly reckless extravagance, their apparent defiance of good sense, Wallace's solution cannot get far. It clearly has enormous potential for explaining why females prefer the strongest or swiftest or best camouflaged. But it stops short at just those very cases – the problematic 'peacocks' tails' – that Darwin's theory was designed to explain.

Is 'good sense' sensible?

This problem remained a stumbling block for sensible choice theories for a century. But modern developments have come to Wallace's rescue. Today's Darwinism can take such theories in its stride. It incorporates several notions that can explain, at least in principle, how a Wallacean good-sense choice might give rise to characteristics that are so luxuriant, so extravagant that they seem intuitively not to make good sense at all. Three interrelated ideas in particular have proved fruitful: markers, conflicts of interest and evolutionary arms races. These concepts were there in classical Darwinism but undeveloped.

We have already come across the idea of markers. Suppose that, say, brightly-coloured plumage and a robust constitution are, in general, closely and reliably connected, perhaps because only robust males have what it takes to keep their feathers bright. Then females could use brightness as a marker for robustness. And the brightest males would come to be preferred, not because brightness is in itself of any use, but because it is a marker of a useful quality. Assuming that both brightness and robustness are heritable, the use of markers obviously opens the way for direct selection on the marker rather than on the sensible quality alone. Bright colours could evolve under the joint pressure of female scrutiny and male attempts to pass muster. Note, by the way, how different this is from Wallace's theory of sensible preferences. A colourblind peahen who wanted to make a sensible Wallacean choice could not take advantage of the information that markers offer.

The second notion is that there could be a conflict of interest between males and females, resulting in an evolutionary chase between males who are cheating over markers, puffing themselves up beyond their true worth, and females who are evolving counter-adaptations to detect deception, so as not to be caught out by dishonest advertising. Both sides would be better off if they could opt out of this extravagant escalation. But they would be locked into the strategic logic of response and counter-response.

Third, there is the idea of an evolutionary arms race going on among the males in their competition for females. This has the explosive potential that is typical of a symmetric arms race – one in which, unlike the arms race between males and females, the competitors are trying to get better at doing the same thing (like building the biggest bomb) rather than using different strategies (like better radar versus better means of avoiding detection) (Dawkins and Krebs 1978, 1979; Krebs and Dawkins 1984; see also Thornhill 1980; West-Eberhard 1979, 1983). All the peacocks are competing to develop ever bigger and better tails. The result is that selection will in general favour males with tails slightly longer than the average, however long that average has become:

Imagine a species in which large size is an advantage in male–male competition but not an advantage from any other point of view. It is entirely reasonable that competition will favour males that are slightly larger than the current population mode, *whatever the current mode may be*. This is a recipe for progressive evolution of the kind that we expect from an arms race. It is a true symmetric arms race ...

(Dawkins and Krebs 1979, p. 502)

Now combine these selective forces and the result is a powerful mechanism for escalation – powerful enough to generate the kind of extravagant exaggeration, the precipitous runaway, that Wallace was unable to account for. During the first half of this century, the climate was uncongenial to the notion that natural selection could indulge, let alone foster, such escalation, such apparent absurdity. We shall see that, for many a decade, Darwinism was influenced by a vague harmony-of-interests, good-for-the-species way of thought. Mating was looked upon as, above all, a cooperative venture. Not until that viewpoint had been shed could this revised version of Wallace's idea become accepted. Nowadays, however, it is flourishing. This line of thinking, unlike the natural selection tradition, has not developed primarily by lineal descent from him. But, from our historical perspective, many modern theorists are revealed as 'Wallaceans' in a new and unexpected guise. Good-sense theories have lost none of the attraction that they had for Wallace: they provide an adaptive explanation not only of the males' characteristics but also of female choice. And, what's more, they do not rely on a notion of adaptation as unorthodox and counter-intuitive as Wallace and many a Darwinian today have felt that Darwin's theory involves.

So how might a modern 'Wallacean' female choose her mate? If the males of her species provide paternal care, then she could obviously go for the best of the providers. She could try to end up with a safe nest for her eggs, a steady supply of food, protection from predators. Now, there may be no genetic difference, no difference in genes for nest-building, between the male who builds the finest nest and the male who builds the worst one. The difference in quality could arise entirely from environmental factors – availability of materials and so on. In that case, female choice would not be acting as a selective force. Her choice could evolve but it would not in turn influence the evolution of male nest-building. Alternatively, differences in nest quality could reflect an underlying heritable genetic difference. In that case, female preference would become a major selective force on male proficiency at building nests.

Now consider a species in which the female protects and nourishes and instructs her offspring without any help from their father, a species in which the male merely meets and mates with the female, contributing nothing but sperm to the reproductive effort. In this case, she has no option, if she is choosing at all, but to pick her mate solely on the grounds of whose genes would contribute most to her offspring's survival and reproduction. The only good-sense factor that could ultimately determine her choice is the genetic endowment that the male is likely to bring to their offspring. Her sole concern will be whether he has good genes – genes for a robust constitution, for example. Of course, genes must be detected indirectly, via phenotypes; females are no better equipped than are any other selective forces to see naked genes. So females might well use just the kinds of phenotypic qualities – vigour, strength and so on – that Wallace suggested.

Unfortunately, in the literature, there are no consistent names for these different kinds of choice and the terminology can be confusing. So, before going any further, I'll briefly sort out the terms that are commonly used. This list is, I fear, too tortuous for instant enlightenment but I hope it can be useful nonetheless; it is easier to follow in conjunction with the accompanying diagram.

The best-nest kind of choice is sometimes called a 'good-resources' choice, and the robust-constitution kind, 'good genes'. 'Good resources' sometimes

FEMALES CHOOSE BY...

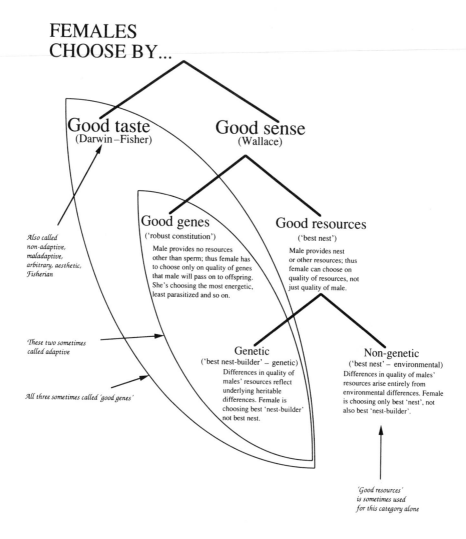

Good taste
(Darwin–Fisher)

Also called non-adaptive, maladaptive, arbitrary, aesthetic, Fisherian

Good sense
(Wallace)

Good genes
('robust constitution')

Male provides no resources other than sperm; thus female has to choose only on quality of genes that male will pass on to offspring. She's choosing the most energetic, least parasitized and so on.

Good resources
('best nest')

Male provides nest or other resources; thus female can choose on quality of resources, not just quality of male.

These two sometimes called adaptive

All three sometimes called 'good genes'

Genetic
('best nest-builder' – genetic)

Differences in quality of males' resources reflect underlying heritable differences. Female is choosing best 'nest-builder' not best nest.

Non-genetic
('best nest' – environmental)

Differences in quality of males' resources arise entirely from environmental differences. Female is choosing only best 'nest', not also best 'nest-builder'.

'Good resources' is sometimes used for this category alone

covers both choice of the best nest and choice of the best nest-maker – that is, choices that do not reflect genetic differences and choices that do. But sometimes good resources is restricted to instances in which the female is choosing only the best nest, not also the best nest-maker – that is, instances in which her choice does not reflect genetic differences (in which case she does not discriminate between different genes and so does not act as a selective force on the males). As for the robust-constitution kind of choice, I have said that it is called a good-genes choice, in contrast to a best-nest or good-resources choice. But some authors use the term good genes to mark off a wider category, pretty much the category that I have called good sense. In that case, good genes covers not only the robust-constitution kind of choice but also genetically differentiated good resources – the whole category that I have called good sense, apart from non-genetic good resources. When good genes is used in this way, the point is to make a contrast between what I have called good-sense choice (at least, good-sense choice that reflects genetic differences) and what I have called good-taste choice (Darwin's notion of choice). In that context, good-taste choice is sometimes called non-adaptive, maladaptive, arbitrary, aesthetic or Fisherian and the alternative (genetic good sense) is called adaptive choice. Finally, good genes sometimes covers not only the category that I have mentioned but also good taste – in other words, all female choice that involves genetic differences. In that case, the good-taste subcategory is often called Fisherian good genes. (Well, I did warn that it would be tortuous.)

For my part, one fundamental distinction that I want to make is between good taste and good sense. And within the good-sense category, we shall find it important to distinguish between species in which the males put resources into the reproductive effort (paternal care) and those in which they are nothing but sperm donors. Where the male does provide resources, the female will be interested in the quality of those resources; where he provides nothing but sperm, she will be interested in nothing but his genes. As we have noted, in paternal-care species the female's concerns might well include resources that do not reflect genetic differences; but at that point our interest on the whole diverges from hers, for we are concerned with evolution and therefore ultimately with genes – so good resources will generally mean choices between genetically different males. Of course, in practice a female might be making more than one kind of choice – and they might be very difficult for us to disentangle.

Terminology apart, there is a serious difficulty with these good-sense theories, particularly the good-genes version; we shall come to it in a moment. On a more positive note, modern Darwinism has clearly transformed Wallace's idea of good-sense female choice in a most fruitful way. But it's startling to see how great the transformations are.

Nothing turns Wallace's idea so much on its head as what are called handicap theories of female choice (see particularly Zahavi 1975, 1977, 1978, 1980, 1981, 1987; see also e.g. Andersson 1982a, 1986; Dawkins 1976, pp. 171–3, 2nd edn., pp. 304–13; Dominey 1983; Eshel 1978; Gadgil 1981; Hamilton and Zuk 1982; Kodric-Brown and Brown 1984; Maynard Smith 1985; Nur and Hasson 1984; Pomiankowski 1987). Indeed, handicap theories in general turn our whole Darwinian world upside-down. We last met them when we looked at adaptive explanations. There we left the zebra potentially

Pelecanus onocrotalus's bump seems to obscure vision. Has it evolved in spite of this fact ... or because of it?

handicapping itself by wearing stripes that would mercilessly betray its underdeveloped muscles or weak legs. That is, if it had any such defects. Of course, if it was strapping and sturdy, then its stripes would tell that more auspicious tale. According to the handicap way of thinking, when the aim is to impress potential mates, the results could make even the zebra's dazzle look sober. Amotz Zahavi, the originator of the handicap principle, tells in his talks of male pelicans (*Pelecanus onocrotalus*) growing large bumps on their beaks during the mating season, bumps so large that they find it hard to see.

Now, one thing that a pelican must have is clear vision, to take an accurate sight before he darts for fish. It appears, then, as if the males are deliberately handicapping themselves. Exactly so, says Zahavi. The point of the exercise is to show off, and to do so reliably. 'Look how well I can feed myself, even with this great big bump in front of my eyes!' The bigger the bump, the more telling the test and the more reliable the claim.

So Zahavian females are exploiting the fact that bumps and bright colours and long tails carry real costs. The handicaps convey the message that the male can support these costs. Of course, the males could try to fake their burden. But Zahavi and others argue that dishonest advertising in the face of female scrutiny is not evolutionarily stable. Thus, burdens will evolve that are hard to fake. And so a female will know that if a male has managed to dodge predators, feed himself and generally keep up the struggle for existence even with his eye-catching plumage or unwieldy tail, then he must surely be of sterling quality. She can rely on male handicaps to guide her to a mate with good genes. Now, all this dramatically changes the rules of Wallace's sensible game. Wallace implicitly assumed that good-sense choices would be low cost. Handicap theories assume that females choose males not in spite of the costliness of a characteristic but because of it.

In Zahavi's original version, the handicap principle was widely thought to be largely unworkable (Bell 1978; Davis and O'Donald 1976; Kirkpatrick 1986; Maynard Smith 1976a, 1978, pp. 173–4, 1978a; O'Donald 1980, pp. 111, 167–74). It seemed – what one would intuitively have imagined until Zahavi shook up our intuitions – that the advantages of a son's good genes would be outweighed by the disadvantages of his handicap. But subsequent models have been more successful (e.g. Andersson 1986; Grafen 1990, 1990a, also in Dawkins 1976, 2nd edn., pp. 308–13; Pomiankowski 1987). Several authors have claimed that variants of the handicap principle – some watered-down, some less so – can, after all, work. Handicap theories, it is increasingly being urged, can be both mathematically and biologically respectable.

I mentioned that modern good-sense theories present a difficulty. It arises from female choice. The stumbling-block is not, as those morose Victorian misogynists would have it, that females are too fickle. On the contrary, it is their obdurate constancy that disconcerts theoreticians. The problem is: Why doesn't the variation among males disappear? The question arises in the following way (see e.g. Arnold 1983; Borgia 1979; Davis and O'Donald 1976; Maynard Smith 1978, pp. 170–1). Natural selection cannot act unless there are differences between which it can choose. Generally, there is enough variation in the population for selective forces to pick the best, winnowing out those that are too slow or too fast, too small or too large. Natural selection

will favour those that come closest to what suits the current environment –
perhaps a tail just four inches long or a running speed of just under 20 miles
an hour. Female choice, however, does not let its standards rest. It exerts a
relentless, ever-demanding pressure, calling not for a tail four inches long, but
for a longer tail, whatever length the males have already managed to reach.
One can readily imagine, and population genetics theory confirms this, that
female choice will rapidly cause the disappearance of the very variation that
is the object of that choice. If females consistently and successfully choose
the heritably best males, there will be no 'best' left to choose: eventually all
the males will tend to become equally good. Female choice requires heritable
genetic differences between males, but the effect of selection on the basis of
such choice is to exhaust these differences, swallowing them up by
ceaselessly exacting more, always more. (Females, let me hastily say, are not
the only gauleiters of selection. The same problem arises with any strong
selective force that pushes consistently in one direction.) It seems, then, that
choice will undermine its own success. And yet it has apparently managed to
mould many a 'peacock's tail'. What is it that rescues female choice from
destroying that which it feeds on?

Parasites. That, at least, is one answer. It is the intriguing theory of one of
the most important Darwinians of the second half of this century: W. D.
Hamilton. His argument, which he developed with Marlene Zuk, runs like
this (Hamilton and Zuk 1982). Of all the threats that an organism has to
contend with – cold, hunger, predation – attack from parasites is among the
most profound. And it is a threat that is ever-renewed. Over evolutionary
time, there is a never-ending arms race: as organisms develop adaptations to
resist their parasites, so the parasites develop counter-adaptations to continue
their plundering, or new parasites take over with new tricks. The hosts then
have to counter these adaptations – and so the cycle grinds on. Thus what are
good genes for resistance at one time might well be ineffective a few
generations on, when the existing parasites have retaliated or fresh
opportunists have seized their chance. There is, then, a constant revision of
what is best, a constant turnover in what constitutes good genes. Today's
most resistant genes may prove a liability to the great-grandchildren of their
current bearers.

So much for the males and their parasites. Now to the females and their
choice. Clearly, a female in search of a mate would be ill-advised to choose
one who has succumbed to parasites or is vulnerable to them. Indeed, if
parasites are so oppressive and so unremitting a threat, a female would be
well-advised to make hereditary resistance her prime criterion for choice of
partner. We can now see why, although the females are constantly selecting

the 'best' males, the genetic variation among the males never runs out. It is because the criterion for 'best' is ever-shifting.

But how can a female detect genes for parasite resistance? She needs some kind of external indicator of genetic quality. A sound procedure would be to pick the healthiest-looking male. A male who is parasitised would be likely to cut a poor figure whereas a male who is resistant would be able to cut a dash with the brightness and glossiness of his fur or feathers, the fine sweep of his tail, the vigour of his display. So it's a fair bet that if a male looks healthy, he will have superior genes to pass on to his offspring. And this indicator is likely to remain reliable, even though the particular parasites that he needs to resist are changing all the time. If females adopt this policy, they will be putting selection pressure on the males to present a healthy appearance. Indeed, the males will be under pressure to try to outdo one another, to try to look just that bit healthier than the healthiest-looking. Over evolutionary time, they will be pushed into advertising their health with brighter and brighter colours, longer and longer tails, more and more flamboyant displays. All the males will get drawn into this escalation, even those who are so sapped by parasites that their ornament betrays the fact. After all, if they didn't even try, then the females would think the worst of them. Males could, of course, try to fake the signs of good health. But selection is busy refining female judgement, too, favouring females who can spot an honest advertisement and weeding out those who are taken in by cheats. Females, then, will force males into adopting crests, colours, and so on that readily reveal their true state. And the most likely way for this to work would be through handicaps, a parasitised male being unable to afford the production costs of a truly spectacular display – or, at least, unable to afford those costs and at the same time keep up with all the other necessities of life. And so, driven by arms races with parasites, with females, and with one another, males evolve their ornaments, glorious through escalation but honest through scrutiny.

As soon as parasites enter the scene, a new set of interests enters with them. We've seen that parasites don't always stop at merely living off ready-made resources. Some also reap adaptive benefit by taking more active control of their hosts' bodies. Remember the manipulative thorny-headed worms and their hapless crustacean hosts? Now think about Hamilton's and Zuk's theory. It assumes that the tell-tale signs of a parasite's presence are an 'unintended' side effect of the guest exploiting its host's body. And, certainly, parasitism typically results in signs of debilitation. But I find it tempting to speculate that when it comes to male ornament and female choice, parasites could sometimes be running the show. Consider a parasite that (unlike the worm in the crustacean) uses its host's reproductive pathway in its own reproductive cycle – their reproductive interests run parallel. It would be in the parasite's interests for its host to mate successfully. In that case, it would be as unfortunate for the parasite as for the host if the side effect of its depredations gave away its presence. Indeed, the parasite would benefit if it could make its host appear the least parasitised of prospective mates – perhaps by putting an extra gloss on its host's plumage or extra brightness in its colour. The external signs of its presence would then no longer be mere side effects; they would be adaptations – adaptations for the benefit of the parasite, the result of parasite manipulation. The male's ornamentation would be the joint product of the selective pressures of female choice and the extended phenotypic effects of a gene in the parasite's body. Admittedly, this seems far-fetched. For one thing, ornamentation is costly even to a healthy host, so parasites would have to bend the rules of physiology quite a lot in their favour. But then it turns out that male fence lizards that are the most heavily infected with malaria are the most strikingly coloured (Read 1988) ... And some parasites are known to make the colours of their hosts more conspicuous to predators that are the parasites' final destination (Moore 1984, p. 82; Moore and Gotelli 1990) ...

Talking of manipulation, why anyway assume that if good sense is prevailing, it's to be found in female choice? It has been suggested that sometimes the selection pressures may be driven not by the females (nor by parasites) but by the males, manipulating female taste; the ornamented males are not mere creations of female fancy but themselves the prime movers. Consider the case of elaborate bird song. It is sometimes thought to have evolved as a marker for good genes or resources, with female choice selecting for elaborate song as a test of potential mates. But it could be the other way round – that female taste has been moulded under selective pressures originating with the male. Take the song of the male canary (*Serinus canarius*). It brings the female into readiness for reproduction; a complex song is more effective than an artificially simplified repertoire. Older males have larger repertoires and it has been suggested that females are using the song as a guide to male vigour because males who hatch early survive better and have more complex songs (Kroodsma 1976). But perhaps the male is manipulating the female to 'choose' him; her behaviour could be an extended phenotypic effect of his manipulative genes (Dawkins 1982, pp. 63–4). The result is likely to be an evolutionary arms race between manipulation and resistance to it. Why, then, hasn't the female been more able to resist? Why is the male apparently winning the race? It may be that the sensory channel that he is exploiting is crucial to her for its original adaptive purpose; so she can defend it to a limited extent but cannot afford to cut it off entirely from invasion. Could the peacock's tail have developed in a similar way? It has been proposed that the peacock is exploiting a standard adaptive response on the part of peahens: a propensity to pay careful attention to eyes (Ridley 1981). Manipulation theories have the advantage of neatly explaining why humans find many sexually selected characteristics beautiful. On other theories this is something of an embarrassment. When manipulation is at work, beauty emerges as a tool of manipulative power. The hold that a dazzling tail or a glorious song has over humans is just a side effect of its hold over members of the species for which it was intended.

We have come a long way from Wallace. On his version of the good-sense view, it was implausible that females who choose the most decorative males could be making a sensible choice. This was because Wallace insisted that female choice would make little or no difference to the effects of natural selection. Modern Darwinism, however, assumes no such thing. It can explain how utilitarian choice, far from making no difference, can lead to ornamentation so costly that it seems to be anything but sensible. This is what Wallace's good-sense theory needed. And, with this new twist, it is finally turning out to have more potential than at first appeared.

Fisher's solution: Good taste makes good sense

We left Darwin and Wallace in the 1880s in an impasse. Darwin's aesthetic theory could explain male ornament adaptively but not female choice. Wallace's utilitarian theory could explain female choice but not the typical extravagance of male ornament. And that is where classical Darwinism left the matter and where the theory of sexual selection rested for its first half century. We have seen how modern Darwinism revitalised Wallace's theory. It was R. A. Fisher who supplied the crucial turning point for Darwin's view. In a paper of 1915 and in his classic book *The Genetical Theory of Natural Selection* (Fisher 1915, 1930, pp. 143–56, particularly pp. 151–3), he underpinned Darwin's theory with the adaptive explanation of female taste that Wallace had rightly demanded. Fisher explained how female choice could be for attractiveness alone, as Darwin claimed, and yet be adaptive, as Wallace insisted it must be. In short, Fisher showed how Darwin's good taste could make surprisingly good sense.

Fisher argued that choosing an attractive mate can be adaptive for a female because she will have attractive sons. In a population in which there is a majority preference for anything whatsoever, a female would do best to follow the fashion, however arbitrary, however absurd, because the next generation of daughters will inherit their mothers' preference whilst her sons will inherit their father's attractive feature. Think of it this way. Imagine you're a peahen in a population in which there's a majority preference among the peahens for males with cumbersome, costly long tails. You could make an apparently sensible choice of mate and go for one with a sensibly short tail. But what would happen in the next generation? Your son will have inherited the short tail but the next generation of females will have inherited the majority preference for long tails. Your son may be better equipped for survival but what evolutionary good is that if he can't get a mate? Natural selection will eventually eliminate both your mating preference and your mate's short tail. It would have been a better strategy if you'd gone for a mate who could have endowed you with attractive sons. You would have lowered your sons' chances of surviving but increased your chances of having grandchildren.

But what reinforces such a fashion, why does it spread? And why does it ever catch on in the first place? The fashion is fuelled by a tie between the preference gene and the ornament gene. Consider a female who has genes for preferring a long-tailed mate. Her offspring will inherit both her preference genes and her mate's long-tail genes, although the preference will be expressed phenotypically only in her daughters and the long tail only in her sons. So her union solders a connection between preference genes and long-tail genes, a closer connection than would arise from random mating. (A measure of this tie is called the coefficient of linkage disequilibrium.) And the same will happen in subsequent generations. It is this connection that fuels the fashion. The more that females exercise a fashionable preference for long tails, the more the fashion is reinforced, each choice of long-tailed mate automatically being likely to select in favour of copies of genes for that very choice.

It is easy to see how readily escalation can take off. The whole process can get started from any majority preference at all, however small; the 'majority' doesn't have to be a large section of the population, just larger than any other. The preponderance could be created initially through nothing more than chance fluctuations. (Remember how the sexes diverge into different reproductive strategies from minimal beginnings merely by self-reinforcement.) Alternatively, as Fisher suggested, the fashion could start from a 'sensible' choice and then cut loose from its utilitarian moorings to soar off into the realms of extravagance. Imagine, say, that longer-than-normal tails help the males to fly better, so female preference for ever-longer tails is favoured by natural selection; eventually they become a downright encumbrance, but female preference for them is by that time sufficiently widespread to take off under its own sexually-selected steam. However it begins, any preference that has more adherents than any other, even if the preponderance is slight, will be favoured by selection because of the 'attractive sons' effect. And then, of course, it will become a larger majority and the advantages of having attractive sons will be even greater – and so on.

Fisher's theory involves a potentially explosive process of positive feedback: success breeds success. The more a preference for long tails is successful, the more in successive generations there will be males with increasingly long tails and females with a preference for even longer ones – and the more successful having and preferring ever-longer tails becomes (until natural selection calls a halt). Success is frequency-dependent and in a self-reinforcing way; the best thing to do is what the majority does, thus the more it is done, the more it becomes an even better thing to do. So selection in favour of long tails and selection in favour of preference for long tails proceed together – male ornament and female taste evolve hand in hand, reinforcing one another, pushing one another in a spiral, to the spectacular extravagance of the peacock's tail. That is what gives the evolution of ornament and taste their typically immoderate, escalating, runaway quality.

Here is how Fisher put it. He pointed out that female preference gives ornament an advantage and ornamented sons give the preference an advantage:

> the modification of the plumage character in the cock proceeds under ... an ... advantage conferred by female preference, which will be proportional to the intensity of this preference. The intensity of preference will itself be increased by selection so long as the sons of hens exercising the preference most decidedly have any advantage over the sons of other hens ... [A]s long as there is a net advantage in favour of further plumage development, there will also be a net advantage in favour of giving to it a more decided preference. (Fisher 1930, pp. 151–2)

And this positive feedback generates a runaway process:

> plumage development in the male, and sexual preference for such developments in the female, must thus advance together, and so long as the process is unchecked by severe counterselection, will advance with ever-increasing speed ... [T]he speed of development will be proportional to the development already attained, which will therefore increase with time exponentially, or in geometric progression. There is thus ... the potentiality of a runaway process, which, however small the beginnings from which it arose, must, unless checked, produce great effects, and in the later stages with great rapidity. (Fisher 1930, p. 152)

Fisher didn't fill out his explanation much further, nor, if he developed it mathematically, did he leave any record of doing so. And neither, for almost half a century, did anyone else. But it's a theory whose time has at last come. In recent years, population geneticists have taken up Fisher's idea with enthusiasm, elaborating it in a variety of formal models (e.g. Kirkpatrick 1982; Lande 1981; O'Donald 1962, 1980; Seger 1985; see also Dawkins 1986, pp. 195–215). And, not surprisingly, Fisher's ingenious revamping of

Darwin's theory has been vindicated: it has been shown that Fisherian runaway is certainly at least theoretically possible.

Darwin's critics can at last be answered. They rightly asked why females should choose ornamentation for its own sake – a choice with no apparent adaptive advantages; and why their taste, if it is not controlled by selection, should not fluctuate arbitrarily. If female choice is Fisherian then Darwin can finally reply. The origin, persistence and escalation of both female preference and male ornamentation can, at least in principle, be accounted for adaptively. But the selection pressures are generated ultimately only by female taste itself; they act on a female only because of what other females happen to be doing. The whole process is based on arbitrary female aesthetics alone, not on sensible, natural selectionist criteria. So, although Fisher's theory is an adaptive theory of female choice, it is radically different from Wallace's. It captures the spirit of Darwin's 'beauty for beauty's sake' without any concession to Wallace's utilitarian leanings. Darwin's idea – that selection could result in females choosing for aesthetics alone, in spite of the cost to the males – has been shown to be theoretically possible. Fisher manages to unite Darwin's good taste with Wallace's good sense: they are conjoined by the choice of the majority, by the consensus of pure fashion.

9

'Until careful experiments are made ...'

What satisfaction Darwin and Wallace would have had in seeing their theories put to the test! But it eluded them. Although they both made suggestions for mate-choice experiments, few were implemented in their lifetimes. It's only very recently that systematic tests have been made at all. So can we now say who was right, Darwin (or, rather, Darwin–Fisher) or Wallace? It would be very gratifying to be able to answer that question, to report that we now have some idea of whether, in any particular species, females predominantly have good taste, good sense or some judicious mixture of the two. But, in spite of extensive probing into the most intimate details of many an animal's mating behaviour, on the whole we still don't know. The difficulties are largely methodological – as a brief trip through the territory will soon reveal.

Darwin and Wallace were both keen to wreak havoc on male ornament (Darwin 1871, ii, pp. 118, 120; Darwin, F. 1887, iii, pp. 94–5; Darwin, F. and Seward 1903, ii, pp. 57–9, 64–5; Wallace 1892). Darwin knew of some cases in which females had rejected male birds after the ornamental plumage had been spoiled, although they had formerly accepted them. He was eager to observe more carefully the effect of removing or damaging ornamental feathers on a bird (particularly a peacock) that had previously been successful in mating. But, whatever the females would have felt, the birds' owners were reluctant to sacrifice their ornaments. He suggested dyeing the tail and crown feathers of young, unpaired white male pigeons to see how the decoration would affect their mating success. And he succeeded in persuading one pigeon owner to dye his birds magenta. But their unnatural splendour apparently passed unremarked by all their companions. He managed to have a dragonfly painted in 'gorgeous colours' but got no further with the experiment. He proposed staining the bright red breasts of male bullfinches with dingy colours to see how they fared with females in competition with normal birds. But this was never done.

Would such experiments have been able to decide between Darwin's and Wallace's views? Clearly, if the females show no preference for the best ornamented males, then Darwin is wrong. But suppose they do go for the most beautiful. This is where the complications begin. If, as Wallace claimed, there is a positive correlation between beauty and many sensible characteristics, then the females might not really be preferring the most beautiful at all. They could just be expressing a Wallacean preference for the sensible, with the ornaments playing no part in their choice. Wallace cites the case of a male hybrid canary-goldfinch (Wallace 1889, p. 300a). This bird was larger and more finely coloured and sang louder and better than the normal goldfinch; he was also highly attractive to wild females. But, asks Wallace, was it his size, colour or voice that attracted them? (And, one might add, which qualities in this case were ornamental and which sensible?) Until we know which features were the attractive ones, female preference for the most ornamented males cannot be taken as evidence in favour of Darwin rather than Wallace.

But suppose we can prise the male's qualities apart and pin down which of them attracts the female. And suppose we find that she really is choosing the best ornamented. Even then, there is still a major difficulty. She could be using markers. As Wallace pointed out, she could be choosing the most decorative males only because she is taking beauty as an indication of sensible characteristics: 'brilliancy of colour in male birds is closely connected with health and vigour, and until careful experiments are made we cannot tell whether it is this health and vigour, or the colour that accompanies it and which therefore becomes an indication of it, that is attractive to the females' (Wallace 1889, p. 300b; see also Wallace 1892). Wallace does not say why he thinks such 'careful experiments' are important (nor, unfortunately, what they would look like). But, clearly, if beauty serves as a marker then it is more difficult to distinguish experimentally between his theory and Darwin's. Females might continue to prefer the most beautiful males even when the sensible characteristics that normally accompany them are removed experimentally; and males might be rejected when bereft of their ornamental plumage, as in the cases Darwin cites, not because females prefer ornament for its own sake but because it is a sign of some advantageous characteristic, such as sexual maturity (Wallace 1889, p. 286).

It seems, then, that there is an asymmetry in what these mate-choice experiments can tell us. They can rule decisively in favour of Wallace. But, however convincing the evidence of female choice and however elaborate the male ornament, can they can ever rule out the possibility that the female is merely using the ornament as a marker for some sensible Wallacean quality? Consider the Australian satin bowerbird (*Ptilonorhynchus violaceus*) (Borgia 1985, 1985a, 1986; Borgia and Gore 1986; Borgia *et al.* 1987; Pagel *et al.* 1988; see also e.g. Diamond 1982, 1987). What at first sight could be more purely aesthetic and non-utilitarian than the male's artistic endeavours? He is pavilioned in splendour in a decorated bower of his own construction. The decorations are predominantly blue and yellow – flowers, shells, snake skins, feathers and, nowadays, the occasional beer can. The female inspects the bower and they mate there but neither male nor female makes any other use of it. Gerald Borgia manipulated the decorations experimentally. And he found that a male's mating success depended on the quality of the ornaments, particularly the number of snail shells and blue feathers in the bower. So far, so aesthetic. But female choice is probably nevertheless Wallacean. One indication is that males attempt to destroy each others' bowers and they accumulate their decorations in part by stealing from others. So the decorative state of a male's bower reflects his ability to defend it and to steal from others. These qualities required by the struggling artist could presumably indicate 'really' useful qualities – strength, stamina, stealth and so on. And, indeed (taking aggressive dominance at feeding sites as a measure of dominance), aggressiveness in bower destruction correlates positively with male dominance status. Another indication of sensible choice is that males seem to prefer scarce objects for decoration. If so, then they probably require ingenuity, memory and endurance to furbish a bower, and the decor could be advertising all this to females. In another species of bowerbird, the Vogelkop gardener (*Amblyornis inornatus*) (Diamond 1988), geographically separate populations have been found to show different colour preferences and it seems that, again, the preferred colours might be those that are least likely to turn up in the different natural environments. What's more, there is evidence, as we shall see in a moment, that females are seeking the males with the greatest resistance to parasites.

Now consider another example: males displaying their finery to the females. Darwin understandably attached great significance to female preference for superior display. It was the most direct evidence that nature provided of females exercising choice. And, what's more, display seemed to have no function other than to show off 'beauty for beauty's sake'. Nevertheless, as Wallace stressed, proficiency in display is likely to go hand in hand with superiority in sensible qualities. In sage grouse (*Centrocercus urophasianus*) (Gibson and Bradbury 1985; Krebs and Harvey 1988), males engage in elaborate strutting: beating their wings and inflating the chest with a pair of orange air sacs set in white feathers, with which they make popping and whistling sounds – a dandified posturing that, to human eyes, bears an unfortunate resemblance to noisy, animated fried eggs. Not surprisingly, this high-impact advertising is very energy-costly, and males vary enormously in how much they strut. Females prefer the males that strut the most. It seems that they are picking the males who are best able to sustain themselves –

perhaps, it has been suggested, because they are the most efficient at finding food. Or, again, as we shall see, perhaps females are influenced by signs of parasites. In ruffs (*Philomachus pugnax*), too, female choice appears to be influenced by the vigour and frequency of male display (Hogan-Warburg 1966); again, the display might well be a marker.

The problem is that 'beauty for beauty's sake' interpretations are always vulnerable to such findings. One can never establish that a preference is pure good taste because one can never close the door on good sense. There is a multitude of sensible qualities that the female could be opting for. So it is impossible to establish that utilitarian forces are not at work alongside the aesthetic. As far as such experiments are concerned, then, Darwin–Fisher explanations seem to maintain themselves by default rather than by right.

But is the position really so bleak for aesthetic interpretations? Are empirical investigations inevitably biased in favour of Wallace? No, they needn't be. Admittedly, in theory Wallacean interpretations cannot be excluded. But in practice, if one drew up a substantial list of likely Wallacean factors, informed by sound and imaginative evolutionary intuitions, and then showed that they were not correlated with male ornament and female choice, that would be a plausible argument against Wallacean good sense and for Darwin–Fisher good taste.

And the plausibility would be strengthened if one could track down predictions from the two theories that went in entirely different directions. Think again, for example, about the suggestion that male Vogelkop gardener bowerbirds prefer colours that are least likely to turn up in their locality. Mark Pagel has pointed out that these geographically separated populations provide a ready means of testing whether female taste really does reflect scarcity value, as would be expected if the males were advertising some utilitarian quality, or whether female taste and scarcity show no correlation at all, either positive or negative, as would be expected if females were guided by an arbitrary Darwin–Fisher choice (Pagel *et al.* 1988, p. 289; see also Borgia 1986, p. 79). If we leave it at that, and there turns out to be no correlation, we must still allow the possibility that the females really are making a utilitarian choice but are using some criterion other than scarcity by which to judge male quality. But an intriguing idea of Darwin's suggests that we needn't leave it at that. 'Arbitrary' choice needn't be unpredictable. As we've noted, Darwin toyed with the idea that female taste in ornaments might be formed by the colours that females are most familiar with from their natural environments (Darwin 1876a, p. 211; Darwin, F. 1887, iii, pp. 151, 157; Marchant 1916, i, p. 270; Poulton 1896, p. 202; see also Wallace 1889, p. 335). In that case, far from expecting no correlation between scarcity and preference, one would expect a negative correlation, with females inclining towards colours that are naturally most abundant – exactly the opposite of the Wallacean prediction.

Perhaps most experiments so far have found only sober, utilitarian criteria of choice because that is all that most experiments have looked for. Why not probe for the typical extravagance and absurdity that are likely to arise among Fisherian followers of fashion? If females are choosing on Darwin–Fisher lines they should, given the opportunity, go for males that are decked out in ornaments even more lavish than nature would normally provide. This is because normal ornaments are a compromise between sexual selection and natural selection – between female taste attempting to push the ornament in runaway selection into ever-greater extravagance and natural selection clamping down on it. So it should be possible to uncover experimentally what is usually a latent, unexpressed preference.

And this, indeed, is what one test has perhaps done. The experiment was on a species of widowbird (*Euplectes progne*) in which the males have strikingly long tails, especially at mating time. Malte Andersson (1982, 1983) clipped the tails of some males to almost a quarter of their length, from about 50 cm (20 inches) to about 14 cm (5.5 inches), and glued the clipped feathers onto other males, increasing their tail-length by about half. So he had one group of under-ornamented males and one group of super-males. He also had two groups of normal males; in case the operation of cutting and glueing affected female preference, one group was left untouched and the others had their tails clipped but glued on again in their entirety. Then he let the females choose. He measured mating success by the number of new nests containing eggs or young in the male's territory – which is a component of reproductive success as well as an indication of female preference. The super-tailed males turned out to be the clear winners. They attracted on average markedly more females than their short-tailed or normal rivals. The short-tailed and normal-unglued groups attracted the same number of females and the difference between them and the super-tailed males was statistically significant. (With the small numbers involved, the difference between the super-tailed and the normal-but-glued groups was too slight to be statistically significant (Baker and Parker 1983).)

All this is plausible evidence that the tails of male long-tailed widowbirds have evolved by a Darwin–Fisher runaway process. But female choice could nevertheless be Wallacean. The preference for super-tailed males could perhaps just be reaction to a super-normal stimulus – like the reaction of the cuckoo's foster-parents to their charge's supernormal gape – with no Darwin–Fisher mechanism behind it. Against such a utilitarian interpretation it was found that mating success was not related to two likely Wallacean possibilities: territory quality or ability to hold territory. However, other equally likely possibilities remain; perhaps, for example, the longest-tailed males are the most resistant to parasites.

It's not only preferences favouring super-ornamentation that could indicate a Darwin–Fisher choice. The discovery of any normally latent preferences that are apparently capricious or quirky is suggestive. Arbitrary preferences are, after all, what Darwin–Fisher choice is about. Darwin may have had some such idea in mind with his whimsical suggestion of dyeing pigeons magenta. He would no doubt have been highly gratified (albeit puzzled) to learn of some apparently bizarre preferences in captive zebra finches (*Poephila guttata*) (Burley 1981, 1985, 1986, 1986a, 1986b; Trivers 1985, pp. 256–60; see also Harvey 1986). When presented with a choice of mates whose legs had been ringed with coloured plastic, the females preferred the males ringed in red to orange- or green-ringed males and the males went for black-ringed females in preference to blue or orange. What is more, the most attractive females (the black-ringed) had greater reproductive success; they reared more young to independence. Presumably these females were not superior; after all, the rings were assigned randomly. More likely, it was the males that made the difference. It seems that zebra finches put more resources into rearing offspring when they secure an attractive mate (Burley 1988a). Further experiments uncovered even more curious predilections among the females. When the males were dressed up in hats of various colours, the females preferred the ones in white.

What is going on here? What is the evolutionary significance of these odd preferences? They could be a sign of the kind of female choice that Darwin posited. But the answer might not lie in sexual selection at all. There's some evidence that these unlikely ornaments tap into signals that the birds would normally respond to – that they perhaps enhance the bright red beaks that signify good health, or coincide with the colours used for identifying members of the zebra finch's own species. In that case the predilections are probably a by-product of a good-sense choice or of species recognition. And if it is interspecific recognition that's at work, then, far from being on Darwin's side, the females' choice leads straight back down the natural selectionist path that Wallace laid out a century ago. At most, it provides a starting point from which Darwin–Fisher preferences could originate and proliferate in the population before Fisherian runaway takes off. Admittedly, the artificiality of the males' ornaments and the fact that the zebra finches were captive add to the difficulties of interpretation: in their natural surroundings, female zebra finches are unlikely to encounter males sporting bright bangles or exotic millinery (although free-living zebra finches have been found to have much the same band-colour preferences as their captive relatives (Burley 1988)). But similar natural selectionist claims have been made for species in the wild that have not been subject to such manipulation. Snow geese (*Anser caerulescens*) choose their mates on the basis of plumage colour, but a study of their choice concluded that this selection within the species had no adaptive advantages and was probably a mere side effect of selection for fine-tuned ability to discriminate between species (Cooke and Davies 1983).

Maybe the search for Darwin–Fisher ornaments should not, anyway, be confined to the showy, the fantastic, the wondrous. After all, in principle, runaway sexual selection could whittle away at a tail just as readily as it could build one up (Dawkins 1986, p. 215). Perhaps inconspicuous ornamentation is commoner than we think, a rich realm still waiting to be explored, hitherto obscured by the grossness of our aesthetic expectations.

Perhaps, too, the search should not be confined to obvious garnishes, to plumes and crests and other embellishments. Even male genitalia can boast a florid architecture. Wherever animals favour internal fertilisation, from fleas to rodents, from snakes to primates, the genitals of the males present an exuberance of forms. Traditionally, these essential organs have been regarded as purely utilitarian – the products of lock-and-key engineering, species isolation and the like. But is a penis just a handy tool? William Eberhard has argued that such profusion, prodigality and arbitrariness, such rapid, divergent evolution bear all the marks of Fisherian runaway: these male genitals owe their typical luxuriance to female whim (Eberhard 1985).

I have concentrated on the difficulties of distinguishing between good-sense and good-taste explanations. Let's now move on to some of the other reasons why empirical questions about sexual selection can be hard to answer. Consider, for example, Hamilton and Zuk's intriguing theory that males owe their glory to parasite detection, to females using ornament as a guide to hereditary parasite-resistance. A highly plausible idea. But as for testing it, there are stumbling blocks – as Andrew Read has documented in detail (Read 1990). We'll look at some of them, for they illustrate the kinds of difficulties that crop up not only on this theory but on any theory of sexual selection. (And, whilst we're on the subject, we'll also see how this particular hypothesis has fared empirically.)

Take, first, the major novel prediction from the Hamilton–Zuk theory, a prediction about cross-species comparisons. According to the hypothesis, species that are the most susceptible to invasion by parasites should be the showiest because, over evolutionary time, males in these species have been under the greatest selective pressure to advertise their hereditary resistance. When Hamilton and Zuk advanced the theory, they surveyed 109 North American passerine bird species and found that their prediction was borne out: there was indeed a positive correlation between chronic blood infection and male showiness, as measured by bright coloration and song complexity (Hamilton and Zuk 1982).

But correlations across species, just like any other correlations, raise problems (e.g. Clutton-Brock and Harvey 1984; Harvey and Mace 1982; Harvey and Pagel 1991; Pagel and Harvey 1988; Ridley 1983). Two of the most notorious are 'the inflation of n' and 'correlation but not causation'. The 'inflation of n' problem arises in the following way. If, say, 100 out of 109 species of birds all have both bright coloration and a high parasite load, we seem to have a respectable 100 supporting cases. But if all 100 of them inherited both features from a common ancestor we really have only one supporting case; we are counting the same thing 100 times over. So we must try to ensure that our data points are independent. After all, if we were to allow those 100 common-phylogeny species to constitute 100 independent bits of data, then why should we stop at counting species? Why not count individuals, and tick off possibly millions of apparently supporting cases? Fortunately there are solutions to this difficulty, at least in principle. The trick is not naively to count species, or any other particular level in the taxonomic hierarchy, but to count independent evolutionary origins of the characteristics of interest (Ridley 1983). Alternatively, one can simply see whether the correlation holds independently across diverse taxa.

The other problem arises from the familiar fact that correlation does not imply causation. Both bright coloration and high parasite load could be caused independently by some third factor, a factor that may or may not be known to us. Remember the correlations that Wallace found between brightness (in females in his case) and nest-type. If, say, being covered also made nests more attractive to parasites or their vectors then that could be the reason for brightness being correlated with parasite load. In principle this second type of problem, too, can be overcome. But in practice the task is dauntingly difficult, so legion are the possible alternative explanations and so profound is our ignorance of them – quite apart from the difficulties of testing even those that we do suspect. Again, one answer is to take correlations from a wide range of groups – mammals and reptiles as well as birds in the hosts, for example, and a similar wide spread for the parasites. After all, widely different groups are unlikely to share the same confounding variables.

How the Hamilton–Zuk theory would fare on a wide-ranging analysis of taxa we don't know, for nothing so ambitious has been attempted. There are, however, some more limited investigations that have tried to remove the effects of phylogeny, or confounding variables such as ecology, or both. On the whole, the conclusions have been fairly favourable, some studies having found a positive association and almost all the rest having found no association rather than a negative one (but see Read 1990).

So, for example, a survey involving 526 species of neotropical birds, in which the effects of phylogeny were removed, showed a positive relationship between male brightness and parasite load (at least in families consisting all or mostly of resident species, though not in families of migrant species – which is perhaps to be expected because resident species, being exposed to the same parasites all year round, are under greater selection pressure) (Zuk 1991). And across ten species of birds of paradise, allowing for variables such as body size, diet and altitudinal range, the brighter the males, the greater the average number of parasites that were found on them (parasite intensity); what's more, promiscuous species, which were brighter than monogamous species, had a higher proportion of males playing host to at least one parasite (parasite prevalence) (Pruett-Jones et al. 1990). A survey covering 79 bird species from Papua New Guinea, which took account of similar ecological variables, found correlations between male showiness and parasite burdens at some phylogenetic levels (though not at others) (Pruett-Jones et al. 1991). In a study of 113 species of European passerines, in which the effects of phylogeny and of ten wide-ranging ecological and behavioural variables were removed, it was found that the brighter the males, the higher the prevalence of blood parasites (Read 1987). And when an enlarged version of Hamilton and Zuk's original data set was controlled for phylogeny, male brightness again correlated positively with parasite prevalence (Read 1987). In addition to this evidence from birds, 24 species of British and Irish freshwater fish, ranging across ten families, showed a similar correlation, this time between parasite load and male–female differences in brightness (the effects of several ecological and behavioural factors having been removed) (Ward 1988).

Against these results, Hamilton and Zuk's second criterion of male showiness – song complexity – showed no such relationship (song duration even showed a negative relationship) when several components of complexity (repertoire size, versatility and so on) were analysed in 131 European and North American passerines, again with phylogenetic associations removed (Read and Weary 1990). And when the enlarged version of Hamilton and Zuk's original data set and the set of 113 European passerines were analysed again, this time using a different (although not necessarily more authoritative) way of scoring brightness, the correlations became less persuasive; in the European birds, although the correlation actually became stronger, it depended heavily on species in which few birds had been sampled, and in the American birds, too, the correlation might have been influenced by small samples and perhaps by phylogeny (Read and Harvey 1989).

These investigations remind us that, even if all the confounding variables that are known to be plausible are controlled for, many difficulties of interpretation still need to be thrashed out (Cox 1989; Hamilton and Zuk 1989; Read and Harvey 1989a; Zuk 1989). How, for example, should elaboration of ornament be quantified and compared across species – bright red with iridescent blue, a long tail with a fancy crest? And are results seriously distorted by including only some of the types of parasite that the hosts are susceptible to?

Cross-species comparisons are, of course, only one way to test the theory. Another is to take correlations within species. The prediction here is that the showiest males will have the greatest hereditary resistance to parasites. And they should also be the most attractive to females. But, again, there are problems. For example, the number of parasites that a male is host to might not be a reliable measure of resistance because differences in numbers will also depend on chance differences of exposure. And the males that happen to have the most parasites might be, as a direct consequence of their unwanted guests' activity, the dullest. What's more, if resistance is costly, as it is very likely to be – a thickened shell, a finely-tuned immune response – then resistant males are paying twice, once for their resistance and once for their ornament; thus, in populations that happen not to have been exposed to parasites, if males can 'choose' their level of display, the most resistant individuals might be the ones with the least well-developed ornaments. And if to develop elaborate ornamentation really is a handicap, there is the problem of finding tests to distinguish between (Hamilton–Zukian) female choice for handicapped males because they are resistant and (Zahavian) female choice for handicapped males because they are displaying more general qualities like strength and vigour; this might involve, for example, discovering the specific mechanisms by which parasites interfere with a male's ability to colour his comb a really deep red or grow antlers that are really large. And there is also the problem, more easily solved, of finding out whether females are simply trying to avoid getting parasites transmitted onto themselves or whether they are trying to mate with males that have hereditary resistance.

These difficulties have not prevented within-species studies from becoming a fast-growing area of investigation. The analyses have covered a wide range of host (and parasite) species. The results so far, like those across species, have been somewhat mixed but largely favourable (mind you, many of them are much as would be expected on any theory of sexual selection). On the whole the findings have been first, that the more flamboyant a male's ornament, the lower his parasite burden; and second, that females favour males with fewer parasites. This is assuming, by the way, that 'ornament' – display rate, colour or whatever – has been correctly identified; in most cases this has been decided by human intuition rather than by experiment or field observations. I'll sketch just a few examples of the results that are coming in.

In barn swallows (*Hirundo rustica*), males that were heavily parasitised by a blood-eating mite had shorter tails than parasite-free males – and, as we shall see, females prefer long-tailed males; unmated males were more often parasitised and more heavily so than mated males; high levels of parasites in the nest reduced breeding success (as shown both by field observation and by experimental manipulation of parasites); and – a key factor in the Hamilton–Zuk theory – parasite resistance is heritable, judging from the fact that when half the nestlings in some broods were switched with half from other broods, the parasite burdens of individual nestlings matched those of their genetic parents more closely than those of foster parents (Møller 1990, 1991). In a study of red jungle fowl (*Gallus gallus*), parasite loads in males were experimentally manipulated (using an intestinal roundworm); it was found that ornamental characteristics were more impressive in unparasitised males; that they were a more reliable indicator than non-ornamental characteristics (such as body weight) that females might have used as cues if they were not followers of Hamilton and Zuk; and that females preferred unparasitised males (Zuk *et al.* 1990). Rock doves (*Columba livia*) were subjected to the same kind of manipulation (taking two parasite species), with similar results; the females were probably using reduced courtship display as their cue because, judging by the effects of parasites on females, parasites did not impair other aspects of behaviour or other characteristics that humans, at least, could detect visually (Clayton 1990). In a study of ring-necked pheasant (*Phasianus colchicus*), half the male chicks were fortified against parasites with the help of anti-parasite drugs and strict hygiene, whereas the other half were left to fend for themselves much as they would under natural conditions. The progeny of the unaided group suffered greater mortality but those that did survive the selection pressure turned out to be more resistant than the progeny of the artificially aided group, suggesting, as the Hamilton–Zuk theory posits, that resistance is heritable. But mate-choice experiments on these offspring were inconclusive, females showing no preference for the sons of 'naturally selected' males over those of featherbedded males (Hillgarth 1990). In sage grouse (*Centrocercus urophasianus*), males with lice are less likely to mate than are louse-free males (Johnson and Boyce 1991); when captive male sage grouse had their air sacs daubed with 'blood marks' to look like those of males with lice, females tended to avoid them, although they had previously accepted them as readily as undaubed males (Spurrier *et al.* 1991).

In satin bowerbirds (*Ptilonorhynchus violaceus*), males that held bowers had the lowest intensity of head lice, although head lice intensity did not correlate with other ornamental characteristics that, as we have seen, females are thought to use when judging potential mates; what is more, nearly all the matings went to bower-holders; and, within bower-holders, those with the lowest intensity of head lice had the greatest mating success (that was in a second study – in an earlier study too few bower-holders were infected that season for any correlation to show up) (Borgia 1986a, Borgia and Collis 1989). An investigation of a small sample of Lawes' parotia, a species of bird of paradise, was equivocal but suggestive: the more intensely that males were parasitised, the less they exhibited display-related characteristics; not surprisingly under the circumstances, females did not mate with highly-parasitised males; nevertheless, females accepted males with low parasite intensity – although, consistent with the Hamilton–Zuk hypothesis, the females could have been sorting out unexposed males from exposed-but-fairly-resistant ones (Pruett-Jones *et al.* 1990). In guppies (*Poecilia reticulata*), display rates were found to be inversely correlated with parasite load, and females preferred less heavily parasitised males (Kennedy *et al.* 1987). Gray treefrogs (*Hyla versicolor*) that were highly parasitised (measured by number of helminth worms) had a lower call rate and lower mating success (females judge males by their call); in lightly parasitised males, however, calls were unaffected and these suitors were as popular with females as unparasitised males – which could be the same phenomenon as suggested in Lawes' parotia (Hausfater *et al.* 1990). In fruit flies (*Drosophila testacea*), parasitised males were less successful at mating, and when females did mate with them the offspring were less likely to be viable; but it is not known how far this was female choice and how far male competition, nor is it known what cues the females use, although the abdomens of parasitised males are often distended, which makes them a lighter colour than usual (Jaenike 1988). In field crickets (*Gryllus veletis* and *G. pennsylvanicus*), the higher the levels of a gut parasite, the lower the number of spermatophores that males produced per unit time (an important component of mating success); females also mated preferentially with less parasitised (and older) males (age rather than ornament perhaps being their cue) (Zuk 1987, 1988).

I have dwelt on the difficulties of testing. Let's now get a more systematic idea of what experiments have been done on theories of sexual selection and with what results.

Darwin and Wallace opened up a promising programme for experimental work, although few of their own ideas were taken up at the time. The earliest detailed attempt to observe the effect of female choice on sexual selection was published shortly after Darwin's death. Two American experts on spiders, George and Elisabeth Peckham, were interested in testing between Wallace's physiological theory that the male's ornaments arose from his greater 'vital force' and Darwin's theory that it resulted from female choice (Peckham and Peckham 1889, 1890; see also Pocock 1890; Poulton 1890, pp.

297–303). To this end, they made detailed observations of spiders in their natural habitats. ('The courtship of spiders is a very tedious affair, going on hour after hour' they remarked acidly (Peckham and Peckham 1889, p. 37).) They concluded that Darwin was right:

> The fact that in the *Attidae* the males vie with each other in making an elaborate display, not only of their grace and agility but also of their beauty, before the females, and that the females, after attentively watching the dances and tournaments which have been executed for their gratification, select for their mates the males that they find most pleasing, points strongly to the conclusion that the great differences in color and in ornament between the males and females of these spiders are the result of sexual selection.   (Peckham and Peckham 1889, p. 60)

Shortly after this, Alfred Mayer, also in the United States, experimentally manipulated several species of sexually dimorphic moths to see whether female choice was affected (Mayer 1900; Mayer and Soule 1906, pp. 427–31; see also Kellogg 1907, pp. 120–3). He cut off the males' blackish wings and glued on the females' reddish-brown ones but was 'unable to detect that the females displayed any unusual aversion toward their effeminate looking consorts' (Mayer 1900, p. 19); females proved equally undiscriminating towards males with wings painted scarlet or green, although (unless they were blinded) they rejected males with no wings at all. Mayer felt that the results told against Darwin's view.

Courtship may be a 'tedious affair' for the observer. Even so, one might have expected studies like these to be among the first in a century-long stream. Not at all. However, bear in mind the influence of Wallace's attempt to replace sexual selection by natural selection, and this neglect after all comes as no surprise. Not that there were no empirical studies of mating behaviour. But until quite recently, most of them, particularly studies in the wild, were geared not to sexual selection theory but to Wallace's natural selection tradition: 'naturalists focused their attention on ... [such problems] as mating signals and behaviour, and reproductive isolation ... With respect to sexual behaviour, an animal was expected to get a mate of the same species (kind) – what else was there?' (Lloyd 1979, p. 293). There were honourable exceptions. One of the most notable was Edmund Selous, a pioneer of the study of bird behaviour in Britain (though a barrister by training). Writing in the early decades of this century, he concluded that his observations of mating birds spoke 'trumpet-tongued' in favour of sexual selection (Selous 1910, p. 264). But look what happened even to his contribution. It was largely Julian Huxley who built on Selous's work and we have seen how squarely Huxley stood within the natural selectionist way of thinking; even his early papers epitomise that approach (Huxley 1914, 1921, 1923). Such studies revealed very little about sexual selection. It was several decades before more systematic experiments were started (on *Drosophila*) to investigate the role of mate choice (see e.g. O'Donald 1980, p. 16). So, for most of the history of the theory of sexual selection, there was very little attempt to investigate it empirically. Only with the recent revival of interest in the theory has female choice begun to be studied in a wide range of species, both in the wild and in captivity. Now, at last, serious attempts are being made to discover how, if at all, such preferences have influenced the evolution of the males' extravagant ornaments (see e.g. Bateson 1983a; Blum and Blum 1979; Thornhill and Alcock 1983; see also Catchpole 1988; Kirkpatrick 1987 for summaries of current knowledge, theoretical as well as empirical). Nevertheless, as we have seen, the strongest influence even now is not Darwin's but the view that we can trace to Wallace. Experiments tend to be undertaken in the spirit of testing between conjectures about good-sense choices: Is the female choosing good genes or good resources? If good genes, is she looking for heritable resistance to parasites? And if good resources then is it food or territory or what? The possibility of testing between any of those conjectures and Darwin–Fisher choice has attracted less attention – although that has begun to change.

The search for Wallacean choices (leaving aside parasite resistance, which we have looked at) has proved to be extremely fruitful. The females of many species apparently make their selection along sensible Wallacean lines. Female moorhens (*Gallinula chloropus*) prefer the males with the largest fat reserves, probably because they are more efficient egg-incubators than thinner males (Petrie 1983). The mottled sculpin (*Cottus bairdi*), a freshwater fish, prefers large males, apparently for their proficiency in guarding eggs (Brown 1981; Downhower and Brown 1980, 1981). And female hangingflies (species belonging to the genus *Bittacus*) select the mate that brings the largest prey insect during courtship feeding, presumably because this offering sustains them during sperm transfer and egg-laying (Thornhill 1976, 1979, 1980, 1980a, 1980b). Many a species has been found that would equally meet with Wallace's approval.

Females who choose incubators, guards and so on obviously value a male for his resources. What about species in which the male leaves all the offspring's needs to the female? Do they furnish any evidence of female preference for good genes, genes for the qualities needed in the struggle for existence? The pheasant *Phasianus colchicus* is one species in which there is no paternal care. Torbjörn von Schantz and his colleagues (von Schantz *et al.* 1989) carefully observed not only pheasants with natural variations in male spur length but also pheasants in which they had manipulated the males' spurs, shortening some and lengthening others with a plastic 'spur' (but keeping all lengths within natural bounds – none super-short or super-long). They found that females prefer the males with the longest spurs. And it turns out, significantly for the good-genes hypothesis, that these males also survive longer than males with less attractive spurs. Spurs apparently play no part in deciding dominance among males (weight and tail length are what matter), so female preference is not simply taking its cue from the males' own hierarchy and letting selective forces among the males set the standards. Neither are females using spur length as a guide to territory quality or age. It seems that female preference is guided by the males' survival qualities – by good genes – alone. Apparently female *Colias* butterflies have a similar preference (Watt *et al.* 1986). Their strategy is not quite the same because the spermatophore carries nutrients as well as sperm – resources as well as genes. But, as far as the genes are concerned, females prefer males with a genotype that is best at fuelling flight and maintaining temperature, related attributes that are both crucial for butterflies. Females are probably guided to such males because these very attributes also enable the males to persist in their courtship.

All these good-sense examples are impressive. Nevertheless, such evidence alone is not enough. There will be no evolution of female choice and male ornament, whether Wallacean or Darwin–Fisherian, unless both the preference and the preferred characteristic are heritable, and unless matings with the preferred males result in greater than average reproductive success (or, at least, if this was true in the evolutionary past). Some evidence of this kind has been found, albeit far from complete. In seaweed flies (*Coelopa frigida*), for example, females with one particular gene make a particular choice of mate and also mate more successfully than females with a different allele – though it is not necessarily that gene itself that is responsible for their behaviour (Engelhard *et al.* 1989). The hybrid offspring of unions between two species of Australian field crickets (*Teleogryllus commodus* and *T. oceanicus*) occur in two types; females prefer the calling song of males of their own type (only males sing), suggesting that their mate preference is genetically coupled with the males' calls (Hoy *et al.* 1977; see also Doherty and Gerhardt 1983). In pheasants, females that mate with the longest-spurred males hatch more chicks than the others and long-spurred males also enjoy greater reproductive success. In mottled sculpins, the larger males seem to have greater hatching success. In hangingflies, discriminating females have greater success in egg-laying and males with the more attractive prey items have greater success in transferring sperm; what is more, both female preference and male prey selection are apparently heritable.

A detailed laboratory study by Linda Partridge set out specifically to test for a connection between mate choice and a component of reproductive success (and therefore perhaps for sexual selection) (Partridge 1980). Fruit flies (*Drosophila melanogaster*) were divided into two groups; in one the females were allowed free choice of mates and in the other they were assigned mates at random. The offspring of these two groups were both pitted against standard competitors, the competition being for access to a limited food supply. It was found that a significantly higher proportion of the offspring of choice pairings survived to adulthood. It seems, then, that by exercising choice the parents could affect at least this component of reproductive success. But other questions remain open (see e.g. Arnold 1983; Maynard Smith 1982c, p. 184). Were the parents able to choose for good genes, for example, contrary to the view that genetic variation affecting fitness would not be heritable? Or was it just a case of choosing those unlike themselves (negative assortative mating)? Was total reproductive success enhanced? Or were the benefits outweighed by losses in some other component of reproductive success – which, as John Maynard Smith (1985, p. 2) points out, might be expected on both theoretical (Williams 1957) and empirical (Rose and Charlesworth 1980) grounds? Did the females choose the superior males or were the superior males more able to obtain access to mates? If there was female choice, was it heritable (or would it have been in the past)? And did the criteria for mate choice involve the kind of exaggerated male characteristics that Darwin was trying to explain? Until such questions are answered, we really don't know what these results can tell us about whether the females are followers of Wallace.

Wallace would have been even more pleased with an interpretation that is now commonly being placed on many of the experimental findings. For him, the idea that females choose their mates, even sensibly, was an explanation of last resort. His first resort, of course, was orthodox natural selection. But his next preference was to explain male ornament as the result of direct competition between males. He would feel well vindicated by many claims that are now being made.

Take birdsong, which Darwin definitely thought was 'to charm the female' (Darwin 1871, ii, pp. 51–68). Darwin was convinced that it was a sexually selected characteristic (although, out of all the evidence he collected, he quoted only one naturalist who claimed to have observed a connection – it was in finches and canaries – between male singing ability and mating success (Darwin 1871, ii, p. 52)). Darwin would have appreciated the results of some recent studies. In two species of flycatcher (*Ficedula hypoleuca* and *F. albicollis*), the females were found overwhelmingly to favour nest-boxes in which dummies 'sang' (courtesy of tape-recordings) in preference to nest-boxes in which they remained silent (Eriksson and Wallin 1986). In sedge warblers (*Acrocephalus schoenobaenus*), it was found that the males with the most elaborate song succeeded in mating the earliest (which probably gave them a reproductive advantage) (Catchpole 1980); this female preference continued to hold under laboratory conditions, when confounding Wallacean factors, such as quality of the male or his territory, were removed (Catchpole *et al.* 1984) (although this does not preclude the possibility that the song is a marker for some other sensible quality). But female preference does not rule out male competition as a selective force; both influences could be at work (Catchpole 1987). In brown-headed cowbirds (*Molothrus ater*), females were found to show a preference for the distinctive songs of dominant males over those of subordinate males (West *et al.* 1981). And in the village indigobird of Zambia (*Vidua chalybeata*), although male song and the highly conspicuous behaviour that accompanies it have been shaped to some extent by female choice, intermale aggression seems to have played an important role (Payne 1983; Payne and Payne 1977). Indeed, some investigators hold that competition between males is frequently the major evolutionary force behind elaborate song. This has been claimed, for example, of red-winged blackbirds (*Agelaius phoeniceus*), because it was found that a large song repertoire aided territory defence, whereas, by contrast, a correlation between repertoire size and female choice (as measured by harem size) disappeared when repertoire size was controlled for male reproductive age (Peek 1972; Searcy and Yasukawa 1983; Yasukawa 1981; Yasukawa *et al.* 1980).

Similar claims have been made about other characteristics, such as splashes of bright colour. Taking male red-winged blackbirds again, it was found that painting out the males' red epaulets had no direct effect on female choice (Peek 1972; Searcy and Yasukawa 1983; Smith D. G. 1972). Females seemed to be influenced primarily by a Wallacean factor – the quality of the male's territory. But males with their epaulets painted out were less able to defend their territory. In three-spine sticklebacks (*Gasterosteus aculeatus*), the females seem at first sight to be making a purely aesthetic choice, which has nothing to do with male competition. In some populations there are two kinds of male: a minority develops a red throat during the mating season, the rest remain drab. In laboratory experiments the females have been found to prefer red-throated males (as measured by the choice of nest in which they lay their eggs). And it appears to be the red throat that they like; when drab males are adorned with an artificial throat of lipstick or nail varnish, the females respond to them as if they were genetically red-throated (Semler 1971). Such choices certainly look as if they would have pleased Darwin. But it transpires that red males are less likely to lose eggs from the nest through predation by other sticklebacks and this is probably because the red throat has threat value. So, not only do the females seem to be making a thoroughly sensible Wallacean choice, but, what's more, male competition might also be at work on those red throats. This was just the type of result that Wallace had hoped for: that characteristics that couldn't be assimilated to the more standard forces of natural selection would be accounted for by competition between males.

Surely it would have been beyond even Wallace's wildest dreams for male rivalry to have captured the tails of the peacock, the Argus pheasant and other really spectacular male birds. And yet, over the last few years, interpretations have swung so far in this direction that even these heights of aesthetic extravagance have been widely regarded as, after all, having far less to do with female choice than with competition among males. Most of these gorgeously-plumed birds belong to lek (or lek-like) species (see e.g. Borgia 1979; Bradbury 1981; Bradbury and Gibson 1983). These are species in which males congregate and display on particular patches of ground, ground that is used only for this purpose – not for food or cover or anything else. The females visit the males there and apparently look them over – though to what extent is one of the points in dispute. Either way, the lek is a meeting-place for mating. Typically, the males in such species provide no parental care. So if the females do choose, they must be going for Wallacean good genes or exercising Darwin–Fisher good taste. Darwin, reasonably enough, regarded leks as strong indirect evidence that females were choosing the flamboyant characteristics on display there – and, of course, that they were choosing them for their aesthetic qualities alone (Darwin 1871, ii, pp. 100–3, 122–4). Some Darwinians nowadays, however, take the view that in several of these species the showiest, most exaggerated of the males' characteristics have evolved above all through competition between the males, and that if females have played any part it is only because they prefer to mate with the victors. One review of empirical findings ends with this comment: 'Existing evidence points to the conclusion that the importance of female choice in the evolution of exaggerated traits has been largely indirect, through female preferences for dominant males, and an importance of exaggerated traits in determining or signalling dominance' (Searcy 1982, p. 80). Linda Partridge and Tim Halliday come to a similar conclusion:

it is common for the consequences of intersexual selection to be exemplified by the peacock and birds of paradise. Evidence that females actually choose their mates in these species is, however, slight or non-existent. Indeed, some recent studies suggest that elaborate male plumage in these birds may be, at least in part, the evolutionary result of inter-male competition; males may be intimidated by the elaborate plumage of rivals in aggressive encounters ... Such field studies as have been carried out on species in which the evolution of elaborate male plumage has classically been attributed to female choice generally fail to support that hypothesis unequivocally.   (Partridge and Halliday 1984, pp. 233–5)

Take the birds of paradise, one of the most stunningly ornamented of all groups of birds. It has been claimed of one species (*Paradisaea decora*) that their extravagant plumage and display result almost entirely from competition between males for dominance and mating precedence (Diamond 1981). The males confine their most gorgeous displays to one another. When the females are present, they put up a relatively meagre show, and female choice anyway amounts to little more than accepting the victor. The Argus pheasant (*Argusianus argus*) has not escaped this downgrading of the female's role in shaping bodies – or, in this case, feathers – beautiful. It has been argued that female choice is not determined by the subtleties of the male's artistic plumage but by the gross effect of his display (echoes there of disagreements between Darwin and Wallace) and, more important, by whether he holds a display site (Davison 1981). The onslaught on female choice doesn't end there. Some authors have suggested that even when there is choice, males sometimes pre-empt it by successfully disrupting the attempts of other males to mate; the golden-headed manakin (*Pipra erythrocephala*) has been cited as an example (Lill 1976). In short, it is argued that there is often little or no female choice in lek species; that even when the female does choose, she may not be going for the male's most flamboyant ornaments; and that even when her choice is for ornament, she may merely be reinforcing the results of male competition.

Some critics have challenged these interpretations of the data (e.g. Cox and Le Boeuf 1977). After all, they have argued, the females could just be getting the males to do the work for them – inciting the males to compete with one another so that they will sort out which are superior. What's more, females can often choose between leks even when they have little choice within them. They also have the option of choosing a low-ranking male on the periphery of a lek. And attempts by other males to disrupt copulations between a female and the male of her choice have a low success rate.

New findings are also undermining this 'male club' view of lek species. It had been urged, for example, that in sage grouse much of the male's showy plumage and display had developed through contests between males over territory boundaries (Wiley 1973). On this view, the female wasn't interested in ornament but in whether the male occupied a central position in the lek – a fashionable address, not fashionable dress. But it has now been found that position is not a major determinant of mating success. What's more, as we have seen, females set their own agenda rather than just accepting the verdict of male combat: they choose males for their strutting display, an energy-costly activity and so presumably a sign of quality (Gibson and Bradbury 1985; Krebs and Harvey 1988).

That reinterpretation restores female choice. However, the female's criterion is still good sense. There are no reports so far in lek species of Darwin–Fisher, beauty for beauty's sake, choices. But perhaps that icon of sexual selection, the peacock, will after all come to Darwin's rescue? Marion Petrie and her colleagues have found that, in the peafowl *Pavo cristatus*, females apparently prefer males with the most eye spots in their train (Petrie *et al.* 1991). What happens is this. As in all lek species, males attempt to secure a display site within the lek area and only those males that manage to secure one put on a display. Females visit males at the lek. They never mate with the first male that courts them, always rejecting some potential mates before deciding. There is enormous variance in male mating success; of the ten males observed at one lek, the most successful copulated 12 times (eight different females) and the least successful not at all. Over 50% of this variance could be accounted for by the splendour of the male's train, in particular by number of eye-spots. It was found, for example, that in ten out of eleven successful copulations the female had chosen the male with the highest spot number of those she had sampled (in the one odd case, the chosen male had only one spot less). Mating success could not be accounted for by factors that have been thought to be important in other lek species (mostly to do with competition between males), such as the male's call rate, his display rate, challenges from intruders and whether his position in the lek was central or peripheral. (All this, of course, applies only to males that manage to obtain a display site at all; it's not known whether they have more

Beauty for whose sake?

Natives of Aru shooting the Great Bird of Paradise (from Wallace's *The Malay Archipelago*)

"I thought of the long ages of the past, during which the successive generations of this little creature [the King Bird of Paradise (Paradisea regia)] had run their course – year by year being born, and living and dying amid these dark and gloomy woods, with no intelligent eye to gaze upon their loveliness; to all appearance such a wanton waste of beauty. Such ideas excite a feeling of melancholy. It seems sad, that on the one hand such exquisite creatures should live out their lives and exhibit their charms only in these wild inhospitable regions, doomed for ages yet to come to hopeless barbarism; while on the other hand, should civilized man ever reach these distant lands, and bring moral, intellectual, and physical light into the recesses of these virgin forests, we may be sure that he will so disturb the nicely-balanced relations of organic and inorganic nature as to cause the disappearance, and finally the extinction, of these very beings whose wonderful structure and beauty he alone is fitted to appreciate and enjoy."

(Wallace: The Malay Archipelago)

eye-spots than males without sites, for these 'floaters' don't show their trains to experimenters any more than to females; it is known, however, that males with sites are heavier and have longer trains.) It's possible that the females are not using eye-spot number as their cue or, at least, as their sole cue. But they certainly seem, on whatever cues, to prefer elaborate trains. The next question is Why? Number of eye-spots and, to a lesser extent, length and colour of train change with age. So females could be using elaboration as an indicator of some good-sense quality that goes with age – perhaps ability to survive. Or is it wrong to assume that females are making a good-sense choice at all? Perhaps, as Darwin thought, they are choosing the most gorgeous males just because they are the most gorgeous. We still don't know.

Whatever tale the peacock eventually has to tell, at present most of the other stock symbols of sexual selection, that is most lek species, are quite commonly put beyond Darwin's (or Darwin–Fisher's) grasp. Won't this perhaps seem ungenerous in a few years' time, as ungenerous as it now seems to have claimed that the function of the peacock's tail and the nightingale's song is species recognition?

I have said, by the way, that it is a victory for Wallace to see Darwin's prime exhibits apparently succumbing to the uncontentious force of male competition. It is, however, a pyrrhic victory as far as classical Darwinism is concerned. Only recently has Darwinism been able to deal with conventional, ritualised aspects of behaviour. Wallace envisaged that male competition would work straightforwardly along the lines of natural selection; the males would engage in head-on battles with weapons that would be useful for other purposes (e.g. Wallace 1889, pp. 136–7, 282–3). There was no place in his theory for the extravagant escalation, so characteristic of lek species, that male competition can generate. Even the less showy examples that we granted as successes for Wallace – the songs of the red-wing blackbird, cowbird and village indigobird, the stickleback's red throat, the red-wing blackbird's epaulets – involve a highly conventional element that lies beyond the scope of classical thinking and requires careful analysis still today. I once asked Britain's most eminent population geneticist and the pioneer of game theoretical explanations of conventional competition what he thought of the rather cavalier assumptions that are often made about the value of conventional threats. 'If I were a peacock and another male flaunted his tail at me, I'd kick him in the balls' was his authoritative reply. What is more, unlike Darwin (e.g. Darwin 1871, ii, pp. 50, 232–3, 269), Wallace assumed that competition between males would almost entirely exclude female choice. But, as we've noted, there can be plenty of leeway for female preference, especially of the sensible kind, even if male competition is the major driving force.

We saw that Darwin, and Wallace far more so, underestimated the likely costliness of male ornaments. And, in species after species, what an underestimate that has now been found to be. To modern Darwinians, one source of costs is obvious. If males are signalling to females, then those signals are ripe for exploitation by a monstrous regiment of scavengers – predators, parasites and competing males. And this is indeed what happens. In polymorphic three-spine stickleback populations, the red-throated males suffer far more predation than the black-throated because of their bright colours (Moodie 1972). The same is true of the males compared with the females in several other species of fish (e.g. Haas 1978). In a species of field cricket (*Gryllus integer*), the males who call longest and most intensely to females suffer a much higher rate of parasitism from a fly that deposits its host-devouring larvae on them (Cade 1979, 1980). In the túngara frog (*Physalaemus pustulosus*), females prefer a chuck sound, particularly a low frequency one, in the mating call rather than a high frequency whine – it gives them more information on the body size of their potential mates; but the low frequency chuck calls are also more attractive to a frog-eating bat (*Trachops cirrhosus*) (Ryan 1985, pp. 163–78; Ryan *et al.* 1982).

And that is by no means the only kind of cost of being attractive. Sexual selection for increased body size in male birds invariably brings with it an increase in bill size, in some cases so great that males are forced to exploit suboptimal food niches (Selander 1972). The energetic costs of the males' display may be so high that they are pushed into abandoning safe foraging options for ones that possibly give higher energy returns but are more risky (Vehrencamp and Bradbury 1984). In the great-tailed grackle (*Quiscalus mexicanus*), not only does the males' bright plumage attract predators but also their long tails impede flight and their large size is above optimum for efficient foraging (Selander 1972). But spare some sympathy for females for they, too, may have to bear the direct costs of sexually selected traits. In some species in which the males are sexually selected to be larger than the females, to produce a son takes a greater toll of the mother than to produce a daughter (Clutton-Brock *et al.* 1981).

Suppose that the males seem to bear up well under their impediment, so well that those with the most extravagant ornaments do best in the struggle for survival as well as in the struggle for mates. Should we then conclude that the burden of being attractive is, after all, no burden, that survival and mating success, rather than pulling in opposite directions, actually agree? Should we assume that, for example, in pheasants, 'survival and reproduction unanimously favour larger spurs' (Kirkpatrick 1989, p. 116)? Wallace, as we have noted, tried to deal with the embarrassment of males being encumbered, on his theory to no purpose, by pointing to 'the great abundance of most of the species which possess these wonderful superfluities of plumage' (Wallace 1889, p. 293); the costs must be pretty minimal, he urged, if the males manage to thrive in spite of them (though his embarrassment was mitigated by the fact that, on his view, ornaments develop only if males are physiologically able to afford the overspill). But one could draw the opposite of a no-cost or low-cost conclusion: that the cost is so high that only the resilient can bear it. Think of those naively idealised monuments, so beloved by artists of socialist realism, showing muscle-bound Stakhanovite workers stoically supporting Herculean loads. Laughable as they are, they do capture reality in one way. Would we expect the hero of the excess quota to look like a seven-stone weakling? No. Surely only someone who is up to it would take on the burden in the first place. And if he is still there to tell the tale at the end of the season, this shows that he really did have those survival qualities, not that it was after all no burden to do far more than the allotted share. Indeed, a handicap view would go further: a male shoulders his burden precisely in order to advertise his quality, to proclaim his ability to take it on and yet cope in spite of it. So, for example, male pheasants with long spurs would survive even better without them, but they grow them because they are accurate indicators of quality (Pomiankowski 1989).

We have seen Darwin's solution to the problem of how sexual selection could act in monogamous species. He suggested that the healthiest females in any season breed the earliest. Having the pick of the males, they choose the best ornamented; and with their early start they have the most offspring. Healthiness is not hereditary but female preference and male ornament are. R. A. Fisher cautioned that 'it would seem no easy matter to demonstrate' (Fisher 1930, p. 153) whether there really is a correlation between healthiness, early breeding and numbers of offspring. Not easy, but it has for some years been demonstrated in several species (O'Donald 1980, pp. 3, 25–7, 41, 136–48, 1987). And Peter O'Donald showed that in one monogamous species, the arctic skua (*Stercorarius parasiticus*), there is an impressive fit between data on breeding dates and the predictions about links between ornament genes and preference genes that can be made when Darwin's conjecture is formalised in genetic models. But O'Donald did not make the kind of mate-choice experiments that would help to show what female preference was really for. Anders Møller has now filled that gap (Howlett 1988; Møller 1988). He took a monogamous species of swallow (the species that he used to test the Hamilton–Zuk hypothesis, *H. rustica*), in which the males' trailing outermost tail feathers are about 16% longer than the females'; males attract females by displaying their tails. Møller subjected the males to the same clip-and-glue treatment as Andersson's widowbirds. Once again, the super-tailed males were an overwhelming success with the females:

on average, they paired in only a quarter of the time of the extra-short-tailed males. Because of this speedier matching, the super-tailed males and their mates were more likely to have a second clutch together. So the super-tailed males ended the season by having an average of twice as many fledglings – and vindicating Darwin's biological intuition, at least on this point if not on the question of why females prefer long-tailed males. By the way, it has been suggested that sexual selection may sometimes work in monogamous species because they're not, after all, entirely monogamous. It turns out that the super-tailed swallows did indeed benefit from non-monogamous pairing – and again, far more so than other males, although not from the others' want of trying. So sexual selection could be getting an extra boost this way.

Finally, let's remind ourselves of just a few of the practical difficulties of testing theories of sexual selection. For one thing, it's no easy matter to judge when animals are choosing at all. Non-random mating, for example, might at first glance seem like good evidence for choice – but not at second glance. Common toads (*Bufo bufo*) practice size-assortative mating, and at one time this was attributed to female choice; but it is now thought to be merely a consequence of the mechanical fact that only in pairs that are well-matched for size can a male grip firmly enough to resist take-overs by other males (Arak 1983; Halliday 1983). Neither is choice necessarily a sign of sexual selection. Assortative mating on the basis of, say, kin relationships may involve choice but may not give rise to, and may even oppose, sexual selection (Bateson 1983a, p. xi). And then there are tasks that present formidable practical difficulties in the wild. To demonstrate that sexual selection is at work, there must be evidence of differential reproductive success (in the past, even if no longer); in particular, this requires measures of life-time reproductive success rather than the short-term periods that most studies have relied on.

In 1890 Wallace remarked that so many more observations were needed to answer problems about sexual selection that 'this most interesting question ... in all probability, will not be finally settled by the present generation of naturalists' (Wallace 1890a, p. 291). One century on, matters are still far from settled. Indeed, 'this most interesting question' has proliferated an abundance of new questions – such an abundance that many of the mysteries of mating are unlikely to be penetrated for several generations yet.

10

Ghosts of Darwinism surpassed

The changing face of sexual selection

For classical Darwinism sexual selection was an oddity, entirely different from natural selection and generally opposed to it. It's not hard to see why. Sexual selection was driven by preferences of members of the male's own species, leading to competition between males; the paradigmatic forces of natural selection were between species, not within them, and asocial. Sexual selection was solely to do with mating success; for natural selection, success and failure covered a vastly broader sweep – survival and all the remaining aspects of reproduction. And sexual selection seemed to favour the ornamental, the pointless, even the downright damaging; natural selection was thought to opt invariably for the efficient and utilitarian. To classical Darwinism such distinctions were of great importance, marking a deep gulf between sexual selection and natural selection. Modern Darwinism takes a different view.

Let's start with the fact that sexual selection is to do with social relations within species. We have seen how classical Darwinism neglected the social aspects of selective forces; even selective forces within species, when they were discussed at all, were often dealt with asocially, in much the same way as inorganic pressures. Such thinking so permeated classical Darwinism that sexual selection was thought to be quite unlike natural selection. Classical Darwinism was forced to recognise sexual selection as social because, for one thing, the theory is quintessentially about forces within a species and competition within one sex. Here is how Darwin contrasted it with natural selection: 'This form of selection depends, not on a struggle for existence in relation to other organic beings or to external conditions, but on a struggle between the individuals of one sex, generally the males, for the possession of the other sex' (Peckham 1959, pp. 173–4). What is more, sexual selection involves what was taken to be an unusual selective force: 'will, choice, and rivalry' (Darwin 1871, i, p. 258). The theory assumed that female preference could mould flamboyant males, in just the same way as the need for efficient foraging could shape the woodpecker's beak or the advantages of wide dispersal could favour plumed seeds.

Generations of Darwinians felt, like Darwin, that all this added up to a major difference between the two theories (although, unlike Darwin, they usually concluded that this difference told against sexual selection). Groos, who was for ousting choice and assimilating sexual selection to natural selection, put it like this: 'The selective principle involved ... is not the mechanical law of survival of the fittest, but rather the will of a living, feeling being capable of making a choice, and is much like that employed in artificial breeding ... [A] fitting designation of this theory of sexual selection would be "a multiplication of the most pleasing"' (Groos 1898, p. 230) – a principle that he found thoroughly unconvincing. Lloyd Morgan drew attention to what he saw as a difference

> between natural selection through elimination and conscious selection through choice. The two processes begin at different ends of the scale of efficiency. Natural selection begins by eliminating the weakest, and so works up the scale from its lower end until none but the fittest survive; there is no conscious choice in the matter. Sexual selection by preferential mating begins by selecting the most successful in stimulating the pairing instinct, and so works down the scale until none but the hopelessly unattractive remain unmated. The process is determined by conscious choice.   (Morgan 1896, p. 219)

The assumption that this kind of difference is fundamental still crops up occasionally even today. Peter Vorzimmer, for example, seems to agree with Darwin: 'Because the individual organism (of the opposite sex to the one being selected), rather than the elements of the environment, constituted the source of the selective standard, Darwin saw that a distinctly different form of selection was involved' (Vorzimmer 1972, p. 189).

Incidentally, this traditional attitude to sexual selection provides a telling counterexample to a widely held view that classical Darwinism (or at least Darwin's own Darwinism) systematically incorporated social pressures into selective forces. It is commonly claimed, for example, that Malthus was important to Darwin and Wallace because he viewed competition as intraspecific and social, in contrast with the prevailing idea (notably Lyell's) that biological struggle was primarily an asocial battle against inorganic forces or members of other species (e.g. Herbert 1971; Kohn 1980; Manier 1978, p. 78; Ruse 1979a, p. 175; Sober 1984, pp. 16–17, 195–6; Vorzimmer 1969). Even if Malthus did provide that starting point, classical Darwinism's standard contrast between natural and sexual selection shows how far Darwin and Wallace travelled from this beginning.

The contrast between non-social and social selective forces was reflected in Darwin's idea that, whereas natural selection would more or less grind to a halt in a constant environment, sexual selection was, in principle, capable of continuing indefinitely on its giddy spiral of ornamental exaggeration:

> In regard to structures acquired through ordinary or natural selection, there is in most cases, as long as the conditions of life remain the same, a limit to the amount of advantageous modification in relation to certain special ends; but in regard to structures adapted to make one male victorious over another, either in fighting or in charming the female, there is no definite limit to the amount of advantageous modification; so that as long as the proper variations arise the work of sexual selection will go on. (Darwin 1871, i, p. 278)

Classical Darwinism offered no explicit theoretical reasons for maintaining this view. It crops up in *Descent* as if Darwin had simply read it off from the data – from the disparity between the parsimonious economy of the woodpecker's beak and the baroque flamboyance of the peacock's tail. Indeed, extravagant escalation was for him a diagnostic feature of sexually selected characteristics. But Darwin's view reflects his recognition of sexual selection as a selective force that was social. Unlike natural selection, sexual selection was seen as internally generated, and, as a result, inevitably changing and dynamic – female demands provoking male competition, and each pushing the other to ever-greater excesses. The result, as Darwin said, was 'no definite limit'. It's significant that in just one sphere Darwin envisaged natural selection as acting in the same way. He took the view that mental improvement in humans could continue indefinitely. (He did, by the way, consider it to be improvement and not merely change.) It is no coincidence that this was another of the rare cases in which he recognised the selection pressures to be social.

Could those social forces push ornaments to such heights of escalation that they might carry a species to extinction, or would natural selection put a stop to such excess before it got out of hand? According to Darwin, natural selection would invariably intervene (e.g. Darwin 1871, i, pp. 278–9). There was, of course, something to be said for this view; natural selection may act as a countervailing force and probably quite commonly does. It may work by moderating the sexually selected characteristic in all individuals; that was how Darwin envisaged it. Or, as we have seen, it may favour variability; this appears to be what is happening with, say, the black- and red-throated stickleback populations, in which the polymorphism is maintained in a frequency-dependent way by the relative costs and benefits of being cryptic-and-unattractive or conspicuous-and-attractive (Moodie 1972; O'Donald 1980, pp. 67, 170, 182). Nevertheless, although natural selection may keep sexual selection within bounds, it does not invariably act as the *deus ex machina* that Darwin supposed: 'It has often been assumed that evolution' – that is, natural selection – 'somehow rescues populations from sexual selection ... Genetic models of the evolution of sexual selection do not confirm this belief. The notion that evolution will necessarily extricate a species from the maladaptive tendencies of sexual selection is unfounded' (Kirkpatrick 1982, p. 10).

Incidentally, it has been suggested (Cohen 1984) that natural selection might sometimes call a halt to ornamentation when advertising is at a threshold of perceptual saturation. Imagine what it would cost a peacock to make an even greater impact on a peahen's perceptions. His ornaments might have become so exaggerated that any increment would have to be extremely large in order for females to be able to appreciate the difference. If so, a greater impact might be so costly that it wouldn't be worth his while to attempt it.

For modern Darwinism, nothing remains of the traditional idea that the intraspecific and social nature of sexual selection sets it apart from natural selection. Nowadays there is nothing unusual in these properties even when it comes to natural selection. It is now routine to regard relations between organisms, particularly members of the same species, as highly significant selective pressures. Mating preferences and intrasexual competition no longer stand out as atypical. Modern Darwinism can also explain why 'selection unlimited' might be expected when female choice is at work. Fisherian escalation is an obvious reason; and we have seen that Wallacean good-sense choices can have similar effects. Indeed, it's now standardly recognised that social competition among members of the same species, not merely for mates but for any resource, can be a powerful force for a co-evolutionary spiral. Again, sexual selection turns out not to be anomalous after all.

Whilst we're on the idea of intrasexual competition, I'd like to stress that *intra*sexual is indeed what the competition is. I mention this because of a widespread habit of referring to mate choice as '*inter*sexual' selection. What on earth could this mean? Consider intra- and interspecific competition. Intraspecific competition means competition within a species – cats competing against cats in sibling rivalry or in scrambling for prey. Interspecific competition means competition between two different species – cats against mice. Now consider intra- and intersexual competition. Intrasexual competition does indeed mean reproductive competition between members of the same sex – males fighting with males or singing the loudest or growing the showiest tail. 'Intersexual' competition, then, if it meant anything, should mean reproductive competition between the two sexes, males and females competing for the privilege of being the sex that does all the mating – surely a dubious triumph in a sexually reproducing species! In fact, of course, so-called 'intersexual selection' is, like intrasexual competition, about males competing with other males. Certainly, they are competing for females. But this does not make their competition 'intersexual'. Two cats competing for a mouse does not constitute interspecific competition, even though the mouse is a different species. How this terminology arose, I don't know. Jerram Brown conjectures that Julian Huxley, although not the outright culprit, nevertheless fostered the muddle by introducing the term 'intrasexual' to cover aggressive struggles between males for mates, thereby inviting the term 'intersexual' to cover the other alternative, female choice of mates (what Huxley called epigamic selection) (Brown 1983). But this is perhaps unfair on Huxley. An invitation it may have been – but an invitation that one should be pleased to refuse.

Now to the second reason why sexual selection was viewed as being outside natural selection. It is to do with the stress that classical Darwinism laid on survival as opposed to reproduction. Natural selection, of course, involves both; in Darwin's words, it comprises 'not only the life of the individual, but success in leaving progeny' (Darwin 1859, p. 62). Nevertheless, being organism-centred, classical Darwinism gives overwhelming priority to individual survival; in comparison, reproduction gets overlooked. But sexual selection deals only with reproduction: it 'depends on the advantage which certain individuals have over other individuals of the same sex and species, in exclusive relation to reproduction ... whilst natural selection depends on the success of both sexes, at all ages, in relation to the general conditions of life' (Darwin 1871, i, p. 256, ii, p. 398). And, what is more, it deals with only one component of reproduction – mating advantage.

It was for this reason that sexual selection was also regarded as being less rigorous than natural selection. To paraphrase Darwin, natural selection involves life and death whereas sexual selection involves only differential mating success:

> Sexual selection acts in a less rigorous manner than natural selection. The latter produces its effects by the life or death at all ages of the more or less successful individuals ... But [with sexual selection] ... the less successful male merely fails to obtain a female, or obtains later in the season a retarded or less vigorous female, or, if polygamous, obtains fewer females; so that they leave fewer, or less vigorous, or no offspring. (Darwin 1871, i, p. 278; see also 1859, pp. 88, 156–7)

'Merely' leave no offspring? If even Darwin could slip into the view that failing to reproduce was less important than failing to survive, then individual survival must indeed have taken precedence over reproduction. (To be fair,

such passages could be interpreted as referring to the reproductive fate of an individual in a particular season; but Darwin makes no such qualification in any of them.)

Classical Darwinism came to lay great stress on the relative roles of differential survival and differential reproduction. There grew up a view that natural selection was to do with the real business of the struggle for existence, whereas sexual selection was relatively unimportant because it was 'only' to do with reproduction and, what's more, 'only' one aspect of that. According to Shull 'the prevalent view' was 'that natural selection need[s to] ... render life-and-death decisions in order to work' (Shull 1936, pp. 152–4) (a view that he criticised, but not because he accepted sexual selection). Julian Huxley, for example, distinguished between what he called 'survival selection' and 'reproductive selection' (which subsumed sexual selection) and claimed that 'survival selection is much the more important: selection ... operates primarily by means of ... differential survival to maturity ... Natural selection clearly may also operate by means of the differential reproduction of mature individuals, but ... this reproductive selection has only minor evolutionary effects' (Huxley 1942, p. xix). The emphasis on survival and relative neglect of reproduction became such an established way of thinking that Darwin has often been attributed with the view that natural selection was concerned almost exclusively with survival. Simpson, for example, said: 'He recognised the fact that natural selection operated by differential reproduction, but he did not equate the two. In the modern theory natural selection is differential reproduction ... In the Darwinian system, natural selection was elimination, death, of the unfit and survival of the fit in a struggle for existence' (Simpson 1950, p. 268). According to Michael Ruse, a greater misapprehension is widespread: 'It is commonly argued today that Darwin was obsessed with the fact of death, to the complete exclusion of the fact of reproduction' (Ruse 1971, p. 348).

For modern Darwinism this is all a storm in a teacup. Nowadays, the distinction between the survival and the reproduction of individual organisms has lost that supreme significance. From a gene-centred viewpoint the question that matters is 'What contribution can either make towards the replication of genes?'

Talking of sexual selection being seen as less rigorous than natural selection, this view had an interesting consequence. It led to a striking contrast between classical and modern Darwinian views about the variability of sexually selected characteristics within a species. Classical Darwinism assumed that, on the whole, variability would arise only when selective forces were not stringent – that under exacting conditions adaptations would generally be uniform. (Remember how Darwin treated the apparently erratic egg-laying habits of rheas and cowbirds as an 'imperfect' instinct.) Darwin noticed that sexually selected characteristics often exhibited marked structural and behavioural differences within a species (e.g. Darwin 1871, i, pp. 401–3, ii, pp. 46, 132–5) and he took this as evidence for his view that sexual selection was less rigorous than natural selection. Speaking of beetles, for example, he says: 'The extraordinary size of the horns, and their widely different structure in closely-allied forms, indicate that they have been formed for some important purpose; but their excessive variability in the males of the same species leads to the inference that this purpose cannot be of a definite nature' (Darwin 1871, ii, p. 371). He concludes that they are sexual ornaments.

Where classical Darwinism saw pointless individual differences, modern Darwinism more often than not finds selection at work. Variability is generally brought about by frequency-dependent selection: if the rarer of two types has an advantage by virtue of being rare, then variability will automatically be maintained. Think of a left-handed boxer. Any left-handed person will testify that it's generally quite a disadvantage to be left-handed in a right-handed world. But, given that all boxers are used to fighting right-handed opponents, then a left-handed boxer will be able to pack an unexpected punch. Now think again of Darwin's horned beetles. There are some species in which the horns, as well as showing the 'excessive' variability that Darwin noted, fall neatly into only two sorts in the largest males. This dimorphism is the typical product of frequency dependence. Darwin may have been right that the horns often have an ornamental function. But W. D. Hamilton has pointed out that something else may be going on here. If the largest males take on more of the out-and-out fighting for females, then 'an uncommon variant may get an advantage similar to that of a left-handed boxer' even though he may be somewhat disadvantaged in other respects (Hamilton 1979, p. 204; see also Eberhard 1979, 1980). It is the same kind of force that may often be at work on sexual selection for male ornament. In this case, the frequency-dependent selection pressure is for different mating tactics. Suppose, say, some males are holding territories and attracting females to them by elaborate song. Then it could pay other males, so-called satellite males, to parasitise their efforts by trying to intercept the females who are en route to their hard-won territory. Variability in sexually selected characteristics is turning out to be quite common across a wide range of species (e.g. Cade 1979), sometimes at extremely high levels (e.g. Harvey and Wilcove 1985). And this is no surprise. Darwinians no longer regard it as the result of weak selection. On the contrary, it is expected because of the very strength of the selective forces; the drive for 'mere' ornamentation is no longer regarded as a lax selective pressure.

Now to the last distinction that classical Darwinism made between natural and sexual selection: the soberly utilitarian adaptations of the one and the ornamental flourishes of the other. One only had to look at the females, Darwin said, to see that sexually selected characteristics were useless in other aspects of life: 'unarmed, unornamented, or unattractive males would succeed equally well in the battle for life and in leaving numerous progeny, if better endowed males were not present. We may infer that this would be the case, for the females, which are unarmed and unornamented, are able to survive and procreate their kind' (Darwin 1871, i, p. 258). What's more, the end results of natural selection and sexual selection typically pulled in opposite directions. Where natural selection favoured camouflage, streamlining and low energy costs, female taste called for dazzling colours, extravagant structures and conspicuous behaviour. This contrast between the useful and the ornamental was considered to be the most salient difference between natural and sexual selection. It was, after all, the baroque extravagance of male ornament that originally posed the problem of sexual selection for Darwin.

The assumptions behind this distinction are typical of classical Darwinism. Sexual selection was acknowledged to involve a trade-off (between mating advantage on the one hand and survival together with the remaining aspects of reproduction on the other); but the costs incurred by the adaptations of natural selection tended to be overlooked. And sexually selected characteristics were generally regarded as of no 'real' use and even damaging to their possessors; but it was taken for granted that the adaptations of natural selection would be useful to the bearer.

Many Darwinians, often under the influence of vague species-level thinking, came to view ornamental lavishness in a very gloomy light. Sexually selected adaptations (along with those of other intraspecific forces) were seen as useful to the individual but so 'selfish' as to be bad for the species, perhaps bad enough to carry it inexorably to extinction. Male ornaments (and weapons developed by males to compete for females) were looked upon as self-interested aids to the reproductive success of their possessors, aids that jeopardised the collective good. These were the feelings of Konrad Lorenz:

purely intra-specific selective breeding can lead to ... forms and behaviour patterns which are not only non-adaptive but can even have adverse effects on species-preservation ... If sexual rivalry ... exerts selection pressure uninfluenced by any environmental exigencies, it may develop in a direction which is ... irrelevant, if not positively detrimental to survival ... [leading to] bizarre physical forms of no use to the species ... Sexual selection by the female often has ... results ... quite against the interests of the species.   (Lorenz 1966, pp. 30–2)

The Argus pheasant, for example, 'has run itself into a blind alley ... [T]hese birds will never reach a sensible solution and "decide" to stop this nonsense at once ... Here ... we are up against a strange and almost uncanny phenomenon ... it is selection itself that has here run into a blind alley which may easily result in destruction' (Lorenz 1966, pp. 32–3). Julian Huxley declared indignantly: 'intraspecific selection is on the whole a biological evil' (Huxley 1942, p. 484; see also p. xx). According to him, 'Inter-specific selection obviously must promote the biological advantage of the species. Intra-specific selection, on the other hand ... may ... favour the evolution of characters which are useless or even deleterious to the species as a whole ... [T]he most extreme examples concern reproduction' (Huxley 1938a, p. 22; see also p. 13); 'display-characters confined to one sex could be ... useless or even deleterious to the species' (Huxley 1942, p. 484; see also 1947, p. 174). J. B. S. Haldane said of intraspecific competition in general and sexual selection in particular (though in his case there was no 'good-of-the-species' implied):

> the results may be biologically advantageous for the individual, but ultimately disastrous for the species ... [I]t is in the struggle between adults of the same species that the biological effects of competition are probably most marked. It seems likely that they render the species as a whole less successful in coping with the environment ... [T]he bright colours and song of many bird species ... serve to attract the other sex ... But ... their value to the species as a whole is dubious.
>
> (Haldane 1932, pp. 120–8)

G. G. Simpson (1950, p. 223) and Verne Grant (1963, pp. 242–3) came to similar conclusions. And some commentators (though generally not biologists) still single out intraspecific competition, particularly sexual selection, as unique in this respect. According to the philosopher Mary Midgley, interspecific competition 'has to be limited sharply by prudence and common sense ... [whereas intraspecific competition] can easily turn out very badly ... Where the motive of competitiveness is strong, it is hard for a species to get out of ... a cul-de-sac' (Midgley 1979, pp. 132–3). Carl Bajema, in his historical commentary on sexual selection, distinguishes between natural selection's adaptations, which are 'beneficial to all members of the species as well as to the individual' and sexual selection's adaptations, which are 'beneficial to the individual but harmful to other members of the species' (Bajema 1984, pp. 111, 113; see also e.g. pp. 110, 146, 262).

It is certainly true that sexual selection's ornaments might make a species more vulnerable to extinction. But this is not peculiar to sexual selection nor even to intraspecific selection in general. Admittedly that is where, intuitively, the most glaring examples seem to come from. But what is so 'prudent' or 'commonsensical' about, say, an interspecific arms race?

Wouldn't 'the biological advantage of the species' be better served if both prey and predator 'decided to stop this nonsense at once' and reach a 'sensible solution'? Suppose the prey species simply agreed to surrender its weakest members to the predators. This would save both sides all their costly investment in more and more powerful muscles, in bigger and better weapons, in protective armour and devices to penetrate it. If one applies the 'extinction' argument systematically, the eagle's talons and the cheetah's sprint turn out to have the same 'dubious value' as the peacock's tail.

More seriously, it is clearly a mistake of woolly species-level thinking to assume that natural selection has any interest in what is good for species – to assume that adaptations will generally be good for both individual and group, and that intraspecific selection is peculiarly 'selfish' if it favours adaptations that are good for the individual but bad for the group. As Fisher said, it is inappropriate to ask

> 'Of what advantage could it be to any species for the males to struggle for the females and for the females to struggle for the males?' ... Natural selection can only explain these instincts in so far as they are individually beneficial, and leaves entirely open the question as to whether in the aggregate they are a benefit or an injury to the species.   (Fisher 1930, p. 50)

And even Fisher has not gone far enough. We must climb down from the individual as well as the species, right down to the gene, in order to find the only entity for which the idea of 'selfishness' is systematically appropriate. (Incidentally, the pervasive atmosphere of species- and group-level thinking was probably a major barrier to Fisher's explanation of sexual selection being properly appreciated in its time.) Genes 'selfishly' have phenotypic effects that further their own replication. Whether those effects will also be 'good' for the individual who carries the gene, for other members of its group, for the species as a whole, for the phylum, even for members of other species, is a contingent matter. Indeed, although it is clear how phenotypic effects can be good for genes, it is not obvious precisely what 'good for' means in the other cases. What's 'good for' an individual's reproductive effort may threaten its survival; what's good for a species' geographical distribution may eventually contribute to its extinction.

Sexual selection, particularly Fisherian runaway, has still not shaken off the suspicion that it is in some way maladaptive. Ernst Mayr has declared that 'various forms of selfish selection (e.g. ... many aspects of sexual selection) may produce changes in the phenotype that could hardly be classified as "adaptations"' (Mayr 1983, p. 324). A highly authoritative textbook – perhaps the most authoritative recent text – on evolutionary biology states: 'Runaway sexual selection is a fascinating example of how selection may proceed without adaptation ... In these models, the evolution of female preference is not an adaptive process' (Futuyma 1986, pp. 278–9). The editors of the papers from an equally authoritative recent conference on sexual selection noted: 'One of the most pervasive controversies [over sexual selection] stems from uncertainty over how "adaptive" we can expect the world to be ... [S]exual selection has become a new battleground over the limits of adaptation ... This issue pervaded nearly all discussions [at the conference] ... and it was brought into explicit focus as the first of our four major topics' (Andersson and Bradbury 1987, pp. 2–3). Stevan Arnold has even urged that this is sufficient ground for preserving the traditional distinction between natural and sexual selection: 'structures that confer mating success may hinder the male in the struggle for survival: sexual selection and natural can be opposing processes' (Arnold 1983, p. 70; see also pp. 68–71). And, as we noted earlier, this feeling is reflected in the standard vocabulary, Fisherian sexual selection often being referred to as 'non-adaptive', 'maladaptive' or 'arbitrary' in contrast to 'adaptive' (good-sense) theories. Of course, one shouldn't read too much into a mere choice of words. They are probably intended as no more than convenient labels, a way of sharpening the distinction between good-taste and good-sense choice. But they are resonant words and it is likely that they capture a tendency in current thinking – and, what's more, that they reinforce it.

An uneasiness over adaptive status had some justification in the nineteenth century. After all, the very core of Darwin's theory, female taste, was not explained adaptively. What's more, sexually selected characteristics violated the nineteenth-century notion of what constituted an adaptation: the woodpecker's beak and plumed seed were elegantly utilitarian and obviously beneficial to their bearers, the peacock's tail and the nightingale's song were not. So it's understandable that nineteenth-century Darwinians, especially committed adaptationists like Wallace, should feel that male ornaments weren't respectable adaptations.

But Fisher changed all that. Darwin–Fisher explanations account for both male ornament and female taste. And they account for them adaptively. Admittedly, Fisherian adaptations might, to some, still look distinctly counter-intuitive. But one of Fisher's contributions was to show just how counter-intuitive the results of selection can be – and to help us revise our intuitions. From that point of view, it is particularly inappropriate that Darwin–Fisher sexual selection has ended up being called 'maladaptive' – not to mention unfair to Fisher!

It is, by the way, odd that Darwinians who are critical of hardline adaptationism should hold that Darwin–Fisher sexually selected characteristics are somehow maladaptive merely because they are not 'sensible'. Such critics standardly accuse committed adaptationists of taking a Panglossian view of adaptations – the view that selection should always result in what is 'best'. But when they question the status of sexually selected characteristics, they are themselves tacitly making Panglossian assumptions about how 'sensible' adaptations ought to be.

In modern Darwinism, Darwin's contrasts between sexual selection's extravagance, its trade-offs, its harmfulness, and natural selection's utility, its efficiency, its benefits all melt away. All adaptations are compromises; a trade-off between mating and predation is no different in principle from a trade-off between foraging and predation. And sexual selection is by no means the only begetter of adaptations that harm the organism that bears them. One of the achievements of modern Darwinism is to have revised our ideas about what constitutes an adaptation and what is the entity that benefits from it. Think not of woodpeckers' beaks but of manipulative parasite genes. The idea that selection, natural selection included, always opts for elegant, utilitarian solutions that are 'best' for their bearers is a nineteenth-century view. Nowadays, even the most thoroughgoing of adaptationists need not feel uncomfortable about sexually selected adaptations.

Sexual selection became an arena for debates about the scope of Darwinism – a telling indication of just how unorthodox it was thought to be. On one side, purists like Wallace viewed sexual selection as an unDarwinian heresy that threatened to usurp natural selection. On the other side, Darwinians who thought of themselves as pluralists welcomed the theory as an alternative to nothing-but-natural-selection. Romanes, for example, declared with satisfaction: 'in so far as any one holds that sexual selection is a true cause of ... modification, he is obliged to believe that innumerable ... characters ... have been produced without reference to utility (other, of course, than utility for sexual purposes), and therefore without reference to natural selection' (Romanes 1892–7, ii, p. 219). He took Wallace's position to be yet another example of his intransigent, narrow vision:

> the objection that the principles of natural selection must necessarily swallow up those of sexual selection ... lies at the root of all Mr Wallace's opposition to the supplementary theory of sexual selection. He is self-consistent in refusing to entertain the evidence of sexual selection, on the ground of his antecedent persuasion that in the great drama of evolution there is no possible standing-ground for any other actor than that which appears in the person of natural selection.
>
> (Romanes 1892–7, i, p. 399)

The century-old prejudices still linger. This may be why, even today, some Darwinians deny that sexual selection is just a special case of natural selection (e.g. Arnold 1983, p. 71). But it is now becoming increasingly recognised that the traditional distinctions between sexual selection and other selective forces have broken down. Sexual selection certainly attracts more than its fair share of odd properties. But, from a gene-centred viewpoint, they all fall well within the scope of natural selection. Sexual selection need no longer be viewed as antithetical to natural selection; modern Darwinism returns it to the fold. Darwin's other theory, it finally emerges, is not so 'other' after all.

A happy ending to the peacock's tale

The theory of sexual selection has had a chequered career. Darwin applied it very liberally. He detected female choice at the drop of a feather. Reaction against the theory went so far that Wallace's programme of reducing sexual selection to natural selection held sway for almost a century. Throughout most of this period, sexual selection remained on the Darwinian sidelines, neglected, distorted or misunderstood. Natural selection suffered a partial eclipse for almost half a century after Darwin's death. Sexual selection suffered an almost total eclipse for almost twice as long. At the turn of the century, for example, Vernon Kellogg's lengthy and largely sympathetic progress report on Darwinian theory dismissed sexual selection as 'now nearly wholly discredited' (Kellogg 1907, p. 3). Twenty years later, Erik Nordenskiöld's *History of Biology* (which was anyway hostile to Darwinism) declared that the 'doctrine of sexual selection ... is nowadays embraced by hardly any true scientists', adding, as tacit proof of its unsoundness, that nevertheless 'popular literature shows traces of it' and citing in particular Strindberg's 'enthusiasm' (Nordenskiöld 1929, pp. 474–5). One of sexual selection's few advocates during this period, Edmund Selous, complained of downright suppression: 'I did everything, within my power, to further scientific truth, and have indeed produced immensely strong evidence in favour of the Darwinian theory of sexual selection. It would seem, however, that, since the theory itself is (officially) out of favour, such evidence is not wanted' (Selous 1913, p. 98); he instanced the entry on blackcocks in *The British Bird Book*, where 'there is no reference to certain facts ... which I have put on record, although these facts quite contradict what is generally stated on the subject ... Nothing is said about the anything but "indifferent" conduct of the hen, showing so clearly her power of choice ... as inferred by Darwin, but still so constantly denied' (Selous 1913, pp. 96–7). With the coming of the modern synthesis, natural selection found its feet again. But sexual selection was still ignored. Look it up in the index of any of the classic texts of that period. If index entries are any measure of perceived importance, then sexual selection was not at the forefront of ideas. In Dobzhansky (1937), Simpson (1944, 1953) and even in a recent historical survey of the period (Mayr and Provine 1980) it is conspicuously absent; in Mayr (1942, 1963) and Rensch (1959) it is mentioned once; in Huxley (1942) it is given the lengthiest treatment – two pages (pp. 35–7). Fisher's contribution should have rescued sexual selection from obscurity. But that took yet another half century. Why did Darwin's theory – which aroused such interest at its inception and has again in recent decades – why, during the intervening years, did it suffer such an inglorious career?

At first sight, a major reason would seem to be that Darwin's own version of the theory left female choice unexplained. And unexplained it remained until Fisher took up the problem:

> If instead of regarding the existence of sexual preference as a basic fact to be established only by direct observation, we consider that the tastes of organisms, like their organs and faculties, must be regarded as the products of evolutionary change, governed by the relative advantage which such tastes may confer, it appears ... that ... a sexual preference of a particular kind may confer a selective advantage, and therefore become established in the species.   (Fisher 1930, p. 151)

With the benefit of up-to-date Fisherian hindsight, Darwin's omission is indeed revealed as an obvious and serious gap. It is surprising, then, to realise that this hardly accounts at all for the rejection of sexual selection. This objection played only a relatively minor part in criticisms of the theory. This is probably one reason why Darwinians failed to appreciate Fisher's theory for the important development that it was. If you don't see the problem, you're unlikely to value the solution. John Maynard Smith is disarmingly frank about how little impact Fisher's analysis made: 'In the extensive publications marking the centenary of the *Origin of Species*, the only explicit treatment of sexual selection was Maynard Smith (1958a); although I did describe a possible mechanism of female choice in *Drosophila subobscura*, it is clear that I had not read or understood Fisher' (Maynard Smith 1987, p. 10).

A more important reason why sexual selection theory got pushed to one side is that it has always aroused fears of anthropomorphism – the idea that human attributes are being unjustifiably foisted on other animals. It was female choice and, worse, female aesthetic taste that proved so upsetting.

Take the notion of choice. As far as nineteenth-century Darwinians, at least, were concerned, it is not obvious why they should have felt so uneasy about birds and insects and fish choosing their mates. Admittedly, few of them followed Darwin in treating humans like other animals. But on the whole they were not averse to dignifying others with being somewhat like us, as long as the analogy wasn't taken too far. And certainly they applied the idea of choice in other spheres. They would talk happily of animals choosing between different foods or nesting materials or habitats. But they became distinctly uncomfortable at the mention of choice of mates. Of course, they could have argued that there were obviously strong selective pressures to evolve discrimination over food, whereas female choice arose for no apparent reason, with no selective forces driving it to improvement, no advantages to the female or her mate in developing finer and finer discrimination. We've seen, however, that this explanatory gap hardly worried nineteenth-century critics at all. So it's baffling to think where exactly the differences between choice of mate and choice of anything else were supposed to lie. Why were judgements about mates viewed as higher accomplishments than a bird's judgements about which egg in the nest is hers and which the cuckoo's, or a chameleon's judgement about matching himself precisely to his background? We find Wallace, for example, doubting whether birds could choose their mates. But this was the Wallace who reported how birds would choose to eat only certain insects, avoiding those they found unpalatable (Wallace 1889, pp. 234–8). And this was the Wallace who stressed that birds would choose mates from their own species rather than others and, what is more, use arbitrary markers to do it. What further mechanisms could he have thought a 'good taste' choice of mates would require? Admittedly, we may not know how an animal manages to discriminate between its own species and others; but, as Fisher said, 'it is no conjecture that a discriminative mechanism exists, variations in which will be capable of giving rise to a similar discrimination within its own species' (Fisher 1930, p. 144). Again, we find Wallace doubting whether birds could make distinctions between very small differences. But this was the Wallace who marvelled at an insect's resemblance to a leaf or flower – a resemblance so close that even a naturalist of his experience could be fooled into examining a 'flower' that would fly away at his touch. And what was the selective force that had brought about such fine-tuning, if not the visual discrimination of birds?

Later Darwinians were more consistent in their uneasiness over female choice. By the turn of the century, 'the main trend in behavioural studies was towards mechanistic explanations and away from anything that smacked of anthropomorphism' (Maynard Smith 1987, p. 10). Choice of mates came under suspicion along with choice of food, habitat or anything else. Indeed, in spite of the protests of a vigorous minority of ethologists, the first decades of the century saw something of a move towards laboratory-based physiology, and away from anything to do with behaviour at all. Not until ethology began to establish itself, from the 1930s to the 1950s, were these restricted views of other animals challenged. 'Animals are emotional people of extremely poor intelligence' ran a favourite slogan of one of the movement's leading figures, Konrad Lorenz (Durant 1981, p. 177). These ethologists saw themselves as still firmly rejecting anthropomorphism. But they did believe 'that it was possible to understand animals in just the same way that we understand our fellow-men' (Durant 1981, p. 186). The two parallel strands in Darwin's principle of continuity were once again being taken seriously; Darwin had not only explained 'human thoughts and actions ... in terms of animal instinct, [but also explained animal behaviour] ... in terms of human thoughts and feelings ... [E]ven as he lowered man's mind into nature, Darwin raised the minds of the other animals to meet it' (Durant 1985, p. 291). Sexual selection was not immediately taken up again. But this was the beginning of a more congenial climate for the theory.

At the risk of appearing to try to have it both ways, I'll now say that sexual selection theory doesn't require such a climate. There's nothing necessarily anthropomorphic about female choice of mates. To talk of female choice is only to say that there has been selection for genes that have the effect of making females behave as if they were choosing. Such talk makes no assumptions – anthropomorphic or otherwise – about what brings the behaviour about, what mechanisms are responsible for that effect. A peahen may go through a process that's like our human understanding of choice; she may not. To say, for example, that females 'prefer' males who can give them attractive sons, is merely to say that there are now, or have been in the evolutionary past, genetic differences in the population that cause, or have caused, differences in behaviour; and that, because of these differences, some females have a greater probability than others of mating preferentially in such a way that they end up with 'attractive' sons – that is, sons who will benefit from the same kind of preferential mating. So, as with any theory involving 'selfish' genes (Dawkins 1981), the theory of sexual selection does not, after all, require a climate of anthropomorphism. The theory is not about discriminating animals but about discriminating genes, and only a pedant (Midgley 1979a) would call that anthropomorphic.

Indeed, anthropomorphic interpretations of mate choice might even hinder our understanding by making us think in terms of individuals rather than genes. Take, for example, the recent discussion that we noted earlier (when examining the notion of epigamic characteristics) about when a choice is really a choice. At what point does, say, competition between males or male coercion turn active female choice into something more passive – so passive that it no longer deserves to be viewed as choice, or certainly not *her* choice? If we think in terms of animals making choices, it's hard to avoid such questions (and very hard to answer them!). But if we think in terms of genes and their phenotypic effects, we might be able to by-pass these problems and look at the issue of choice more fruitfully. Consider a model of courtship in which males coerce females. Looked at from the point of view of individuals it's surprising to find that one sex ends up being systematically manipulated; what on earth is in it for them? But from a gene's-eye view, there is no surprise. A gene 'for' male manipulation of female choice exerts its (extended) phenotypic effects in both male and female. If those effects confer a selective advantage over the available alternatives – a not implausible assumption on this model – then that gene will proliferate. In general, it might well prove more illuminating to ask how such genes exert their power through their phenotypic effects on either sex than to ask which sex is 'really' making the choice. Surely that is the question that is of more interest to Darwinians. It's difficult enough, after all, to work out what constitutes free will in humans; why burden ourselves unnecessarily with the metaphysics of peahen preference?

Given that the notion of choice doesn't commit us to anthropomorphism, Darwinians are finally free of the associated notion of aesthetic taste, which has been the other traditional objection to choice. Darwin's insistence that mate choice was aesthetic provoked a chorus of criticism from his day to this. For nineteenth-century naturalists, particularly, the notion of 'lower' animals sharing an experience so elevated as aesthetic sense offended their feelings (largely aesthetic!) about what was proper to 'them' and 'us'. Aesthetics was generally seen as one of those areas, like moral sense and rationality, that could prevent other animals from getting too close for comfort. So one of the very reasons that attracted Darwin to his theory of sexual selection – that female aesthetic taste soldered a link in the continuity chain – made it repugnant to many of his contemporaries (as we've seen in the case of Wallace). Throughout the history of the theory, well-meaning defenders of Darwin have argued that he did not assume aesthetic sense (or, at least, did not mean to), that the idea that he did was an invention of his critics and that the assumption is anyway irrelevant to his theory. Lloyd Morgan, for example, said that 'Darwin occasionally expresses himself unguardedly in the matter' but that his theory could be '[s]tripped of all its unnecessary aesthetic surplusage' (Morgan 1896, pp. 218, 263). And some modern commentators have said much the same (e.g. Ghiselin 1969, p. 218; Morgan 1896, p. 263; O'Donald 1980, pp. 2–3, 5). With a non-anthropomorphic interpretation of choice, this debate loses its interest. It reduces to a question of what we call things. And we all know that such questions aren't important – that names don't matter and that it's pointless to argue about words.

Having dutifully said that, I'm nevertheless about to make an issue of it, just this once – to suggest that we should, after all, go along with Darwin and call a Darwin–Fisher 'good taste' choice of mates an aesthetic choice. Why?

Because this helps us to bear in mind its similarities to human aesthetic taste, judgement and fashion, and its differences from Wallace's good-sense choice. Fisher's model resembles most models of aesthetic choice in humans in that the criteria for choice have a self-governing, autonomous, whimsical, for-their-own-sake quality; it is taste itself and taste alone that sets the standard, without reference to utilitarian considerations. A particular tail is popular just because it is popular, and it is solely through this self-reinforcement that the popularity is maintained from generation to generation. On Darwin's own version of his theory there was less justification for calling choice aesthetic. But Fisher vindicated Darwin's analogy; when Darwin's theory is augmented with Fisher's analysis, the description becomes entirely apt. Admittedly, beauty may not be in the eye of the beholder; but it's certainly in the genes of the beholder. All this is in sharp contrast to Wallacean good-sense choice. There's nothing aesthetic about a utilitarian preference for a plentiful food supply or immunity to disease – no arbitrary standards there.

Incidentally, whilst we're on the topic of how to describe female choice, note that it has become standard to talk of 'coy females' (and 'eager males') (e.g. Bradbury and Andersson 1987, p. 4). I can't resist wondering what words would be used if the sex-roles were reversed. Would a (male) investor or business executive be called coy for not rushing headlong into the first option? If males were choosy about mates, would they be 'coy' – or discriminating, judicious, responsible, prudent, discerning? (And, by the way, would females be 'eager' – or would they be wanton, frivolous, wayward, brazen?)

Now to a final influence that worked against sexual selection. It is one that we have already examined: the popularity of Wallace's natural selectionist alternative. This elbowing-out of sexual selection reflects – and itself contributed to – a growing preoccupation within Darwinism with the question of the origin of species. We've seen how the problems faced by Darwin could be grouped into two: adaptation and speciation. Adaptation had been of great importance to both pre-Darwinian natural theology and early Darwinism. But this interest waned with the eclipse of Darwinism and, in particular, with the rise of orthogenesis and saltationism – both theories that stressed the supposedly non-adaptive aspects of organisms. This shift of emphasis was incorporated into the revival of Darwinism. When Darwinism was reborn with the modern synthesis, it was the problem of speciation that was central. In this light, questions about mate choice reduced to questions about species recognition marks, ethological isolating mechanisms and other such means by which species are established and preserved (e.g. Lack 1968, pp. 159–60; Mayr 1942, p. 254; Rensch 1959, pp. 11–12).

Although the rise of ethology proved in some ways congenial to the study of sexual selection, it also reinforced this tendency to focus on differences between species rather than differences within them. Much of the success of the ethological tradition lay in treating behaviour not as variable but as stereotyped within the species. But sexual selection is about intraspecific variation – slightly longer tails, slightly stronger preferences. As the ethologist Peter Marler has said:

> We are still in debt to our ethological progenitors for the insight that what appears to the uninitiated observer as a series of continuously varying movements, too chaotic to be scientifically manageable, typically proves to have, at its core actions that are stereotyped and species-specific ... [But sexual selection is] about the extent to which behaviour varies between members of the same species, and even within the same population. This variation ... is the raw material upon which the forces of sexual selection can operate.   (Marler 1985, pp. ix–x)

Darwin conceded that to conceive of the peacock's tail as a product of female choice was 'an awful stretcher' (Darwin, F. and Seward 1903, ii, p. 90). But he never wavered in his insistence that we should stretch our belief. 'My conviction of the power of sexual selection remains unshaken' (Darwin 1871, 2nd edn., p. viii), he said in the Preface to the second edition of *Descent*. Contrast these words with his second thoughts on natural selection and they are all the more telling. His conviction remained unshaken to the end. As Romanes remarked,

> his very last words to science – read only a few hours before his death at a meeting of the Zoological Society – were:
> 'I may perhaps be here permitted to say that, after having carefully weighed, to the best of my ability, the various arguments which have been advanced against the principle of sexual selection, I remain firmly convinced of its truth.'
>   (Romanes 1892–7, i, p. 400; see Darwin 1882)

Darwin was confident that eventually 'the idea of sexual selection [would] ... be much more largely accepted' (Darwin 1871, 2nd edn., p. ix). It has taken more than a century. But his prediction has at last proved true. A happy ending to the peacock's tale. Or perhaps this is really just the beginning ...

PART THREE

The Ant

11

Altruism now

The problem with altruism

Natural selection is demanding, exacting, relentless. It is intolerant of weakness, indifferent to suffering. It favours the hardy, the resilient, the healthy. One might expect organisms shaped by such a force to bear its stamp, to suffer in its own image – expect them to be locked in struggle, pursuing their own interests, uncaring of others. Natural selection would surely see off chivalrous self-sacrifice. Selfishness should win the day.

But look carefully at nature and you will find that it doesn't always seem like that. You might well see animals that are apparently strikingly unselfish, particularly with their own species – giving warning of predators, sharing food, grooming others to remove parasites, adopting orphans, fighting without killing or even injuring their adversaries and conducting themselves in numerous other civilised ways. Indeed, in some respects they behave more like the moral paragons of Aesop – working dutifully for the sake of the community, noble in spirit and generous in deed – than the hard-bitten, self-seeking individualists that natural selection would seem to favour. Such behaviour poses a problem for the Darwinian view of nature. It has become known as the problem of altruism.

And a problem it certainly was for classical Darwinism. But no longer. There is now no difficulty, at least in principle, in explaining why an organism might look like an altruist, why it might give up its time, its food, its territory, its mate or even its life to help others.

The problem solved

A bird gives an alarm call. This seems a highly altruistic thing to do: warning others of danger but perilously alerting the predator to its own presence. How can we explain it? If we take an organism-centred view, we shan't be able to. Worse, if we take a group- or species-level view, we might be able to 'explain' it all too easily! And we shall end up in the kind of muddle that, as we shall soon see, permeated Darwinism for several decades. But if we hold steadily to a gene-centred view, the problem dissolves, gratifyingly, before our eyes.

Consider, first, the idea of kin selection (Fisher 1930, pp. 177–81; Haldane 1932, pp. 130–1, 207–10, 1955, p. 44; Hamilton 1963, 1964, 1971, 1971a, 1972, 1975, 1979; see also Dawkins 1979; Grafen 1985). This is the principle that natural selection can favour the act of an organism giving help to its relatives, even though the help is costly to the organism itself. How does it work? Imagine a gene that has the effect of making the organism in which it is housed behave in such a way that it helps copies of itself in other organisms – a bird that gives alarm calls to help other birds but only when those other birds also have a gene for behaving in the same way. A gene that made its bearer operate such a policy of differential aid could, all other things being equal, obviously prosper. But the problem is that the aid must be differential; if birds that don't bear the alarm-call gene are helped as much as those that do, natural selection won't favour that gene. It isn't easy for a gene to 'recognise' copies of itself in other individuals. One way of increasing the likelihood that the altruism will reach only its intended target is, to put it roughly, to keep it in the family. If I bear a gene for behaving altruistically in some way then my kin are more likely to bear it than is anyone picked at random from the population. The closer the kin, the more likely we are to share that gene; the more distant the kin, the more the likelihood approaches that of a random member of the population. If we think about the probable distribution of such an altruism gene, then we can see what natural selection's preferences might be. If, for example, I had the choice of saving my own life or the lives of two brothers or those of eight cousins, then (all other things being equal – a crucial proviso) natural selection would be indifferent as to which I should do. And if I could save, say, three brothers or nine cousins then natural selection would favour this self-sacrificial altruism, favour saving my kin rather than my skin. A gene for such altruistic life-saving would, on average, proliferate more copies of itself than would an alternative gene for clinging non-altruistically to one's own life. But all other things do have to be equal. The reason why we don't find individuals risking their lives for hordes of second cousins is that it's unlikely to be practical to assemble the hordes. And the reason why help so often goes only one way, even though the genetic relationship is symmetrical, is that there are practical asymmetries; parents are in a better position to suckle offspring or teach them to fly than offspring are to attempt the reverse, and the same goes for older sibs helping younger ones – which is just as well, or each item would be no sooner given than punctiliously handed back!

We can see from all this that, in spite of the name 'kin selection', there is nothing magical about helping kin rather than anyone else. It's just that kin selection can be an efficient and practicable method by which a gene for altruism could practise discrimination. The discrimination rules need not require brothers and sisters and nieces and nephews to be identified as such. They could be very simple indeed: 'Help those reared in the same nest as yourself' or 'Help those with the same smell as yourself' or (in species that tend to stay in one place) 'Help your neighbour'.

Real-life examples abound. In several species of ground squirrel (such as *Spermophilus beldingi*), the females, unlike the males, live near to close relatives. This creates leeway for kin selection. And it transpires that when the females give alarm calls, a highly dangerous activity, they discriminate in favour of mothers, daughters and sisters (Dunford 1977; Sherman 1977, 1980, 1980a). In the Tasmanian native hen (*Tribonyx mortierii*), a breeding group sometimes consists of two males and a female. It turns out that, when this occurs, the males are brothers; when there is an excess of males it is of greater selective advantage to the mated males to share their mate with a brother than to drive him out (Maynard Smith and Ridpath 1972). Kin selection might sometimes be involved in 'helping at the nest' – forgoing breeding and assisting with rearing the offspring of others – which is known to occur in over 150 bird species (Brown 1978; Davies 1982; Emlen 1984). In the majority of cases (though by no means all (e.g. Ligon and Ligon 1978; Stacey and Koenig 1984)) the young in the nest are sibs or other close relatives of the helper. Under certain conditions (for instance, when there are very few breeding sites in the territory) helping to rear kin might pay better than attempting to breed oneself.

But what if the beneficiaries are not the animal's kin? How might we explain altruistic behaviour then? Reciprocity is one answer. What looks like altruism might really pay the participants: they could be exchanging altruistic favours in such a way that each does better from cooperating than it would from failing to cooperate. The costs of a good deed are compensated for by a good deed in return. But how could such a cosy, mutually beneficial arrangement come about? To a selfish Darwinian strategist it is ripe for exploitation. Certainly, cooperation pays. But wouldn't reneging pay the reneger even better? Far from evolving, surely the cooperation would degenerate into cheating, with defectors seizing unrequited good turns from an ever-dwindling source. If only everyone would cooperate, everyone would be better off; but the best course for any individual is to pursue its own self-interest; and so everyone will inevitably end up worse off.

Or so it seems at first glance. A second glance shows, however, that such pessimism is unwarranted. The right way to look at the problem is by using game theory, as did the American political scientist Robert Axelrod together with W. D. Hamilton, foreshadowed by Robert Trivers (Axelrod 1984,

Prisoner One

The payoffs are to Prisoner One

	COOPERATE	DEFECT
COOPERATE	Fairly good **R**: Reward for mutual cooperation 2 year sentence	Best **T**: Temptation to defect 1 year sentence
DEFECT	Worst **S**: Sucker's payoff 10 year sentence	Fairly bad **P**: Punishment for mutual defection 6 year sentence

Prisoner Two (rows)

The Prisoner's Dilemma

particularly pp. 88–105; Axelrod and Hamilton 1981; Trivers 1971, 1985, pp. 361–94, particularly pp. 389–92; see also Dawkins 1976, 2nd edn., pp. 202–33). Axelrod and Hamilton turned to a well-analysed model in game theory, the Prisoner's Dilemma, because it captures just that problem: the rational pursuit of individual self-interest driving everyone into an outcome that nobody prefers. Imagine two partners in crime, awaiting trial. Each is faced with two options: cooperating with the other one by refusing to confess or defecting from their alliance by confessing. If both cooperate – keeping their lips sealed – then the authorities can't pin much on them and each gets a very light sentence (R: the Reward for mutual cooperation); if one defects by turning King's Evidence whilst the other refuses to talk, then the defector is rewarded with an even lighter sentence (T: the Temptation to defect) whilst the other gets the stiffest sentence going (S: the Sucker's payoff); if both squeal then each gets a lighter sentence than he would have got from maintaining a lone, staunch silence but still heavier than the sentence that each would have reaped from mutual cooperation (P: the Punishment for defection). Both, then, have a scale of preferences T, R, P, S; T is the best and S is the worst outcome. Note that this is a non-zero-sum game. In a zero-sum game, my loss is your gain – as if a banker were dividing up a fixed sum between the two of us; in a non-zero-sum game, I can benefit without your losing – working together, we can both profit at the expense of the banker.

The prisoners have to make their decisions without knowing what the other will do. How would a rational prisoner act?

He would defect. Whatever the other prisoner does, defection pays better than cooperation. His argument would run as follows: 'Suppose my partner in crime cooperates. I could do fairly well by cooperating, too (R). But I could do even better by defecting (T). Suppose, alternatively, that he defects. Then if I cooperate I'll end up worst of all (S). So, again, I should defect (P). I am hoping for the best (T) and avoiding the worst (S).' And, because both prisoners are reasoning in this way, both will end up defecting. So they are landed with the lower-ranking preference P rather than the higher-ranking preference R. That is the dilemma: it pays each of them to defect whatever the other one does, yet if both defect, each does less well than if both had cooperated; 'what is best for each person individually leads to mutual defection, whereas everyone would have been better off with mutual cooperation' (Axelrod 1984, p. 9).

But the dilemma has a solution. We have been talking about a one-off game. Suppose, however, that the participants play the game repeatedly, suppose that each knows that the two of them are likely to meet an indefinite number of times. Suppose, to use Axelrod's potent metaphor, that the future can cast a long shadow backwards onto the present. Under such conditions, cooperation can evolve. Consider, for example, the strategy Tit for Tat: cooperate on the first move and after that copy what the other player did on the previous move. Tit for Tat is never the first to defect; it retaliates against defection by defecting on the next move but subsequently lets bygones be bygones. It turns out that this highly cooperative strategy can evolve, even when initially pitted against exploitative, readily-defecting strategies. And it can be stable against invasion by them. If it is to get off the ground, a critical proportion of its encounters must be with cooperators like itself; otherwise the strategy Always Defect will evolve and be stable instead. In short, to use a concept we touched on earlier, Tit for Tat pretty well amounts to an evolutionarily stable strategy (ESS): once it, or something very like it, exceeds a critical frequency in the population then (not strictly but to all intents and purposes) such a strategy will be stable against invasion from any other.

What accounts for this success? Axelrod identified several properties – in particular, being 'nice' (never the first to defect), 'provokable' (retaliating against defection) and 'forgiving' (letting bygones be bygones and resuming cooperation). Niceness generates the rewards of cooperation; provokability discourages persistent defection; and forgivingness heads off long, reverberating bouts of recrimination and counter-recrimination. The reason why a strategy with these properties can be so successful is that, when it plays against another such strategy, not least itself, then both players can win the reward for mutual cooperation (R) on every encounter; they can take full advantage of playing a non-zero-sum game to help one another to attain a high average score for each of them. Unlike less cooperative strategies, they never scoop a spectacular payoff for defection (T); but neither do they drop to those low payoffs of lone sucker or mutual defection (S or P) that less cooperative strategies are particularly likely to inflict on one another, often back and forth, in amplifying recrimination. In evolution, a strategy is represented in any generation in proportion to its success in the previous generation. So, the more a Tit-for-Tat-like strategy is successful, the more likely it will be to encounter itself and the more it will be able to reap the rewards of mutual cooperation. And so it is that out of Darwinian self-interest cooperation can evolve; out of selfishness comes forth altruism.

Such cooperation apparently occurs among female vampire bats (*Desmodus rotundus*) when they regurgitate scavenged blood to certain roost-mates that have failed to find a meal during their night-time search (Wilkinson 1984; see also Wilkinson 1985). Sometimes the recipients are offspring or other kin but sometimes they are unrelated. It turns out that, in such transactions, there is plenty of scope for Tit-for-Tat-like cooperation. The future casts a long shadow; the same females (kin and non-kin, without males) often roost together for many years. The cost of regurgitation is relatively low when a bat is a donor but the benefit is relatively high when it is a recipient (because the value of a meal rises markedly with the time since the last meal – a well-fed bat has usually eaten surplus to requirements but weight-loss, once started, increases so rapidly that it takes only three days for a bat to die of starvation). Unsuccessful feeding trips are common and equally likely to befall any member of the roost (apart from young bats, who fail more often), so the roles of donor and recipient are likely to alternate frequently. Individuals can recognise one another; and the more closely any two bats associate, the more likely it is that each will favour the other when it regurgitates. Gerald Wilkinson carefully investigated these and other conditions that would be expected if the bats were engaged in a game of Prisoner's Dilemma with a Tit-for-Tat-like solution. He came to the conclusion that this was, indeed, what they were doing.

Tree swallows (*Tachycineta bicolor*) have perhaps evolved a Tit-for-Tat relationship between breeding adults and non-breeders (neither kin nor helpers) who hang around nests opportunistically hoping to take them over (Lombardo 1985). The two groups generally practise mutual restraint rather than engaging in out-and-out aggression; parents are possibly gaining help with defending the nest and non-breeders possibly gaining information about suitable nest sites. When Michael Lombardo simulated defection by non-breeders – making it appear as if two stuffed birds that he placed near the nest had killed two nestlings – the parents retaliated by attacking the stuffed birds; but they quickly 'forgave' the apparent defectors when their live nestlings were restored. The same forces might well be at work when male olive baboons (*Papio anubis*) join up in temporary (non-kin) coalitions against single opponents (Packer 1977); when vervet monkeys (*Cercopithecus aethiops*) are more willing to aid others (again, non-relatives) if the individual that is soliciting help has recently groomed the helper (Seyfarth and Cheney 1984); when dwarf mongooses (*Helogale parvula*) 'baby-sit' for non-relatives (Rood 1978); when sticklebacks (*Gasterosteus aculeatus*) together undertake the dangerous task of approaching a stalking predator (Milinski 1987); and when pairs of a hermaphroditic coral reef fish, the black hamlet (*Hypoplectrus nigricans*), take turns, during egg-laying, at being the 'male' (low reproductive investment) partner and 'female' (high investment) partner (Fischer 1980).

Reciprocal altruists have to have some means of recognising one another, a means of discriminating in favour of those who do good turns and against those who do not. But they don't need a highly developed brain, or any brain at all, to manage this; as we noted with kin selection, any functional equivalent to intelligent discrimination will do. It could be constant contact between two mutually dependent species, such as a hermit crab and its sea-anemone partner. Or it could be a unique meeting place, such as the reliable locations adopted by fish that need their parasites removed and those that remove them. So games of Prisoner's Dilemma need not be confined to bats and swallows and monkeys. Even microbes and their hosts could play. Axelrod and Hamilton have speculated that a Prisoner's Dilemma type of analysis might explain why microbes that are normally benign can suddenly turn virulent when their host is severely injured or terminally ill. The shadow of the future has suddenly shrunk. If the microorganism needs to be infective in order to spread to other hosts, then this is the time for it to seize the opportunity. And perhaps, they have suggested, chromosomes in a woman's reproductive cells do much the same when she gets towards the end of her reproductive life. This could explain the increase of certain kinds of genetic defects in offspring with increasing maternal age. An offspring suffering from Down's syndrome, for example, has an extra copy of chromosome 21. As the shadow of the future shortens, chromosomes that have previously cooperated in the fair lottery of cell division could do better by defecting so as to avoid the dead-end polar body and install themselves instead in the egg nucleus. But defection could breed defection. And the end result – unfortunate for the human victim as well as for chromosomes that get caught in double defection – would be an extra chromosome in the offspring.

'Strange freak' or selfish handicap?

Sentinel guanaco (huanaco) (from Hudson's *The Naturalist in La Plata*)

"While the herd feeds one animal acts as sentinel, stationed on the hillside, and on the appearance of danger utters a shrill neigh of alarm, and instantly all take to flight ... They are ... excitable, and at times indulge in strange freaks. Darwin writes:- 'On the mountains of Tierra del Fuego I have more than once seen a huanaco, on being approached, not only neigh and squeal, but prance and leap about in a most ridiculous manner, apparently in defiance as a challenge'." (Hudson: The Naturalist in La Plata)

Kin selection and reciprocally altruistic cooperation are two well-established explanations of altruism. A characteristically more maverick account is Amotz Zahavi's handicap theory. We have already met this as a counter-intuitive explanation of sexually selected flamboyance. When applied to altruism, handicaps turn the world just as disconcertingly upside down. Consider a bird – Zahavi studied the Arabian babbler (*Turdoides squamiceps*) – that is acting as a sentinel. Is it doing so, in spite of the danger, to help its kin or to reciprocate favours? Not at all, says Zahavi (1977, 1987, particularly pp. 322–3, 1990, pp. 122, 125–9). It is doing so to help itself – and *because* of the danger! 'Look at what I can manage' the babbler is saying to its companions. 'I am strong and robust and alert enough to bear the burden of sentinel duty, to take on the costs and still be able to thrive. And you can rely on that; only an individual of high quality could afford to handicap itself so much.' So babblers positively 'compete to ... replace other group members as sentinels instead of letting others waste their time and energy' (Zahavi 1987, p. 323). One can almost see them jockeying with one another for guard duty, vying for the most dangerous post, the longest vigil, the hottest hour of day! Difficult as this is to swallow, we have already seen, in the context of sexual selection, that there can in theory be substantial benefits to showing off reliably, even though reliability imposes severe costs.

Finally, a more sinister explanation of self-sacrificial behaviour. We must consider the possibility that the behaviour really is self-sacrificial, that of a victim, a pawn, the instrument of another. We have already met the idea of one organism manipulating another to the manipulator's advantage – the hapless shrimp surrendering itself to predators, the female canary drawn irresistibly to the male's song. Perhaps some altruists really are altruists, really are acting against their own best interests, under the influence of genes that are in another organism's body, dancing to another's evolutionary tune. If so, their altruism is the extended phenotypic expression of those genes. And it is those genes that are reaping the selective benefits.

Consider a cuckoo's unwitting hosts, sacrificing themselves and their own offspring to satisfy their demanding foster-child. We could look on their behaviour merely as a mistake, an adaptation with its purpose perverted, a ready-made niche that the cuckoo is using for its own ends rather than the ends 'intended' by natural selection. On this analysis, the cuckoo's behaviour is explained adaptively but the hosts' is not. Their altruism is no more than a temporary aberration, the result of an inevitable time-lag in the arms race between cuckoos and their victims; eventually the host species will probably evolve defences against this parasitism and their exploiters will have to improve their deception or find a new, naive species to take on the burden of parental care (Brooke and Davies 1988). That is one view.

We could, however, look on the behaviour of the hosts as an adaptation, an adaptation that benefits the cuckoos, the adaptive phenotypic effect of a manipulative gene in the cuckoo's body (Dawkins 1982, pp. 54, 55, 67–70, 226–7, 233, 247). On this analysis, too, there could be an arms race, with the hosts struggling to take more control of their own destiny and the cuckoos tightening their grip or moving on to easier prey. Indeed, it might seem that on this view we should positively expect the hosts to hit back. After all, there's nothing in it for them – in fact, it's a downright sacrifice, all give and no take. It seems little short of a Darwinian scandal for natural selection to allow the cuckoos their success. But our indignation would be misplaced. We shouldn't look on the hosts as systematic losers, even if they are condemned never to shake off their oppressors. There may well be an asymmetry in the strength of the selective forces acting on the cuckoos and their hosts. On the hosts' side, it may not be worth the costs to invest in counter-adaptations against manipulation; spending a season rearing a cuckoo needn't be fatal to reproductive success and might anyway be a rare event for any individual member of the host species. By contrast, we can expect the cuckoos to put up an impressive evolutionary fight because for them this race is a matter of life and death. 'The cuckoo is descended from a line of ancestors, every single one of whom has successfully fooled a host. The host is descended from a line of ancestors, many of whom may never have encountered a cuckoo in their lives, or may have reproduced successfully after being parasitized by a cuckoo' (Dawkins 1982, p. 70). So the cuckoos probably owe some of their victory to the 'life–dinner principle': 'The rabbit runs faster than the fox, because the rabbit is running for his life while the fox is only running for his dinner' (Dawkins 1982, p. 65).

The 'life–dinner principle' illustrates a more general point about arms races and manipulation. If there is any asymmetry in the strength of the selective forces acting on the two sides, if the forces affecting the manipulator are more critical, more stringent than those affecting the manipulated, then natural selection will be unlikely to rescue the exploited from their exploitation. 'If the individual manipulator has more to lose by failing to manipulate than the individual victim has to lose by failing to resist manipulation, we should expect to see successful manipulation in nature. We should expect to see animals working in the interests of other animals' genes' (Dawkins 1982, p. 67).

Let's think again about that bird giving an alarm call. Perhaps it is manipulating its fellows. Admittedly, it is drawing attention to itself by alerting others. But at the same time it may be providing itself with cover by rousing its companions to accompany it on the dangerous flight to safety (Charnov and Krebs 1975; Dawkins 1976, pp. 182–3). In this example the other birds could be gaining some advantage, even though they are also being used. But manipulation can be unremittingly selfish. The apparently altruistic guard may be raising a false alarm that gets others to do it a good turn although they are doing themselves no good at all. In the Amazon forest at least two species of bird (*Lanio versicolor* and *Thamnomanes schistogynus*) have been found to do this (Munn 1986). They forage for insects in flocks of mixed species; individual members of these two species act as sentinels in their respective flocks. They feed largely on the insects flushed out by the rest of the flock. If the sentinel gives a fake alarm call when it is scrambling for the same insect as a member of another species, then the other bird is distracted and the sentinel is more likely to end up with the food. Why do these other birds let themselves be duped? Again, the answer probably lies in an asymmetry in the selective forces – the useful gains from occasional cheating versus the possibly fatal danger of not taking every alarm call at face value.

Manipulation could be what is behind the Bruce effect – the ability of a male mouse to prevent implantation in a female who has recently been impregnated by another male and to bring her rapidly back to oestrus so that she is ready to mate with him. Darwinians have long been puzzled about the adaptive significance of this behaviour (e.g. Wilson 1975, p. 154). The benefit to the male is obvious. But what advantage does the female get from her apparent self-sacrifice? Well, perhaps none (Dawkins 1982, pp. 229–33); perhaps the adaptive advantage is to genes in the male mouse, genes that have their extended phenotypic expression in the female's ready compliance. Perhaps this is an arms race that she is doomed to 'lose'.

Note, by the way, that in one respect these examples of manipulation all happen to be unlike that of the parasite and shrimp, in their intimate proximity, though like that of the male and female canary, in their physical separation. The cuckoo, the bird giving the alarm call, the male mouse – all these manipulators work through genetic action at a distance. They do not dwell inside their victims; they do not control their bodies by direct physical contact. They exert their power by remote control, tapping into the sensory organs of those that they manipulate, into their central nervous systems, their brains. The cuckoo, for example, unlike the parasite in the shrimp, does not live within its host's body:

so it has less opportunity for manipulating the host's internal biochemistry. It has to rely on other media for its manipulation, for instance sound waves and light waves ... [I]t uses a supernormally bright gape to inject its control into the reed warbler's nervous system via the eyes. It uses an especially loud begging cry to control the reed warbler's nervous system via the ears. Cuckoo genes, in exerting their developmental power over host phenotypes, have to rely on action at a distance.   (Dawkins 1982, p. 227)

It could be said that with manipulation we have at last found true, albeit involuntary, altruism. With kin selection, reciprocal cooperation or handicapping advertisement, there is a benefit to the altruist or to copies of its altruistic genes. With manipulation, the altruist experiences only cost (albeit perhaps a cost that would itself be too costly to eliminate). But this way of looking at it turns out to be unduly organism-centred. What benefits from altruism in every case is a gene for altruism. Whether that gene happens to be carried by the organism performing the altruistic act is of no interest to natural selection. All that matters as far as natural selection is concerned is that the phenotypic expression of the gene should be of selective advantage to the gene itself (as compared with its alleles, the other alternatives that could have been selected). So, from a gene-centred point of view, manipulation turns out not to be a special case. This can easily (though, I must admit, rather tediously) be seen by spelling out exactly how any gene for 'altruism' works. In kin selection, the gene for 'altruism' is helping copies of itself in near relatives; in handicapping advertisement and reciprocal cooperation, it is helping itself through phenotypic expression in the organism that bears the gene; and in extended phenotypic manipulation, it is helping itself through phenotypic expression in another organism. On this analysis, it's hard to see why one would want to single out manipulation at all.

'Altruism' reanalysed

I promised that on a gene-centred understanding the problem of altruism would dissolve. So powerful a solvent has modern Darwinism proved to be that it is as well, in order to appreciate what has been accomplished, to remind ourselves of what all the fuss was about, of how the difficulty arose in the first place.

The problem originated with a central tenet of classical Darwinism: 'every complex structure and instinct ... [should be] ... useful to the possessor'; natural selection 'will never produce anything in a being injurious to itself, for natural selection acts solely by and for the good of each' (Darwin 1859, pp. 485–6, p. 201; see also e.g. pp. 84, 86, 95, 199, 233, 459, 485–6). This rules out altruism, self-sacrifice for the sake of another. But what should altruism include? Maternal care? Or is reproductive success so much a part of Darwinian self-interest that mothering counts as 'useful to the possessor'? And if help to offspring is not altruistic, why insist that help to other kin is? What's more, the question was not only who gets helped but how. Some animals are cannibalistic; should refraining from eating your neighbours be regarded as altruism? Some birds eject their sibs from the nest; should altruism include not doing so? Clearly, the problem of altruism, as it was first seen, was far from precise. It was fuelled by organism-centred thinking; by more or less unarticulated ethological traditions of what was considered 'normal' (suckling offspring, yes – eating them, no ... well, generally not); and by uncritical Hobbesian expectations that Darwinian organisms would bludgeon their way to evolutionary immortality by naked brute force.

Only once it was solved was the problem clearly seen. Only with gene-centred hindsight were Darwinians able to formulate sharply what really should have been considered altruistic, and why. And the results reveal just how misleading the old intuitions could be:

> Consider a pride of lions gnawing at a kill. An individual who eats less than her physiological requirement is, in effect behaving altruistically towards others who get more as a result. If these others were close kin, such restraint might be favoured by kin selection. But the kind of mutation that could lead to such altruistic restraint could be ludicrously simple. A genetic propensity to bad teeth might slow down the rate at which an individual could chew at the meat. The gene for bad teeth would be, in the full sense of the technical term, a gene for altruism, and it might indeed be favoured by kin selection.   (Dawkins 1979, p. 190)

Tooth decay as altruism? That's hardly how saintly self-sacrifice was originally envisaged! And yet the logic is unassailable.

The diversity of solutions to the problem of altruism has roused the suspicions of several critics that there is something shifty afoot in the explanatory enterprise (e.g. Midgley 1979a, p. 440; Sahlins 1976, p. 84). This feeling is presumably based on the idea that there is a single characteristic unifying the phenomena and that it follows from this that there should be a single, unified solution. There isn't and anyway it doesn't.

Inspired by the recent interest in altruism, biologists have begun to detect a wealth of apparently altruistic behaviour previously invisible to Darwinian eyes. Indeed, there are mutterings about the emperor's new clothes:

> There has grown up in biology the comforting supposition that nature is not really red in tooth and claw; that animals behave, in general, in an altruistic or at least a polite fashion to other members of their own species, and that animals belonging to the same species rarely do serious damage to each other. Just how far this is from the truth is revealed ... [by the extent] of cannibalism in natural populations.
> (Jones 1982, p. 202)

Such a complaint would have been unthinkable throughout most of the history of Darwinism. For a century, altruistic behaviour was hardly discussed at all; until recently most Darwinians did not even appreciate that altruism posed a problem. What was happening during this long time?

12

Altruism then

Nature most cruel

We are faced with a puzzle. There is a discrepancy between nature red-in-tooth-and-claw and the willing self-sacrifice that many an animal displays. The puzzle is why Darwinians took so long to recognise this discrepancy. Why is it only in recent decades that altruism has been widely taken up as a problem? Darwinians have typically expected animals to be cruel, remorseless and selfish rather than gentle, meek and mild. At the same time, even nineteenth-century naturalists were familiar with an impressive repertoire of altruistic acts. Why, then, did altruism not become a notorious anomaly for Darwinism?

As a first step to the solution, let's look at the background assumption that nature is red-in-tooth-and-claw. It's understandable that Darwinians should expect the natural order to be more nasty than nice. After all, genes are cut for themselves; and even thinking just of organisms, as Darwinians traditionally did, then – give or take a gloss of niceness from a bit of calculated help to others (particularly offspring) – Darwinism is still about a universe of self-interested individuals, individuals that make their way in the world largely at the expense of other living creatures. Still, it's worth remembering that, even without going into the sophisticatedly altruistic-looking strategies that we examined in the previous chapter, Darwinian organisms struggling for existence need not necessarily be locked together in relentless, unremitting combat. True, that very phrase 'struggle for existence' evokes images of gory encounters, fights to the death, the strong triumphant and the weak trampled underfoot. But, even on the simplest view of natural selection, it also refers to ways of making a living, to the use of resources, to tactics for survival and reproduction in the face of constraints; and the means to these ends need not appear, at least on the face of it, ruthless and selfish. The struggle may be a matter of how best to exploit resources rather than of how to monopolise them; 'a plant on the edge of the desert is said to struggle for life against the drought' (Darwin 1859, p. 62). Or the struggle may be conducted not primarily by armed encounters or peremptory seizure but by subtle camouflage, nocturnal feeding or just lying low. So self-interest need not look brutal and harsh; it could come in a multitude of styles. This wider interpretation was certainly what Darwin intended his theory to encompass: 'I use the term Struggle for Existence in a large and metaphorical sense, including dependence of one being on another' (Darwin 1859, p. 62). He gives his desert plant as an example. Indeed, he chose the term in preference to his original phrase, 'war of nature', as being more likely to convey this wider sense (Stauffer 1975, pp. 172, 186–8, 569).

Nevertheless, even for Darwin and Wallace, connotations of war understandably won the day. Both introduce the idea of the struggle for existence by stressing that we are sadly misled if we think that nature has a smiling disposition:

> To most persons nature appears calm, orderly, and peaceful. They see the birds singing in the trees, the insects hovering over the flowers, the squirrel climbing among the tree-tops, and all living things in the possession of health and vigour, and in the enjoyment of a sunny existence. But they do not see ... the means by which this beauty and harmony and enjoyment is brought about. They do not see the constant and daily search after food, the failure to obtain which means weakness or death; the constant effort to escape enemies; the ever-recurring struggle against the forces of nature. This daily and hourly struggle, this incessant warfare, is nevertheless the very means by which much of the beauty and harmony and enjoyment in nature is produced ... The general impression of the ordinary observer seems to be that wild animals and plants live peaceful lives and have few troubles ... [This] view is, everywhere and always, demonstrably untrue ... [T]here is a continual competition, and struggle, and war going on in nature ...
>
> (Wallace 1889, pp. 14, 20, 25; see also Darwin 1859, p. 62)

In this spirit they both (rightly) rejected Lyell's idea of a happy balance or equilibrium in the number of species, emphasising struggle as the process that Lyell should have proposed. 'When the locust devastates vast regions and causes the death of animals and man, what is the meaning of saying the balance is preserved? [Are] the Sugar Ants in the West Indies [as well as] the locusts which Mr Lyell says have destroyed 800,000 men an instance of the balance of species? To human apprehension there is no balance but a struggle in which one often exterminates another' (Wallace, writing in about 1856, *Species Notebook (1855–9)*, pp. 49–50, manuscript, Linnean Society of London; quoted in McKinney 1966, pp. 345–6). Darwin did feel that in some instances 'equilibrium' was more appropriate than 'struggle' but he came to the same conclusion about the word as did Wallace: 'to my mind it expresses far too much quiescence' (Stauffer 1975, p. 187).

Darwin's and Wallace's interpretation, not surprisingly, became the standard Darwinian view. One indication of just how standard it did become is the voice of dissent raised by a minority of Darwinians who strongly objected to such a ferocious image of nature and who wanted greater emphasis placed on the communal aspects of the struggle for existence (Montagu 1952 documents this movement). These critics amounted to something of an alternative tradition, repudiating the concentration on competition and stressing the role of cooperation in evolution. An early representative of this way of thinking, now better known for his political activities but also a geographer and enthusiastic naturalist, is Peter Kropotkin. The following comment from his *Mutual Aid* (1902), a book still widely regarded by members of this school as a classic (e.g. Montagu 1952, pp. 37–8), is typical of this view:

> It may be objected to this book that both animals and men are represented in it under too favourable an aspect; that their sociable qualities are insisted upon, while their anti-social and self-asserting instincts are hardly touched upon. This was, however, unavoidable. We have heard so much lately of the 'harsh, pitiless struggle for life', which was said to be carried on by every animal against all other animals ... and these assertions have so much become an article of faith – that it was necessary, first of all, to oppose to them a wide series of facts showing animal and human life under a quite different aspect. (Kropotkin 1902, p. 18; see also 1899, ii, pp. 316–18)

The writings of Kropotkin and those like him make clear that they saw themselves as explicitly dissenting from the dominant position. Again and again they contrast their view with what they describe as the 'orthodox canon' or the 'received doctrine' (Montagu 1952, pp. 43, 49) – which, by the way, many of them attribute to an unwarranted acceptance of Malthusian assumptions (e.g. Kropotkin 1902, p. 68), an influence that we shall touch on below. Some of these thinkers have been unclear and unsophisticated in their Darwinism. But their complaint does accurately testify to how the majority were interpreting Darwinian theory.

Another indication that harsh struggle became the standard interpretation is the fact that Darwinism came to be closely associated with a harsh ethical outlook, so closely associated that some Darwinians felt the need to deny the link. Darwin and Wallace accepted that the struggle for existence regularly entailed violent and sudden death, fighting and pain, but they were nevertheless anxious to dispel any impression that natural selection was a cruel force. They explicitly repudiated such ethical overtones, rightly claiming that they were foisted on the theory quite unjustifiably. Darwin takes care in the *Origin* to end the chapter on the struggle for existence with this reassurance: 'When we reflect on this struggle, we may console ourselves with the full belief, that the war of nature is not incessant, that no fear is felt,

Nature most cruel
South American bird-eating spider with its prey
"We behold the face of nature bright with gladness ... we forget, that the birds which are idly singing round us mostly live on insects or seeds, and are thus constantly destroying life; or we forget how largely these songsters, or their eggs, or their nestlings, are destroyed by birds and beasts of prey." (Darwin: Origin)

that death is generally prompt, and the vigorous, the healthy, and the happy survive and multiply' (Darwin 1859, p. 79). Wallace felt that the ethical aspect was so far misunderstood as to warrant detailed discussion along the same lines (Wallace 1889, pp. 36–40). He concluded that 'the poet's picture of "Nature red in tooth and claw ..." is a picture the evil of which is read into it by our imaginations' (Wallace 1889, p. 40). The attempt to paint a rosy glow on the unacceptable face of nature's struggle was, by the way, a standard exercise within pre-Darwinian natural history and utilitarian-creationist theodicy (see e.g. Blaisdell 1982; Gale 1972; Young 1969); Darwin and Wallace were treading well-worn paths.

Several influences combined to reinforce so very unremitting a red-in-tooth-and-claw view of nature. One was classical Darwinism's failure to do justice to social behaviour, for this predisposed Darwinians to interpret the struggle for existence as nothing other than lone individuals pitted against a merciless environment. Remember that, within classical thinking, other organisms, even conspecifics, tended to be viewed more as a static part of the background than as social beings. So the struggle for existence was somewhat like the struggle against the elements. Selfish organisms gained their ends by chasing, avoiding or eating others, not by sharing or cooperating with them. Classical Darwinism more readily conjures up images of fierce predators dismembering hapless prey than of one member of a social group peaceably grooming another.

A second reinforcing influence was nature's own apparently irresponsible fecundity, its 'superfecundity'. According to Darwinian theory, individuals multiply and their numbers are kept down by the onslaught of selection. But this principle alone does not suggest the startlingly prodigious fertility that is almost universal among organisms. For example, Darwin made some calculations (Darwin 1859, p. 64) concerning elephants, which were believed to be the slowest breeder of all animals (what would now be described as K-selected as opposed to r-selected – that is, among other things, adopting a reproductive strategy that goes for quality rather than quantity); he concluded that a single pair would, at a conservative estimate, if unchecked, populate the earth with 15 million elephants in 500 years (for revised figures see Darwin 1869, 1869a; Peckham 1959, p. 148). The idea that nature could opt for a prodigious rate of reproduction received striking empirical support from the work of the German biologist C. G. Ehrenberg in the 1830s, during the period when Darwin was developing his theory (Gruber 1974, pp. 161–2). His findings, which were on micro-organisms, made a great impression on Darwin. In his notebooks he commented: 'When one reads in Ehrenberg's Paper on Infusoria on the enormous production – millions in few days – one doubt[s] that one animal can really produce so great an effect' (de Beer *et al.* 1960–7, 2 (3), [C] 143) and 'One invisible animalcule in four days could form 2 cubic stone' (de Beer *et al.* 1960–7, 2 (4), [D] 167). Darwin was also well primed to notice how widespread superfecundity was from the writings of Thomas Malthus and several other authors whom Darwin admired (some of them in the Malthusian tradition) such as his grandfather Erasmus Darwin, Charles Lyell, and the explorer and polymath of the natural world Alexander von Humboldt (Gruber 1974, pp. 161–3, 174). Of course, a high rate of reproduction does not necessarily imply fierce, relentless behaviour, particularly given that much of the destruction is in the very early stages of life. Nevertheless, it inevitably threw an even harsher light on the cull that is inherent in Darwinian theory. What is more, it seemed to Darwinians that the severe consequences predicted by Malthus for humans applied with even greater force in the non-human world because the rate of increase was generally so much higher. As Wallace said: 'animals [possess] ... powers of increase from twice to a thousand-fold greater than [humans; thus] ... the ever-present annual destruction must also be many times greater' (Linnean Society 1908, p. 117). Nineteenth-century Darwinians were also familiar with the idea of superfecundity from natural theology (see e.g. Grinnell 1985, p. 61) – although in that school of thought it was, of course, associated with the beneficence of nature, mopping up, for example, the apparent evil of predation (e.g. Paley 1802, pp. 476, 479–81) or supplying evidence of plenitude (supposedly a sign of God's unified scheme).

Third, the legacy of Malthus also fed into the grim view that Darwinians took of nature. Grimness permeated the Malthusian outlook and it found its way into Darwinism along with the idea of a struggle for existence. Malthusian thinking had a powerful influence on the development of Darwinian theory. Darwin and Wallace attributed their awareness of the importance of the struggle to Malthus more than to any other thinker. And Malthusian struggle was undoubtedly cruel and harsh. Of course, Darwinism could have incorporated the idea of struggle whilst rejecting quite such uncompromising connotations. But there seemed to be plausible reasons not to do so. Indeed, again, in the light of Malthusian theory the natural world was revealed as even more cruel and harsh than the human society that Malthus described. In Darwin's view, Malthus's description of checks on human population growth served to emphasise even more starkly the extreme severity and inevitability of checks on other organisms – organisms that, unlike 'civilised' humans (and to a lesser extent 'savages'), are powerless to mitigate the effects of such checks by improving harvests, housing or hospitals. The struggle for existence, Darwin said, 'is the doctrine of Malthus applied with manifold force to the whole animal and vegetable kingdoms; for in this case there can be no artificial increase of food, and no prudential restraint from marriage' (Darwin 1859, p. 63). And so, as Wallace says: 'famine, droughts, floods and winter's storms, would have an even greater effect on animals than on men' (Linnean Society 1908, p. 117). According to Malthus human society could be grim; according to Darwin and Wallace 'uncivilised' nature was grimmer still.

Incidentally, this is only one of several ways in which, for Darwinians, Malthusian human society was unlike the non-human sphere. This point bears on the claim that Darwinians took over Malthusian theory intact, replete with political overtones (e.g. Young 1970, 1971). Many commentators, most notably Marx (Meek 1953, p. 25), have rightly claimed that Malthusian theory diverted attention from the political causes of human suffering because it was seen as attributing them to an 'inevitable' 'law of nature' (although Malthus's whole point was that we had it within our powers to avert the 'inevitable'). Nevertheless, it seems that Darwin and Wallace found in Malthusian theory not social factors masquerading as 'natural' but a social theory that they had to 'naturalise'. Take, for example, the passages that Wallace picked out from Malthus as having impressed him most (admittedly sixty years later). It is striking to see how little the checks on human populations are natural and how much they are man-made. Famine, disease and infant mortality may appear to be natural events but in the cases Malthus cites they are invariably caused by human intervention. Shortages of food and water, for example, are caused by enemies plundering, burning fields and stopping up wells or by the breakdown of the irrigation system when oppressive and tyrannical government engenders insecurity over property; infant mortality is partly a result of patriarchal oppression, women killing their daughters to save them from a fate as dire as their own. So the checks on population are enemies rather than the elements, social and political forces rather than the economy of nature. Wallace says that after reading these passages from Malthus, 'I then saw that war, plunder and massacres among men were represented by the attacks of carnivora on herbivora, and of the stronger upon the weaker among animals' (Linnean Society 1908, p. 117). As these examples suggest, the very properties that determine who shall flourish and who shall go to the wall are entirely different in the Malthusian and Darwinian worlds. Indeed, there is a view (e.g. Bowler 1976; Hirst 1976, pp. 20–1; Manier 1978, pp. 77–8) that the idea of systematic selection on the variation that occurs among individuals, which is fundamental to Darwinian theory, is absent from Malthusian theory, where its counterpart is a largely indiscriminate cull.

It should be said that historians have questioned Darwin's and Wallace's own claims about the influence of Malthus on their thinking (see e.g. Bowler 1984, pp. 162–4; Herbert 1971; Manier 1978; Schweber 1977; Vorzimmer 1969) and it is certainly possible that he did not play the direct role that they attributed to him. Nevertheless, one should remember how profound was the impact of Malthus's pessimism on early nineteenth-century thought (Young 1969). Even the cloying optimism of Paley and the *Bridgewater Treatises* was tempered in response to his view. And Tennyson's 'red in tooth and claw' was not a description of the Darwinian outlook; the poem, published in 1850, was pre-Darwinian and reflected a view of nature that was common both inside and outside science at that time (Gliserman 1975). Darwin and Wallace, no less than their contemporaries, were heirs to this bleak tradition.

As a possible fourth influence, it is commonly asserted that laissez-faire thinking in economics pushed Darwinism into an uncompromisingly harsh interpretation of nature. Perhaps economic philosophy did influence Darwinian theory (e.g. Schweber 1977, 1980). But it is not obvious that this influence accords particularly well with a view of nature as harsh. Most laissez-faire economists stressed the benevolence of the 'hidden hand'. In their eyes the final outcome of competition was benign rather than cruel and their model of society was fundamentally optimistic. Indeed, they have been widely criticised for drawing such rosy conclusions. Marx sardonically compared the honesty of Malthus with the evasiveness of those economists who claimed that there is no real conflict of class interests under capitalism (e.g. Meek 1953, pp. 124, 164).

Finally, it has less commonly been suggested that Darwinian nature reflects not only contemporary theories about human society but also the living model provided by Victorian capitalism itself (see e.g. Bernal 1954, pp. 467–8, 748; Bowler 1976, 1984, p. 164; Gale 1972; Harris 1968, pp. 105–7; Ho 1988, pp. 119–20; Montagu 1952, p. 31; and, probably more ironically than seriously, Marx in Meek 1953, p. 173). But this assumes that Darwinians saw the face of capitalism as ugly. There is a view that, on the contrary, the prevailing spirit of the privileged classes, to which on the whole they belonged, was one of optimism – the assumption that struggle was crowned by progress, and that progress ameliorated suffering (Schweber 1980, pp. 271–4).

Altruism unseen

Against this red-in-tooth-and-claw background one might well expect altruism to be viewed as a serious problem. It would seem that even a rather cursory glance at nature would raise doubts. After all, naturalists were well aware of behaviour that appeared to be altruistic – grooming, food-sharing, defence. But the problem of altruism was hardly discussed and certainly not dealt with systematically. Why was this?

If we look back at earlier Darwinian theory, one factor emerges immediately: its failure to appreciate costs. As we have seen, classical Darwinism was well geared to detecting adaptive advantage but relatively poor at spotting disadvantage. But in the case of altruism, unless the disadvantages to the individual performing the altruistic act are recognised, there appears to be no problem. What makes altruism anomalous is that it involves, or seems to involve, net costs to the altruist. When these costs are overlooked or seriously underestimated, altruism does not show up as a difficulty. To add to this, classical Darwinism, as we have noted, paid scant attention to social behaviour. But it is in the social sphere, rather than in the structural adaptations that were for so long Darwinism's prime concern, that one would intuitively expect the most striking forms of altruism to be found (even though the idea of what constituted an altruistic act was vague). It has been claimed that 'the vexatious problem of altruism was ... the greatest stumbling block to a Darwinian theory of social behaviour' (Gould 1980a, p. 260). It could equally be said that classical Darwinism's weak theory of social behaviour was a stumbling block to altruism's being recognised as a problem at all.

We have seen these features of classical Darwinism illustrated in a general way earlier in the book, and in the final chapters we shall see in detail what impact they had on the treatment of altruism. For now, I want to bring to light one important development in the history that we have so far barely touched on. This development played a crucial role first in concealing, and then eventually in revealing, altruism as a problem that needed to be solved. It was the idea of appealing to a higher level at which selection was working, appealing to a greater good.

Adaptations are for the good of ... what? We have seen that according to modern Darwinism, they are for the good of the genes of which they are the phenotypic effects. And according to classical Darwinism, they are typically for the good of the individual that bears them. Typically, but not always. During the twentieth century, another strand of thinking gradually wove its way through the standard view. This was the idea that adaptations could be for the good, not of the individual, but of the group or population or species or some other level higher than the individual. Consider once again that bird giving an alarm call. For organism-centred Darwinism, such behaviour is altruistic, non-adaptive and problematic. For gene-centred Darwinism, the altruism is merely apparent, the behaviour is adaptive and it poses no problem. Now look at it from the point of view of the 'greater-good' interpretation. The alarm call is altruistic but nevertheless still adaptive, for the adaptive benefit accrues not to the individual altruist (or to the alarm-call gene) but to the group or population or species of which the altruist is a member.

This higher-level, greater-good way of thinking can be found in Darwinism from the beginning; we shall see occasional examples of it even in Darwin and Wallace. But it became more common only with later generations of Darwinians, from around the 1920s until about the mid-1960s. During this time, then, Darwinian theory was shadowed by a doppelgänger, a higher level at which selection was thought to be working, a level above the interests of mere individuals. And in the generous embrace of this greater good, the problem of altruism was readily absorbed.

Or so it was thought. Nowadays, such a conclusion seems to us surprising. In the light of the current Darwinian understanding of the problem of altruism – the solutions we have just surveyed – it is immediately obvious that, far from solving the problem, this greater-good view only raises it more acutely. Saintly self-sacrifice is wide open to invasion by selfishness; it is the selfish beneficiaries that will survive, prosper and be represented in future generations, not those that relinquish life's necessities or even life itself for the sake of others.

Why was the same point not equally obvious to the many Darwinians (by the 1960s, very many) who held this view? The reason is curious. Although greater-good thinking was influential, it was rarely more than a vague, background assumption, often only barely explicit, often not articulated at all, often not even consciously acknowledged. Far from being a carefully worked-out alternative to individual-level selection, it was frequently so diffuse and nebulous as hardly to deserve the status of an alternative theory. As we shall see in more detail in the next four chapters, appeals to the good of the group or the species or some other higher level could be so loose, so equivocal, that it is often hard to tell exactly what their authors had in mind.

Taken at face value, these greater-good theories are making some bold claims. They envisage natural selection as able to act not only on organisms (or genes) but on whole groups, selecting not between alternative alleles but between alternative populations, preserving or driving into oblivion entire clusters of organisms, with most changes in gene frequencies occurring only as 'unintended' side effects. Such selection acts on adaptations that are properties only of groups, properties that cannot be reduced to those of their members. And these adaptations, even if they sometimes happen to favour both individual and group, can also oppose what natural selection would favour at the individual level. Individuals will risk danger so that others can live, risk hunger so that others can eat. Natural selection can ride roughshod over the petty adaptations of individuals, impervious to individual struggle, fixing its mind instead on higher things, pronouncing judgement on adaptive harmony at more elevated levels, rewarding groups that promote 'the greatest good' and penalising those whose members pursue only their own selfish ends. At face value, then, bold claims.

But we should not take terms like 'good of the group' (or species or whatever) as signs of an explicit challenge to orthodoxy. More often, they were used in blithe innocence. Sometimes, indeed, they amounted to no more than a turn of phrase, a form of words that was intended to mean nothing but 'good of the individual'. Sometimes they were to do with selection at some higher level; but it is often apparent that this assertion was naive, devoid of any notion that selection of this kind would involve a radical departure from the normal workings of natural selection and would require a radically different mechanism to drive it.

For several decades, then, Darwinians unwittingly displayed an odd mixture of orthodoxy and heterodoxy. Sometimes, all too rarely, they both preached and practised the individual-level orthodoxy of Darwin, Wallace and their contemporaries. Sometimes they merely paid lip-service to it, whilst relying heavily on notions of the 'greater good'. And much of the time they made cavalier, unapologetic use of higher-level explanations, apparently unaware that they were violating orthodox principles – indeed, apparently unaware of what the orthodoxy really was.

> From about 1920 to about 1960 a curious situation developed where the models of 'Neo-Darwinism' were all concerned with selection at levels no higher than that of competing individuals, whereas the biological literature as a whole increasingly proclaimed faith in Neo-Darwinism, and at the same time stated almost all its interpretations of adaptation in terms of 'benefit to the species'.
>
> (Hamilton 1975, p. 135)

We should not underestimate the influence of these views just because they were often ingenuously, even unconsciously, held; such beliefs can be all the more insidious:

> does a species have ... a collective will to avoid extinction or anything at all similar to such a collective interest? *No modern biologist has explicitly proposed that such factors are operative in the history of a species, but I believe that biologists are unconsciously influenced by such thinking, and that this is true of some distinguished and capable scholars.* (Williams 1966, pp. 253–4; my emphasis)

It may seem odd, in the light of all this, for me to be claiming that the problem of altruism was barely recognised. Surely that is what 'greater-goodism' was all about. Why invoke a higher level unless natural selection seems to be giving at least some individuals an inexplicably raw deal? So one might think. But, as we shall see in detail later, these higher-level accounts rarely show much appreciation of the costliness of altruism, of why it poses a problem. What they are really concerned with is the 'social'. Having turned its back on social adaptations for its first half century, Darwinism gradually began to show a sympathetic interest in them. Unfortunately, however, the concern was not with individual adaptations for social characteristics but with the collective characteristics of whole societies. Social adaptations of individuals were of interest only as building blocks in a grander edifice. The question, typically, wasn't 'How do they benefit the bearer?' but 'How do they benefit the group?' The idea of 'social' seemed to trigger a vague, deferential feeling that the good of something more weighty than the mere individual must be at stake. Social characteristics were seen as characteristics that must be selected at a societal level. 'Greater-goodism', then, was not primarily to do with altruism, with apparent conflicts between the individual and the group. Insofar as altruism was acknowledged and referred up to a higher plane, it was not because self-sacrifice was seen as problematic but because it was seen as 'social' and the higher plane was what befitted all social characteristics. Time and time again, altruistic behaviour was assimilated to adaptations that are straightforwardly good for their bearers; the dangers of giving an alarm call were lumped in with the obvious benefits of huddling together for warmth or sticking safely with the rest of the pack. So it was only as an afterthought that higher-level accounts were pressed into service to deal with traits that benefit the group as a whole but not some of its more self-denying members. We shall see, then, that the routine use of greater-good explanations was not a sign that the problem of altruism was routinely appreciated. On the contrary, these views acted a barrier, obfuscating the issues, obscuring questions that should have been asked.

Behind all this was a cosy assumption, rarely explicit and perhaps often not even recognised, that there is, by and large, no conflict between the welfare of the group and that of its individual members, that, by and large, 'true self-love and social are the same'. It's the kind of rosy reassurance that looks more at home in an optimistic natural theology or a crude apologia for capitalism than in organism-centred (or gene-centred) Darwinian theory. Added to this, it was casually assumed that if ever there was conflict between individual and group, the group would generally win. Remember, for example, the dismay with which some Darwinians noted the 'selfishness' of sexually selected ornaments as compared with the 'good-for-all-ness' of most adaptations. In many ways, it was this selfish sexual ornament, which conflicted with the good of the group, rather than selfless altruism, which promoted it, that was the odd one out.

An influential source of simplistic greater-goodism was a group of ecologists based in Chicago, centred on W. C. Allee and Alfred E. Emerson (e.g. Allee 1938, 1951; Allee *et al.* 1949; Emerson 1960; see also Collins 1986, pp. 264–8, 279–83; Egerton 1973, pp. 343–7). Indeed, it was in part the woolly world view of some early work in ecology that fuelled this episode in Darwinian theory. Many an ecologist, equipped with no more than a flimsy analogy, marched cheerfully from the familiar Darwinian territory of individual organisms into a world of populations and groups. Populations were treated as individuals that just happened to be a notch or two up in the hierarchy of life – larger, longer-lived and possessing emergent properties not to be found in individuals, but nonetheless fundamentally just like the organisms familiar to Darwinian theory: 'Populations, like organisms, exhibit self-regulation of optimal conditions of existence and survival (homeostasis)'

(Emerson 1960, p. 342); like an organism, a population has 'structure, ontogeny, heredity and integration, and forms a unit in an environment' (Allee *et al.* 1949, p. 419). All too often, in a laudable attempt to place adaptations in their full context, to see how they were shaped by and in turn shaped their environment, ecologists were led to see adaptations everywhere, at all levels of the organic hierarchy: 'There seems to be no reason to suppose that the unit of selection must be exclusively confined to a single system of organization, either at the individual, sexual, family, or social level of integration' (Emerson 1960, p. 319); 'All living systems exhibit evolutionary adaptation – adaptation for reproduction, adaptation for maintaining metabolic function in the living state, and adaptation of the whole system to its physical and biotic environment' (Emerson 1960, p. 309). The higher the level, the more important the impact was thought to be on evolutionary history. And so ecology could bask in the satisfaction of working on a grander canvas, encompassing a wider sweep, than traditional Darwinian preoccupations: ecology 'tends to be holistic in its approach' and holism 'adds a certain dignity to synthetic sciences' (Allee *et al.* 1949, p. 693) – a 'dignity', by the way, that a Darwinian of more gene-centred persuasion has termed 'holistier than thou self-righteousness' (Dawkins 1982, p. 113). This style of thinking has not quite died out in ecology, even to this day. And in popular natural history, notably television documentaries, it is flourishing. It is epitomised by the idea – apparently intended as no mere metaphor – that the whole world is one gigantic super-organism (Lovelock 1979).

But my concern here is not to promote more violence on television. Let us return to Allee, Emerson and their associates. *Principles of Animal Ecology* (Allee *et al.* 1949), the major textbook of this school, typified higher-level thinking – generally hazy about who or what an adaptation is good for, often coming down on the side of the group, invariably taking for granted that natural selection will prefer group interests if they conflict with those of individuals, never specifying the mechanism by which all this is achieved. Take, for example, this conclusion about the evolutionary fate of selfish genes: 'At the species level, genes that tended to mutate excessively would be deleterious to the population system, even though some of the characters produced by such genes might be advantageous to the individual. One might, therefore, expect selection to exert a control over the rate of mutation' (Allee *et al.* 1949, p. 684). Conversely, altruism would be favoured: 'If the sacrifice of the emigrating individuals [lemmings] had survival value to the population as a whole, emigrating behaviour might well evolve under natural selection of the whole system' (Allee *et al.* 1949, p. 685); 'If whole populations are adaptive, it seems possible that adaptations producing *beneficial death* of the individual – death for the benefit of the population – might evolve' (Allee *et al.* 1949, p. 692); and ageing, senescence and cannibalism could be 'adaptations for the benefit of the species' (Allee *et al.* 1949, p. 692).

Incidentally, it is ironic to find greater-goodism invoked to justify why mutation rates are as low as they are. This very same principle was generally invoked to justify why mutation rates are as *high* as they are; too low a mutation rate, it was argued, would reduce the evolutionary plasticity of the species (see Williams 1966, pp. 138–41 for criticism). And this way of thinking led down even more tortuous paths. Combine the idea that group interests vanquish selfish ones with a survival-of-organism (rather than replication-of-genes) view, and even parental care can become a sacrifice for the good of the group; after all, it brings 'hazards to the individual parent ... [resulting in an] increase in homeostasis at the group level but often involving a decrease in individual homeostasis. *It would be extremely difficult to explain the evolution of the uterus and mammary glands in mammals or the nest-building instincts of birds as the result of natural selection of the fittest individual*' (Emerson 1960, p. 319; my emphasis).

These ecologists usually identified themselves with a strand in Darwinian thinking that we have already come across – an opposition to red-in-tooth-and-claw Darwinism and an emphasis instead on nature's cooperative aspects. According to them, 'the general tone of Darwinism has taken colour from the extreme individualism of Darwin's time' (Allee 1951, p. 10); Darwin himself 'did not ... adequately apply natural selection to whole group or population units in contrast to his theory of natural selection of individuals' (Emerson 1960, p. 309); although, by the 1880s, the 'idea of the existence of natural cooperation was apparently in the air despite the preoccupation with the egoistic phase of Darwinism' (Allee 1951, p. 11), nevertheless, 'the new century opened with the emphasis still centered upon the individual and his problems rather than upon the group ... [T]he turn toward present day emphasis on the importance of natural cooperation did not come until about the beginning of the 1920's' (Allee *et al.* 1949, p. 32). What is surprising about that 'present day emphasis' is how little, since its naive beginnings, it had seen the problems involved in apparent self-sacrifice for the good of the group. In this respect, the faults and omissions of what Allee calls 'the remarkable if uncritical book on mutual aid' (Allee 1938, p. 11) by Kropotkin is not very different from those of many mid-twentieth-century Darwinians who in most other respects were incomparably more sophisticated. Take, for example, Kropotkin's list of the enviable benefits enjoyed by insects in well-organised societies:

The ants and termites have renounced the 'Hobbesian war', and they are the better for it. Their wonderful nests, their buildings ... ; their paved roads and underground vaulted galleries; their spacious halls and granaries; their corn-fields, harvesting and 'malting' of grain; their rational methods of nursing their eggs and larvae, and of building special nests for rearing the aphids ... described as the 'cows of the ants' ...

(Kropotkin 1902, p. 30)

Kropotkin has no Darwinian compunctions about this civic good being achieved by 'self-sacrifice for the common welfare' and 'a battle during which many ants perished for the safety of the commonwealth' (Kropotkin 1902, pp. 30, 31). Discussing burying beetles, he describes how, on finding a dead animal, they 'bury it in a very considerate way, without quarrelling as to which of them will enjoy the privilege of laying its eggs in the buried corpse' (Kropotkin 1902, p. 28). He has no Darwinian compunctions about some beetles labouring for the reproductive success of others. Glaring mistakes. But they are all too reminiscent of sentiments in many subsequent works that were far more scientifically respectable.

We'll leave these examples of greater-goodism-uncritically-assumed and altruism-unseen with a telling illustration from *The Major Features of Evolution*, a classic text, by the highly respected American palaeontologist George Gaylord Simpson (1953). Prompted in part by J. B. S. Haldane's discussion of altruism (which we shall come to soon), Simpson raises the possibility of a 'contrast between individual and group advantage in adaptation' (Simpson 1953, p. 164) (a possibility not considered in the first edition of the book (1944 – which, by the way, has the different title *Tempo and Mode in Evolution*)). In Simpson's view, individual and group interests generally coincide. He claims that earlier Darwinism (what he calls Darwinian selection), concentrating on individual adaptations (for the good of the species), did not appreciate the possibility of a divergence; current Darwinism (genetical selection), being concerned with whole populations, does appreciate it – but still considers it unlikely.

That contrast [between individual and group advantage] is, indeed, usually absent. An adaptation advantageous to the individual is also likely to be advantageous to the species. It used to be assumed that this was always true – or the question was not raised at all. This was when selection was understood and discussed in purely Darwinian terms, and Darwinian selection usually (but even it not always) acts for the advantage of the species by favoring individuals of some sorts and eliminating those of other sorts. Even selection on social aggregrates generally favors the individual, his integration into the group being favorable to survival and adaptive for him, as well as the group, its social structure being favorable for continuing reproduction of the whole unit. Genetical selection as well as Darwinian selection produces no contradiction between individual and specific adaptation in such cases.

(Simpson 1953, p. 164)

The ricochet between individual and group advantage, the confused distinction between earlier and later Darwinism, the sanguine assumption of 'no contradiction', all demonstrate that Simpson no more recognises the problem posed by altruism than he did in the first edition of his book. In the very next passage, he lumps altruistic characteristics together with 'the opposite effect, i.e., individual adaptation deleterious to the group, exemplified by development of bizarre ornamentation and overelaborate weapons by intragroup selection' (Simpson 1953, p. 165). For him, altruism and selfishness are on a par, equally unusual, equally atypical of the harmony that natural selection normally engenders between individual and group – so unusual and atypical that, he says, 'I must confess to a little skepticism regarding some of the examples on both sides' (Simpson 1953, p. 165).

But the idea of selection acting at a higher level was not always casually assumed. In one case, at least, it was proposed as an outright challenge to individual-level orthodoxy. And along with this challenge went a greater appreciation (albeit not great enough) of the problems posed by altruism. This was the position adopted by V. C. Wynne-Edwards, Professor of Natural History at the University of Aberdeen, in his hefty book *Animal Dispersion in Relation to Social Behaviour* (Wynne-Edwards 1962; see also 1959, 1963, 1964, 1977). As that title suggests, the altruistic behaviour that ultimately concerned him was how populations disperse in relation to their resources, particularly their food supply. Dispersion may not sound particularly altruistic. But according to Wynne-Edwards, animals generally disperse themselves at a density that is close to an optimum for the group as a whole, an optimum well below out-and-out exploitation. Orthodox Darwinism, he says, cannot explain this (and it is to his credit that he at least recognises what the orthodoxy should be); on the standard model, every member of the population is out for itself, exploiting resources to their limit, even to the point of 'overfishing'; individual-level selection is powerless to act against such short-term, selfish interests in favour of the longer-term, collective interests of the group. Dispersal, he claims, must be achieved by group selection, by selection favouring whole populations over other populations; only in that way can the interests of the group override those of its members. Group selection will favour populations in which some members forgo selfish striving – by emigrating, refraining from breeding, denying themselves food – so allowing the population as a whole to thrive: 'control of population density frequently demands sacrifices of the individual; and while population control is essential to the long-term survival of the group, the sacrifices impair fertility and survivorship in the individual' (Wynne-Edwards 1963, p. 623). Wynne-Edwards, then, is explicit and uncompromising: standard Darwinian forces cannot explain the evolution of altruistic social adaptations; a supplementary mechanism of group selection must have played a crucial role.

So what does this mechanism look like? At this point, Wynne-Edwards's forthright tone evaporates. His book claims to be setting out a radically new view. And yet it is exceedingly hard to prise any serious theory out of those pages. The voluminous text is almost exclusively devoted to a detailed exposition of data, to cataloguing purportedly altruistic behaviour and reinterpreting a host of social adaptations as mechanisms for population regulation (no mean task, for Wynne-Edwards's ambition was to account for the origins of all social behaviour and he was inclined to construe even the most apparently self-interested social interactions as public-spirited self-abnegation). As Wynne-Edwards himself later graciously admitted: 'I freely invoked group selection without being precise about its mechanics' (Wynne-Edwards 1982, p. 1096); 'what was conspicuously absent from my promotion of it was a credible model or theory of how, in practice, group selection would take place' (Wynne-Edwards 1977, p. 12).

But was a credible model possible? This was the question that was raised by appeals to higher-level selection. With all their inadequacies – or, rather, because of them – such appeals played a useful role. They came to act as a goad, a provocation, a challenge to which orthodox Darwinism responded – and with fruitful results.

Altruism levelled down

Before going into that response, I should make clear that there were honourable exceptions to the greater-good way of thinking. The most notable, not surprisingly, came from R. A. Fisher (particularly 1930) and J. B. S. Haldane (particularly 1932). Admittedly, their writings sometimes look more organism-centred than gene-centred. But what matters is that they are not higher-level; whether they are gene- or organism-level is, from the point of view of their difference from greater-goodism, relatively unimportant. With their anti-higher-level analysis Fisher and Haldane undoubtedly wheeled out the right vehicle for natural selection. Subsequent Darwinians didn't so much miss the bus entirely as keep hopping on the wrong one.

Consider the career of that paradigmatically gene-centred theory, kin selection. Fisher and Haldane pointed the way, albeit sketchily, as early as the 1930s. But it was not until three decades later that this potential was realised. The fundamental idea, that natural selection may favour help to relatives, was familiar even to classical Darwinism; it is, after all, what lies behind parental care. But it was not recognised as a general principle. Fisher took one step in

The Genetical Theory of Natural Selection (Fisher 1930, pp. 177–81). His problem was how distasteful insect larvae could have evolved their protection. What advantage can distastefulness have for an individual that does not survive a predator's abandoned attempt to eat it? Fisher pointed out that the unpleasant experience would teach the predator to avoid similar prey in future; so, if the rest of the parent's brood lived close to the victim, these sibs could benefit. Haldane, too, sketched out the concept of kin selection (Haldane 1932, pp. 130–1, 207–10, 1955, p. 44) and applied it to maternal care and the social insects. Admittedly, neither Fisher nor Haldane spread his net widely; Fisher did not apply his conclusions to kin other than sibs and Haldane suggested that such altruism would on the whole be restricted to species that are reproductively specialised (Haldane 1932, pp. 130, 131) or live in small family groups (Haldane 1932, pp. 208–10, 1955, p. 44). Nevertheless, the elements of the theory were there. But it was not until W. D. Hamilton's classic papers of the early 1960s (Hamilton 1963, 1964) that the notion of kin selection was made explicit, generalised, tightened up and properly incorporated into Darwinian theory.

Incidentally, as John Maynard Smith brought to my attention, even Haldane could be tempted into a greater-good explanation of altruism (Haldane 1939, pp. 123–6; Maynard Smith 1985b, pp. 135–7). 'Animals and plants are not quite such ruthlessly efficient strugglers as they would be if Darwinism were the whole truth' he stated in one of his popular essays (Haldane 1939, p. 125); 'it does not always pay a species to be too well adapted. A variation making for too great efficiency may cause a species to destroy its food and starve itself to death. This very important principle may explain a good deal of the diversity in nature, and the fact that most species have some characters which cannot be accounted for on orthodox Darwinian lines' (Haldane 1939, p. 126) – orthodox lines being the survival of the fittest individuals (Haldane 1939, p. 123). It's probably safe to assume that Haldane was succumbing to the temptation of conveying an uplifting message to his *Daily Worker* readers rather than trying to propagate a genuine Darwinian unorthodoxy.

I did not include the other major founding father of modern Darwinism, Sewall Wright, among the honourable exceptions. On this issue, he perhaps confused matters more than he clarified them, especially in his own country, the United States. This was largely no fault of his own. Most of the confusion probably arose from his use of the term 'intergroup selection' for a process that bore no resemblance to what Wynne-Edwards and others came to mean by group selection (e.g. Wright 1932, 1945, 1951). His theory was about how random drift (working in conjunction with natural selection) could contribute to adaptation. Imagine a population that gets broken up into smaller, inbred sub-populations. The individuals in the original population would not all have been equally well adapted. By sheer chance (random drift) a sub-population could consist of some of the least well-adapted individuals and it could also be so small and so inbred as to lack the variation that natural selection would need if it were to recover the adaptive successes of the original population. But the impoverished genetic legacy of random drift could prove to be a bonus. Being unable to take the sub-population back to where the parent population had been, natural selection would be forced into other options. Selection could set the population on a new path that would eventually lead to an even higher peak of adaptedness. Natural selection is constrained to prefer local optima, the adaptations that are the best at the time; it cannot ignore them in favour of adaptations that would be better in the long run. Random drift could allow a population to escape local optima and thereby move to higher adaptive peaks. What does all this have to do with what Wright called 'intergroup selection'? Wright pointed out that some of these sub-populations would end up better adapted than others. Because they would be more successful, they would eventually swamp the others in what Wright called 'intergroup competition'. The species would come to be dominated by the members of the better adapted groups. Sewall Wright's intergroup selection thus had nothing to do with what the term 'group selection' generally suggests – group welfare opposing individual welfare. His intergroup selection was a means by which individual welfare is promoted because random drift frees individuals from the restrictions that natural selection's dedication to individual welfare would normally impose on them; random drift liberates individuals from the unremitting opportunism, the concentration on their immediate good, that natural selection exercises on their behalf. Nevertheless, Wright's use of the term 'intergroup' was somewhat ambiguous and could have led people to think of selection at group level (Provine 1986, pp. 287–8). What's more, he did use the term 'group selection' for an entirely different mechanism – what is now familiar to us as kin selection – to explain the evolution of altruism (e.g. Wright 1945; see also Provine 1986, pp. 416–17, who, however, adds to the confusion by miscalling this 'the modern theory of group selection').

Darwin once remarked that in the progress of science 'false views, if supported by some evidence, do little harm, as every one takes a salutary pleasure in proving their falseness; and when this is done, one path towards error is closed and the road to truth is often at the same time opened' (Darwin 1871, ii, p. 385). Whatever other contributions Emerson, Wynne-Edwards and their fellow higher-level selectionists made to science, it has often been noted that on the issue of altruism and levels of selection their most distinguished contribution lay in stimulating their critics, in finally persuading those Darwinians who had long deplored these 'false views' that they called for more systematic treatment than the odd remark over coffee or the occasional unfavourable review. One classic work for which we are largely indebted to these unwitting catalysts is *Adaptation and Natural Selection,* by the distinguished American evolutionist George C. Williams (1966, particularly pp. 92–250). Goaded by airy appeals to the welfare of groups, populations and species, Williams retaliated with two types of argument. He spelled out why genes are suitable candidates for units of selection whereas organisms, groups and so on are not (e.g. Williams 1966, pp. 22–3, 109–10). And he showed that if evolution were to proceed by the extinction of whole groups then several highly unlikely conditions would have to be satisfied (such as groups having to consist overwhelmingly of altruists and not being invaded by selfish individuals). Another set of criticisms was levelled by John Maynard Smith (1964, 1976). He set up population genetics models, explicit mathematical models, of group selection to see what assumptions would have to be made if such selection were to work. He, too, concluded that the conditions required (such as small group size combined with extremely low migration rates) were so stringent that they were likely to be realised only very rarely, so rarely that group selection would have little, if any, impact on evolution – a conclusion subsequently confirmed by others in many detailed models. As Wynne-Edwards later conceded: 'The general consensus of theoretical biologists ... is that credible models cannot be devised, by which the slow march of group selection could overtake the much faster spread of selfish genes that bring gains in individual fitness' (Wynne-Edwards 1978, p. 19). (Though even later still, he seems to have reverted to his earlier attachment to the group (Wynne-Edwards 1986).) As for the task of reinterpreting purported evidence of higher-level selection, David Lack, leading ecologist and doyen of British ornithology, demonstrated how that could be done. Taking cases that Wynne-Edwards had used to illustrate population regulation, he showed that group-selectionist explanations were inaccurate and unnecessary; all the examples could be better explained by individual selection alone (Lack 1966, pp. 299–312). This is how Lack's work converted at least one of today's well-known Darwinians when he was grappling with the competing claims of Wynne-Edwards and individual-level selection:

To help me through this crisis, my teacher had me read the work of Wynne-Edwards and his chief opponent, David Lack. I read them for three straight days, one after the other. At first, Wynne-Edwards reconvinced me every time I reread him. But as I continued, his grasp on my thinking began to weaken. Finally, Wynne-Edwards let go completely and slipped off into the surrounding gloom. The evidence was clear: natural selection refers to differences in *individual* reproductive success.   (Trivers 1985, p. 81)

Rather than discuss in detail the many criticisms of higher-level selection that have built up over the last twenty years, I shall go straight to a recent analysis that clarifies the logic behind them. This analysis, which was arrived at independently by Richard Dawkins and the philosopher David Hull, rests on a distinction between vehicles and replicators (Dawkins 1976, pp. 13–21, 2nd edn. pp. 269, 273–4, 1982, pp. 81–117, 134–5, 1986, pp. 128–37, 265–9, 1989; Hull 1981). (Hull chose the less telling word 'interactor' in place of 'vehicle'.)

Consider the properties that anything would need in order to be a unit of natural selection, a unit of which we could say that adaptations – phenotypic effects – are for their good. First, it must be able to reproduce itself (more strictly, of course, copies of itself); it must be self-replicating. Second, it must have the good luck not to replicate absolutely faithfully but to make occasional slight mistakes; mistakes introduce changes, and thereby differences, into the population, and these differences are the material that selection works on. And third, these self-replicating entities must have properties that influence their survival and reproduction, their probability of further self-replication. Anything with such properties we can call a replicator. Genes are replicators. They reproduce copies of themselves, on the whole faithfully, but with occasional mutations; and they have phenotypic effects that influence the gene's fate. So natural selection can act at the level of genes; genes can be units of selection.

There are other candidates for units of selection – organisms, groups, species. Organisms are the most likely candidate. But an organism does not replicate facsimiles of itself: its offspring cannot inherit its acquired characteristics, the accidental changes that it has undergone during its lifetime. Similar considerations hold, though even more strongly, for groups and other higher levels. Although in some loose sense they renew themselves, divide, bud off, persist, nevertheless they cannot be true replicators. They have no reliable means of self-propagation, no more or less automatic mechanism for churning out generation after generation of facsimiles. Genes, then, can be replicators whereas organisms, groups and other levels in the hierarchy cannot. Natural selection is about the differential survival of replicators. So genes are the only serious candidates for units of selection.

If organisms are not replicators, what are they? The answer is that they are vehicles of replicators, carriers of genes, instruments of replicator preservation. Replicators are what get preserved by natural selection; vehicles are means for this preservation. Organisms are well integrated, coherent, discrete vehicles for the genes that they house; but they are not replicators, not even crude, low-fidelity replicators. Groups, too, are vehicles, but far less distinct, less unified.

What light does all this throw on adaptations? Adaptations must be for the good of replicators, for the good of genes. But they are manifested in vehicles. Genes confer on vehicles properties that influence their own replication. So adaptations could, in principle, turn up at any level – at the level of organisms (either in the organism that bears the gene or in another), at the level of groups and even higher. There is no rigid rule as to where they will be manifested, in which vehicle (nor how). They are, however, most likely to occur in the organism that bears the gene. This is not only because the closest vehicle is the most amenable to physical influence. It is also because genes that share a body are likely, to a large extent, to 'agree' over which phenotypic effects are adaptive. As we saw when discussing the gene-centredness of modern Darwinism, conflicts of interest among same-body genes are dampened down by a common interest in the survival and reproduction of that body. Any gene in a genome will have been selected, among other things, for its compatibility with other genes in that genome, its contribution to their joint endeavour. Above all, the genes in any one body, but not in any other body, have the same hoped-for route into future generations. So we should expect to find adaptations at the level of the organism; and, although any adaptation will be for the good of the gene of which it is the phenotypic effect, we should also expect, by and large, that it will be good for the other genes in that organism because of their need to submerge their differences in pursuit of their common purpose, the organism's survival and reproduction.

And yet – remember outlaw genes, remember the speculation about Down's syndrome – warring factions can arise even among genes that share a body. So how much more likely, and how much more acute, conflicts of interest will be among the looser assemblages of genes that make up higher-level vehicles – groups, populations, species. In these more unwieldy aggregations there is no common aim that binds as tightly, nothing that brings divergent interests into such close harmony, as in organisms. The genes within higher-level vehicles are not fettered by a self-interested commitment to the survival and reproductive success of one particular organism. This allows their conflicting priorities to surface. So to find an adaptive property, a more or less universally satisfactory property, at the level of groups and above would obviously be a trickier feat for natural selection to perform.

Suppose, however, that there was such a property. Would it be a sign that group selection was at work? Imagine, for example, that sexually reproducing species evolved more rapidly than asexual species and so flourished at their expense. It could perhaps be argued that this would be a species-level property, for it is species not individuals that evolve. If so, natural selection would be acting on properties not of individuals but of groups, acting on phenotypic effects at group level. Such selection could, then, be called group selection in some weak sense. But one shouldn't let this terminology obscure the fact that it is still just straightforward replicator selection. The groups are not themselves units of selection, not replicators. They are vehicles, units of adaptation for replicator selection. In this weak sense of group selection, just as with classical Darwinism's organism-centred selection, evolution is still about changes in proportions of replicators (genes) as a result of the influence of their own phenotypic effects on their own replication. Like 'organism selection', 'weak group selection' is about the alterations in relative frequencies of alleles in gene pools that occur because of the phenotypic effects of those genes, effects that are manifested at the level of the organism and the group respectively. As ever, selection is to do with differential survival, and the units that survive over evolutionary time are not groups or individuals but replicators. In this weak sense, then, 'group selection' could occur. Wherever genes are gathered together, emergent properties could arise from their individual strivings, properties that manifest themselves only at the level of groups and higher. But even if they did arise – which as we've seen is unlikely – they would in no way undermine the status of genes as the only units of replicator selection. This does not mean that higher level entities are unimportant in evolution. They are important, but in a different way: as vehicles.

There is a telling analogy – and an equally telling disanalogy – here with a long-standing controversy in the social sciences. Reductionists hold that society consists ultimately of nothing but individuals; against this, holists claim that the social world cannot be understood without resort to higher levels. But these positions need not conflict. It is possible to be a reductionist with respect to objects but happily allow holistic characteristics – a combination of entity reductionism and property holism (Ruben 1985, pp. 1– 44, 83–127, particularly pp. 3–6, 83–86). According to this view, individual human beings are ultimately the only constituents of society; nation-states, parliaments, clubs and so on are all strictly reducible to them without remainder (philosophers call this reductive identification). But at the same time, the objects of the social world can have, even usually do have, irreducible, societal properties. This is analogous to what we have just imagined in the biological world, although the issue there, of course, is about what natural selection acts on rather than what exists. Indeed, at the level of the organism the parallel is close: 'The reason I may sound reductionist is that I insist on an atomistic view of units of selection, in the sense of the units that actually survive or fail to survive, while being wholeheartedly interactionist when it comes to the development of the phenotypic *means* by which they survive' (Dawkins 1982, pp. 113–14). But there is also a crucial disanalogy between social properties and biological adaptations. In human societies, the properties that emerge may be good, bad or indifferent for individuals, social groups, nations or whatever. But group selectionism (weak group selectionism) makes claims about adaptations, about characteristics that satisfy the fragmented purposes of all the genes in the group and, what's more, confer an advantage on that group over other groups. Group-level adaptations, then, are a very special case of emergent properties – so special that it would be rash to expect them to have played any significant role in evolution. Of course, the question of what role they have actually played is an empirical, not a conceptual, issue. It is a factual matter about which adaptations happen to have arisen at levels higher than organisms, about the extent to which groups and other higher-level vehicles happen to have been roadworthy.

With the replicator–vehicle distinction clearly in mind we can now see what it was that higher-level selectionists like Wynne-Edwards must have been claiming if they were not merely confused but were indeed posing the bold challenge to orthodox natural selection that we set out earlier. They would not have been championing group selection in the weak sense that we have just spelled out – the idea that groups can manifest adaptations, adaptations that could have an impact on evolution. They must have been proposing that groups were not only vehicles but were also sometimes replicators, and such powerful, successful replicators that natural selection would give them precedence over the feebler efforts of genes (or, as they generally saw it, organisms) to replicate themselves. If this was indeed their theory then we can now see that it stemmed from a conceptual confusion, a confusion between vehicles, the entities that manifest adaptations, and replicators, the entities that adaptations are 'for the good of'. Put in this way, it becomes even harder to understand how altruistic adaptations were supposed to evolve, how higher-level entities were supposed to create and sustain such individual coherence, such copying fidelity and such unity of purpose. (Incidentally, this is quite different from a view that has more recently, in the United States, been called 'group selectionism'; for a sort-out of the differences see Grafen 1984; see also e.g. Maynard Smith 1982b, 1984a.)

I have said that Fisher and Haldane pointed the way to a Darwinism that dispensed with higher-level selection. But the distinction between replicators and vehicles is characteristic of the more recent revolution in Darwinism. And it carries their analysis further. Only in the light of that distinction is it clear that the gene (sometimes, to Fisher and Haldane, the organism) is not merely at a different level of selection from entities further up the hierarchy, but is an entirely different kind of entity. So higher-level selectionism rested on a misplaced categorisation, an assumption that higher-level entities were, as far as natural selection was concerned, on the same ladder as genes but merely on a more elevated rung. A systematically gene-centred analysis helps to clear up this confusion. It shows that, as far as natural selection is concerned, there is a sharp discontinuity between genes and entities at all the higher levels: genes are replicators, all the higher levels are vehicles. For most purposes it did not matter that Fisher and Haldane sometimes talked of organisms rather than genes. But on this issue of levels of selection, a stauncher loyalty to the gene might have pre-empted at least some misunderstandings on the part of other Darwinians.

Incidentally, it is ironic that many of those individuals who are most hostile to adaptationism at the level of the organism are the very same individuals who, in pursuit of 'pluralism', are most readily seduced into detecting adaptations when they encounter entities at higher levels. They are the first to challenge the adaptive significance of the bands on a snail's shell. But they also are the first to spot adaptations popping out all up and down the hierarchy of life – 'species selection', 'holistic properties' and the like.

There is a further irony. As we saw when we looked at adaptive explanations, these pluralists use the term 'Panglossian' to describe Darwinians who are, in their view, over-eager to provide adaptive accounts of the characteristics of organisms. The irony is that Haldane originally adopted this Voltairian term to describe Darwinians more like these very pluralists themselves: Darwinians who, failing to find adaptive advantages at the level of the organism, look instead to the level of the group, population or species, assuming that misfortunes at one level will be redressed at another; 'the phrase "Pangloss's theorem" was first used in the debate about evolution not as a criticism of adaptive explanations, but specifically as a criticism of "group-selectionist", mean-fitness-maximising arguments' (Maynard Smith 1985a, p. 121).

Enlightened by modern Darwinism's understanding of altruism, we are now ready to enter the muddier waters of earlier discussions. We shall look at four cases: the sterility of social insect workers; conventional contests; human morality; and sterility in first or second generation interspecific crosses. The first two are of interest because although, with historical hindsight, they are among the most egregious examples of altruism, this was not in the least obvious to Darwinians until quite recently. The other two are cases that, unusually, classical Darwinism interpreted as in some way to do with altruism.

13

The social insects: Kind kin

For Aesop, the social insects were a source of inspiration. For Darwinians, they were a source of aggravation. Darwin declared that they posed 'by far the most serious special difficulty, which my theory has encountered' (Darwin 1859, p. 242). A century later, when the problem was at last yielding, George Williams declared: 'there is no more important phenomenon [as a challenge to gene-centred theory] than the organization of insect colonies' (Williams 1966, p. 197).

What was this puzzle that worried Darwinians for so long? Among the Hymenoptera (the group that includes ants, bees and wasps) and the Isoptera (the termites) there are species in which sterile castes work for other members of their community – helping with the care of their offspring, defending the colony and performing numerous other civic duties that benefit their fellows. They devote their lives to the survival and reproduction of others. And yet they leave no offspring of their own. This clearly raises a difficulty. How could natural selection, which works on heritable adaptations, have given rise to such behaviour (so-called eusociality)? How do sterile workers benefit from their self-sacrifice? And how do they pass on their characteristics?

The answer, one way or another, probably lies in kin selection. We tend to think of reproductive success in terms of offspring. But kin-selection theory reminds us that a brother or sister can be as valuable as a son or daughter. If I possess a gene for altruism (or for anything else) then my siblings are just as likely to bear copies of it as are my offspring. So, all other things being equal, an animal's mother is as good a potential reproductive source as is the animal itself. It is not difficult to see, then, why animals might opt for being sterile and caring for their mother's offspring rather than having their own. Indeed, in the light of kin-selection theory, our original problem is reversed: Why, one wonders, isn't the practice widespread? Well, I did say 'all other things being equal'. And it turns out that some animals – the Hymenoptera and Isoptera – are more equal than others.

Let's start with the termites. One oddity that might have smoothed the way for kin selection is a quirk of their diet. Their staple food is wood, which is highly indigestible without help. The help comes from microorganisms that live in the termites' guts. The insects need to reinfect themselves with their digestive aids in every generation and in some cases after every moult (because the lining of the gut is lost in moulting). This they achieve by coprophagy: the precious gift of fæces is passed from one generation to the next. Coprophagy requires close proximity. Perhaps this started the termites on an evolutionary path from proximity to sociality and from sociality eventually to kin selection – not an inevitable path but one in which each step made the next that little bit more likely (Wilson 1971, p. 119). Coprophagy also offers a further avenue for kin selection: pheromonal manipulation. In all social insects it is pheromonal suppression of workers by the queen that provides the mechanism for inducing sterility. The termites' indigestible diet might well have given their queens a ready-made channel during the early evolution of their social behaviour.

The termites' diet also introduces a whole new set of interests: those of the microorganisms. Their genes make up as much as a quarter of the DNA in a termite mound – and an essential quarter, for these tiny creatures and their termites are mutually interdependent. What's more, because the microorganisms reproduce asexually, they are usually a genetically identical clone. With such bulk and such singlemindedness, they might well command the biochemical clout for manipulation. Perhaps, then, these valued guests can take some credit for the termites' highly sociable ways?

> is it not almost inevitable that the symbiont genes will have been selected to exert phenotypic power on their surroundings? And will this not include exerting phenotypic power on termite ... bodies, on termite behaviour? Along these lines, could the evolution of eusociality in the Isoptera be explained as an adaptation of the microscopic symbionts rather than of the termites themselves?
>
> (Dawkins 1982, pp. 207–8)

W. D. Hamilton spotted that the termites' cycles of inbreeding within a colony and outbreeding when founding a new one offer a further curious opportunity for kin selection (Hamilton 1972, p. 198). His insight, by the way, was characteristically *en passant*, so much so that it has commonly been attributed to Stephen Bartz, who discussed the same theory later in more detail (Bartz 1979, 1980; see e.g. Myles and Nutting 1988; Trivers 1985, pp. 181–4); even more characteristically, Hamilton himself forgetfully attributed it to Bartz when I asked him about it – his idea of a priority-dispute! Hamilton's reasoning was as follows; I'll put it in idealised form – the real picture is rarely this neat. Established termite colonies can become highly inbred. In some species, the winged kings and queens commonly get replaced by secondary reproductives from within the colony. The new incumbents are likely to be brother and sister, and their offspring can become so inbred that they are largely homozygous (each individual's alleles occurring in identical pairs). This means that they are more closely related to fellow colony members than they are to members of other colonies (who are also becoming increasingly inbred but in different directions). This in itself could predispose termites to altruistic self-sacrifice. But a termite in an inbred colony is no closer to its siblings than it would be to its hypothetical offspring. If only it were closer to its sibs, there would be an even more powerful reason for self-sacrifice to the point of sterility.

And it turns out that there is, indeed, such a reason. This is how it comes about. In the long history of a lineage, perpetuated through many descendant colonies, the intensive inbreeding is regularly punctuated by outbreeding. This happens when a young winged reproductive flies off and founds a colony of its own. Outbreeding would normally be expected to cut the ties of resemblance dramatically. But not so. Our young breeder's mate, coming from another colony, is also likely to be inbred, and homozygous *but for different alleles*. So their offspring (the first generation) will be heterozygous, but *identically* so. The ties of resemblance among siblings are still holding. Such ties will not, however, survive the next generation, when king or queen or both are replaced. Mendelian shuffling will finally break up the genetic identity. This means that a heterozygous offspring of the founding king and queen is genetically closer to its (identically heterozygous) siblings than it would be to its own offspring, if it had any. So here we have an additional reason for the evolution of worker sterility in termites. Perhaps, then, the high ideal of altruism springs from a base alternation of incest and elopement.

In some termite species, kin altruism may have been fostered by another source of genetic uniformity that makes siblings closer than offspring. In this case, the uniformity is among sibs of the same sex and it comes about through many genes being linked together on the sex chromosomes (Lacy 1980, 1984; Syren & Luykx 1977; but see also Crozier and Luykx 1985; Leinaas 1983). Termites, like ourselves, determine sex by the so-called XX XY system. Females have two X chromosomes, males an X and a Y. A male inherits his father's Y chromosome (the only one his father has to give) and one of his mother's two X chromosomes. A female inherits her father's X chromosome (again the only one he has to give) and one of her mother's two X chromosomes. So, with respect to their paternal sex chromosome, males are effectively identical twins to their brothers and females are effectively identical twins to their sisters. The same is not true with respect to their maternal sex chromosomes, or any of their ordinary chromosomes. As far as a gene on a sex chromosome is concerned, then, siblings of the same sex are genetically closer than potential offspring would be. In particular, a gene for

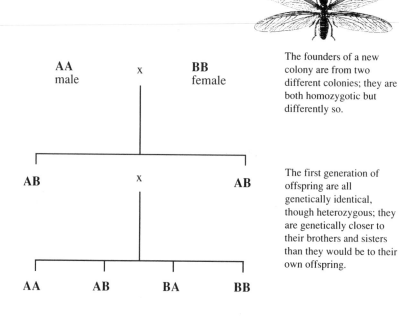

AA	x	**BB**	The founders of a new colony are from two different colonies; they are both homozygotic but differently so.
male		female	

AB   x   **AB**

The first generation of offspring are all genetically identical, though heterozygous; they are genetically closer to their brothers and sisters than they would be to their own offspring.

AA   **AB**   **BA**   **BB**

One path to altruism in termites

individual sterility in the service of same-sex sibling care could be favoured if it happened to be sitting on a sex chromosome.

Theoretically, such 'sex chromosome altruism' is a possible route to worker sterility in any animals with sex chromosomes, even mammals. Hamilton noted this some time ago (Hamilton 1972, p. 201). But he thought that sex chromosomes constitute so small a fraction of the genome (only 5% in some mammals, for example) that such altruism is unlikely. As we shall see, he had bigger things in mind; he had already unravelled the analogous case of sterile worker castes in the Hymenoptera, where full sisters are especially close to one another with respect to the *whole* genome. More recently, however, it has emerged that termites also have something big to offer. In some termite species, a massive portion of the genome (even as much as half of it) is linked to the sex chromosomes, forming what is effectively a giant sex chromosome. Here, then, the objection to 'sex chromosome altruism' melts away. Genes in termites really do have a significantly higher chance of being shared by same-sex siblings than by offspring, because the group of genes that functions as a sex chromosome is so unusually large. Perhaps, then, far back in the evolutionary past, this linkage of large parts of the genome to sex chromosomes occurred in all termite species and helped to get kin altruism off the ground, even in species in which there is no such linkage today. Perhaps. But there are problems. The main one is that termites don't seem to behave as predicted. They appear to dispense altruism to kin regardless of sex; both males and females act as sterile helpers and no evidence has yet been found that either sex biases its good deeds in favour of its own sex. There's also the problem that only some termites have 'giant sex chromosomes' and there is nothing special about the social behaviour of those that do. What's more, these species are placed rather sporadically on the termite evolutionary tree. So possibly the giant chromosomes are too recent an innovation to have played their suggested role. To round off this list of misgivings, Hamilton himself offered me the following reflection: 'The other group notorious for its ring chromosomes is evening primroses! What a Florida mangrove swamp termite has in common with a wasteland evening primrose is totally obscure. At the moment it seems just one of God's jokes'.

Now to the Hymenoptera. One condition that might have favoured kin selection is the fact that we have just noted: full sisters are more closely related to one another than they would be to their offspring. This occurs because of the highly unusual arrangement of their chromosomes – their haplodiploidy (Hamilton 1963, 1964). The females develop from fertilised eggs and are diploid (they have a double set of chromosomes, half from each parent); the males develop from unfertilised eggs and are haploid (they have only a single set of chromosomes). Details of social organization differ from species to species but a colony typically consists of a single queen, who has been fertilised by one male, and her offspring. It is her daughters that are the sterile workers. Why? These female offspring all share identical copies of half their genes, the paternal half, because their father is haploid and so all his sperm are the same; on average, they also have half their maternal genes in common. So, as far as their paternal genome is concerned, they are identical twins (just like same-sex termites, as we've seen was later realised, except that in the termites' case the 'identical twinning' is only for sex chromosomes). The result is that these daughters are more closely related to one another than they would be to their own hypothetical offspring of either sex. So they do better by caring for their potentially fertile sisters – those who will become queens – than by having sons or daughters of their own.

Another condition that favours kin selection in some Hymenoptera is that the mother's monogamy is guaranteed (Dawkins 1976, 2nd edn., pp. 295–6). If a mother is monotonously monogamous then all sibs will have the same father, and the mother becomes as valuable as an identical twin. This point owes nothing to haplodiploidy. It is potentially true of any monogamous animals. The trouble is that in most animals 'monogamy' is highly unreliable. There are some species in which males and females pair up so convincingly for an entire mating season or even for life, that naturalists had long thought them to be monogamous; at one time it was believed that most birds practised such exclusivity (Lack 1968, p. 4). But on closer observation, one after another, these species are revealing a very different picture: in eastern bluebirds (*Sialia sialis*), 9% of all broods examined had multiple parentage (Gowaty and Karlin 1984); in indigo buntings (*Passerina cyanea*), over 22% of the females' copulations were not with their partners and at least 14% of offspring had not been fathered by them (but bore a suggestively close genetic resemblance to neighbouring males) (Westneat 1987, 1987a); in white-fronted bee-eaters (*Merops bullockoides*), between 9% and 12% of all offspring sampled were not genetically related to one or both of their 'parents' (Wrege and Emlen 1987); in mountain white-crowned sparrows (*Zonotrichia leucophrys oriantha*), between 34% and 38% of chicks were probably not the child of their 'father' (Sherman and Morton 1988). Egg-dumping by other members of the species could account for some of these 'illegitimates' but by no means all. The Hymenoptera, however, order things differently. In at least some species, monogamy is unusually reliable. This is because the queen mates just once, sealing her entire reproductive fate on a single nuptial flight, storing the sperm from that sole union and doling it out over the rest of her long life.

Finally, there is the suggestion that the social insects' eusociality originally got off the ground through mutual help among the females (see e.g. Brockmann 1984). Females of the same generation, or mothers and daughters, or both, might well have begun by merely sharing nests. Gradually, with the help of other predisposing conditions, they could have evolved increasingly cooperative behaviour until reaching the extremes of altruism that have so taxed the ingenuity of Darwinians.

So much for what we know today. Now back to Darwin. He was perturbed by the problem of the neuter insects, as his declaration about this 'most serious special difficulty' suggests (although several of his difficulties are his 'most serious'!). Taking the case of ants, he goes over the issues at length in the *Origin* (pp. 235–42). But what exactly is his 'special difficulty'? Surprisingly, the anomaly is not, as we should expect, the ants' sterility and their devotion to the welfare of others. 'How the workers have been rendered sterile is a difficulty; but not so much greater than that of any other striking modification of structure ... I can see no very great difficulty in this being effected by natural selection' (Darwin 1859, p. 236). Indeed, elsewhere – as we shall see when we discuss interspecific sterility – Darwin carefully distinguishes the sterility of the social insects from sterility in non-social organisms as being relatively unproblematic (Darwin 1868, ii, pp. 186–7). And so he decides to 'pass over this preliminary difficulty' (Darwin 1859, p. 236). No, what worries him is that the neuter castes are so very different both from their parents and from one another; how has natural selection been able to work on these individuals, who cannot reproduce, to create such *diverse* characteristics?

> one special difficulty ... at first appeared to me insuperable, and actually fatal to my whole theory ... [The] neuters often differ widely in instinct and in structure from both the males and fertile females, and yet, from being sterile, they cannot propagate their kind ... [And] the climax of the difficulty [is] ... the fact that the neuters of several ants differ, not only from the fertile females and males, but from each other, sometimes to an almost incredible degree ...   (Darwin 1859, pp. 236–8)

So the problem posed by sterile workers is their extraordinary difference in structure and instincts from other members of the species; if only they bore a closer resemblance to their parents and to one another there would be no serious anomaly. His difficulty, then, is the latency of the sterile workers' characteristics in parents that differ so much from them. This was certainly a problem, given the lack of an adequate theory of heredity. But it is hardly what would now be thought of as the 'climax' of the difficulty of sterile insects.

Although for Darwin the anomaly was not the sterile workers' altruism, commentators today commonly take for granted that it was – so commonly that I shall give some additional evidence for my view that this is mistaken and that for him latency was the paramount problem (a view, I am gratified to say, that is also held by Hamilton (1972, p. 193)).

First, there is a telling analogy that Darwin inserts into the fourth (1866) and subsequent editions of the *Origin*. He likens his explanation of the evolution of neuter workers to Wallace's explanation of certain butterfly species occurring in two or even three distinct female forms and to Müller's explanation of two distinct male forms in certain crustaceans (Peckham 1959, pp. 420–1). In these cases there is no sterility, no sacrifice, no altruism. But in Darwin's view the proliferation of distinct forms makes these cases 'equally complex' and the explanations 'analogous' (although unfortunately he doesn't elaborate any further) (Peckham 1959, p. 420). Second, when he summarises his theory in the final chapter of the *Origin* and picks out the sterile workers as 'one of the most curious' 'cases of special difficulty' (Darwin 1859, p. 460), it is the differentiation of castes that he mentions. Third, if Darwin thought that self-sacrifice for the benefit of others was a salient feature of the sterile workers' behaviour he would surely have mentioned it when discussing human sociality and morality in *Descent of*

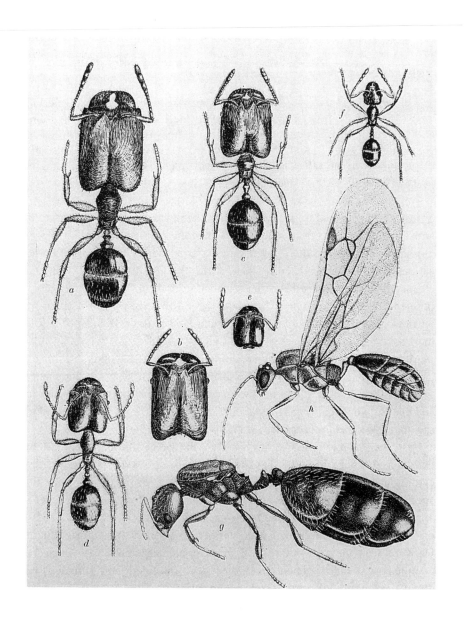

Man; but, although he deals there with proto-morality in other animals, he barely touches on the social insects at all.

Darwin's problem, then, was not the same as that of today's Darwinians. Nevertheless, he had to work out what natural selection was up to, since it couldn't be acting through the differential reproduction of the sterile ants. So, whilst answering his question, he was obliged to answer the questions posed by altruism: Who benefits? And how?

Darwin's solution involves two stages. He begins by making the point that, although the sterile insects are unable to breed, nevertheless their close relatives are likely to share heritable characteristics with them, so the relatives can pass them on, even if the characteristics aren't manifested in the relatives:

> This difficulty [of sterility in insects], though appearing insuperable, is lessened, or, as I believe, disappears, when it is remembered that *selection may be applied to the family, as well as to the individual*, and may thus gain the desired end. Thus, a well-flavoured vegetable is cooked, and the individual is destroyed; but the horticulturist sows seeds of the same stock, and confidently expects to get nearly the same variety; breeders of cattle wish the flesh and fat to be well marbled together; the animal has been slaughtered, but the breeder goes with confidence to the same family ... (Darwin 1859, pp. 237–8; my emphasis)

Strictly, these examples do not quite capture Darwin's point. (I'm admittedly being a bit fussy but we need to pin down exactly what he is saying; we shall see later that others have read all kinds of theories into Darwin's statements.) He is trying to explain cases in which some members of a 'family' produce offspring with phenotypic characteristics that are latent in themselves whereas other members who do manifest the characteristics are incapable of producing offspring. But in Darwin's analogies the fertile members would, in the normal course of things (even if not at the time when the breeder selects them), manifest the same phenotypic characteristics as the cooked and the slaughtered. Darwin's next example is more appropriate: 'a breed of cattle, always yielding oxen [castrated bulls] with extraordinarily long horns, could

---

'By far the most serious special difficulty'

Pheidole kingi instabilis, a small myrmicine harvesting ant of Texas: The worker caste, made up of continuously varying subcastes, from the major worker (a), to media workers (b–d), to the minor worker (e, f); the queen (g); and the male (h).

"One special difficulty at first appeared to me ... fatal to my whole theory ... the fact that the neuters of several ants differ, not only from the fertile females and males, but from each other, sometimes to an almost incredible degree." (Darwin: Origin)

be slowly formed by carefully watching which individual bulls and cows, when matched, produced oxen with the longest horns; and yet no one ox could ever have propagated its kind' (Darwin 1859, p. 238). But he could have sharpened the analogy if he had explicitly stated that the parents of the oxen with extraordinarily long horns were not themselves phenotypically extraordinarily long-horned. In later editions of the *Origin* he came up with the following 'better and real illustration' (Peckham 1959, p. 416) – and it really is a pleasing one. Some varieties of stock produce double as well as single flowers but the double flowers are always sterile; nevertheless the line does not go extinct because it continues to produce the single, fertile stock. Darwin aptly likens the single stock to the fertile relatives and the double stock to the sterile workers (Peckham 1959, p. 417). In sum, then, Darwin is saying that relatives have characteristics in common (whether manifest or latent) and that an individual's manifested characteristics may be latent in its relatives but perpetuated through their germ line alone. A passage in *Descent of Man* supports the view that he is saying nothing more than that. Darwin states that in a human tribe, even if the more 'ingenious' members do not leave offspring, their characteristics might be passed on by other members of the tribe because they are related – and he makes the same reference to the cattle breeder: 'Even if they left no children, the tribe would still include their blood-relations; and it has been ascertained by agriculturalists that by preserving and breeding from the family of an animal, which when slaughtered was found to be valuable, the desired character has been obtained' (Darwin 1871, i, p. 161).

The second stage of Darwin's argument makes the point that selection has been able to act on the characteristics of the sterile insects because these very characteristics affect the success of fertile relatives: 'by the long-continued selection of the fertile parents which produced most neuters with the profitable modification, all the neuters ultimately came to have the desired character' (Darwin 1859, p. 239). By analogy, the cattle breeder chooses which bulls and cows to match using as a guide not their own horns but those he most admires in the ox; and the horticulturalist chooses which single flowers to cultivate using as a guide those of the double variety that take his fancy.

Darwin's general line of thought is clear. (The detailed workings of the mechanism are vague but he did the best that could be done without an adequate theory of heredity.) His idea is twofold. First, even though the sterile insects do not reproduce, their characteristics can be reproduced by others. Second, selection can act on the sterile insects' characteristics through insects in whom they are latent because the characteristics affect the success of those insects. So the continuity of the germ line is maintained through the fertile members of the community, even though it reaches a dead end with the sterile workers, because their characteristics affect the success of their fertile relatives and so can be shaped by natural selection.

So far, so good. But unfortunately this account of the second stage of Darwin's argument, which sounds so like our modern view, is idealised. Confusingly, he also says that the selective advantage is to the community:

> a slight modification of structure, or instinct, correlated with the sterile condition of certain members of the community, has been *advantageous to the community*: consequently the fertile males and females of the same community flourished, and transmitted to their fertile offspring a tendency to produce sterile members having the same modification ... We can see how useful their production may have been to a *social community of insects*, on the same principle that the division of labour is useful to civilised man. (Darwin 1859, pp. 238, 241–2; my emphases)

Similarly, he says that natural selection, by working on the parents, could have produced other forms, such as uniformly small sterile workers or just two highly divergent castes, if they had been useful to the *community* (Darwin 1859, pp. 240–1). And elsewhere he states: 'With sterile neuter insects we have reason to believe that modifications in their structure and fertility have been slowly accumulated by natural selection, from an advantage having been thus indirectly given to the *community* to which they belonged over other *communities* of the same species' (Darwin 1868, ii, pp. 186–7; my emphases). In *Descent of Man*, he goes so far as to cite the social insects as a prime example of natural selection acting on characteristics that benefit the social group but not their bearers:

> With strictly social animals, natural selection sometimes acts indirectly on the individual, through the preservation of variations which are beneficial only to the community ... [M]any remarkable structures, which are of little or no service to the individual or its own offspring, such as the pollen-collecting apparatus, or the sting of the worker-bee, or the great jaws of soldier-ants, have been thus acquired. (Darwin 1871, i, p. 155)

But Darwin's words should not be taken as a sign that he is intentionally adopting a higher-level explanation. He frequently switches between individual- and community-language, apparently feeling free to use them interchangeably. In the fourth (1866) edition of the *Origin*, for example, he omits the reference to the community from the following passage, so that it refers only to the parents, whereas previously it had read: '... the extreme forms, from being the most useful to the *community*, having been produced in greater and greater numbers through the natural selection of the *parents* which generated them' (Peckham 1959, p. 420; my emphasis). But this is not the recantation of a former group-selectionist who has now seen the individual-selectionist light, for in other cases his alterations go exactly the opposite way. In the fifth (1869) edition he changes the following passage: '... by the long-continued selection of the fertile *parents* which produced most neuters with the profitable modification ...' (Peckham 1959, p. 418; my emphasis) to read: '... by the survival of the *communities* with females which produce most neuters ...' (Peckham 1959, p. 418; my emphasis) (although admittedly this is ambiguous; he could be referring to selection on the females). Darwin's level-blind revisions are not confined to the social insects. Discussing natural selection in the *Origin*, he says in the first edition: 'In social animals it will adapt the structure of each individual for the benefit of the community; if *each* in consequence profits by the selected change' (Darwin 1859, p. 87; my emphasis). Yet in the sixth (1872) edition this passage reads: '... for the benefit of the whole community; if the *community* profits by the selected change' (Peckham 1959, p. 172; my emphasis). Darwin, then, seems blithely indifferent to how he expresses himself; he moves back and forth between the language of two levels, apparently without differentiating between them.

So what exactly was Darwin saying? Let's first deal with what he has been thought to have said. Here we find little consensus and enormous confusion. The confusion stems in part from the misidentification of his problem. Bear in mind that, with hindsight, most commentators assume that Darwin's discussion is about the problem of altruistic sterility. So one of the few points on which most are (mistakenly) agreed is that 'the apparent altruism of neuter insects ... seem[s] out of line with ... the struggle for existence. Darwin fully realised the importance of this problem' (Ghiselin 1974, p. 216).

But does Darwin offer an individual- or a higher-level solution? And does it succeed or fail? Here there are even more views than there are commentators. A few examples will be more than enough; I'll confine them to a paragraph (albeit a dauntingly long one!). We'll start in the nineteenth century, with Weismann. He praised Darwin's contribution, apparently without noticing that on Weismann's own account it appeals to selection on two different entities, the parents and the community. Darwin's explanation of the origin of neuter ants, he says,

> must still be regarded as the only possible one – namely, that they arose through *selection of the parents* ... [A] selection of the fruitful females must have taken place, inasmuch as females which produced sterile offspring in addition to fruitful issue were of *special value to the state*; for the existence of members that were workers only was a gain to it and strengthened it ... [A]ll the variations among the workers arose, to make them more fit to be of *service to the state*.
> (Weismann 1893, p. 314; my emphases, original emphasis omitted)

Other commentators have had stronger views on Darwin's level of explanation. Phillip Sloan (1981, p. 623), for example, takes Michael Ruse (1979a) to task for claiming that Darwin is an individual selectionist; this is unconvincing, he says, in the face of Darwin's discussion of the social insects. Turning to Ruse we find that he does view Darwin as an individual selectionist but that he himself claims that Darwin 'saw a whole hive of bees as one individual, rather than seeing individual hive members as competing rivals' (1979a, p. 217). Similarly, he claims elsewhere that Darwin saw the case of the sterile workers as problematic because he was a thoroughgoing individual selectionist (Ruse 1982, p. 190); nevertheless he interprets Darwin's solution as applying to the 'supra-individual' and explicitly contrasts this with what he takes to be Hamilton's individualistic explanation (Ruse 1982, pp. 193, 205). (However, in a more detailed defence of Darwin's individual selectionism (Ruse 1980), he makes no mention of Darwin's 'supra-individualism'. Ruse, by the way, does at one point (1980, pp. 618–19) make the distinction between the problem of altruism and that of the enormous differences between the sterile castes and their parents.) Michael Ghiselin, too, says approvingly that for Darwin selective advantage was always to the individual – and then allows social units to count as individuals (Ghiselin 1969, p. 150). Ghiselin seemed at one time to be under the impression that Darwin's is a kin-selectionist explanation (Ghiselin 1969, p. 58) but subsequently he contrasted the mechanism of kin selection with what he described as Darwin's mechanism of selective advantage to the family as a unit (Ghiselin 1974, p. 137). What is more, Ghiselin objected to families being treated as superorganisms (Ghiselin 1974, p. 218) but nevertheless preferred to treat insect societies as 'integrated wholes' rather than accept kin selection (Ghiselin 1974, pp. 137, 228–33). Elliott Sober (1984, pp. 218–19, 1985, p. 895) also claims that Darwin gives an individual-selectionist account, the individual that benefits being a parent who adopts the reproductive strategy of producing some sterile offspring; Darwin's 'group selectionist phrase "profitable to the community"' he dismisses as 'a verbal slip' (Sober 1984, p. 219) (although Sober asks whether there isn't anyway only a terminological difference between individual and group selection – perhaps because he confusingly interprets 'group' in this context as meaning 'kin group' (Sober 1985, p. 895)). By contrast, Alfred Emerson (1958) specifically praises Darwin for positing a 'supraorganismic social unit' and says that Darwin recognised 'the necessity of treating the societal system as an entity' (Emerson 1958, p. 315). Arthur Caplan, however, criticises Darwin for adopting a group-selectionist solution (Caplan 1981) and Bowler, too, seems to regret that Darwin was 'forced back on a kind of group selection in this case' (Bowler 1984, p. 312). E. O. Wilson (1975) at one point (p. 117) states that Darwin's solution is kin-selectionist and, indeed, goes on to discuss Hamilton's classic kin-selection solution in the same breath (p. 118). But Wilson habitually uses the expression 'kin selection' to mean 'group selection involving only family members' (Wilson 1975, pp. 106, 117–18) so it is possible that he, too, intends to characterise Darwin's solution as group-selectionist. And indeed he does claim that Darwin introduced the concept of group selection to account for the sterile castes (Wilson 1975, p. 106). Either way, Wilson seems to see no inconsistency in his conclusions, for he unequivocally praises Darwin's solution for its 'impeccable logic' (Wilson 1975, p. 117). Ruse assumes that Wilson interprets Darwin as adopting a 'supra-individual' theory and appears to agree with Wilson's interpretation (Ruse 1979a, p. 217); but elsewhere he argues strenuously against the claim that Darwin's solution is group-selectionist (Ruse 1980, pp. 618–19) (although his notion of group selection is also unclear). Robert Richards, too, understands Darwin's solution to be kin-selectionist but does not indicate why and, moreover, equates it with 'community selection' (Richards 1981, p. 225).

I have said that some of this confusion stems from Darwin's problem being misidentified. But some of the responsibility surely lies with Darwin's own ambiguity (though he is not responsible for confusions over group and kin selection!). At times Darwin reads like a group selectionist, at times like an individual selectionist and at times even like an individual selectionist with a kin-selection solution. But it is wrong to attribute any of these views to him unconditionally. Sadly, it seems that there can be no definitive answer as to what Darwin really had in mind. His position certainly comes tantalisingly close to a kin-selection solution. Kin selection explains the evolution of altruistic characteristics by the fact that a gene for altruism can spread because it enhances its own replication through its effects on close relatives. Darwin's problem was the difficulty of latent characteristics. So he laid great stress on two points. First, the sterile workers' characteristics can be reproduced through the germ line of their fertile relatives in whom they are latent. Second, those characteristics increase the reproductive success of the relatives. But this does not amount to a clear-cut kin-selection explanation. We cannot ignore the fact that Darwin paid little attention to the sterile workers' altruism, even if his solution turns out to be relevant to that problem. More important, we cannot ignore the fact that Darwin drags in the idea of benefit to the community (or, at least, drags in community-level language); community benefit, even when that community is a family unit, definitely has no part in a kin-selection explanation. His references to the community equally undermine the claim that he was unambiguously a thoroughgoing individual selectionist. Nevertheless, it is also mistaken to hold that he unequivocally adopted a group-selectionist view. It is not clear what he envisaged by benefit to the community. But in the light of his apparently cavalier revisions such talk certainly cannot be taken as amounting to a fully-fledged higher-level explanation. Perhaps, presumably unwittingly, he conflated these two entirely different kinds of explanation.

There remains the question of why Darwin didn't see the workers' altruism as being at least as important a problem as that of latent characteristics, even though such extreme self-sacrifice, such total commitment to the welfare of others, seems most unDarwinian. Of course, if he was indeed cushioning his individual-level view with an appeal to higher-level benefits then the altruism would be happily absorbed into the good of the community; altruistic sterility would present a difficulty, but a difficulty that would rapidly fade in the light of a greater good. But we cannot assume that Darwin did resort to a higher-level explanation.

Fortunately, however, we don't have to decide what level of explanation he saw himself as proposing in order to explain why he 'passed over' the workers' sterility as merely 'a preliminary difficulty' (Darwin 1859, p. 236). The reason is one that we have already met and one that we shall meet again: Darwin viewed altruistic behaviour as relatively unproblematic in general for individuals in highly social communities. We saw him saying, for example, in the quotation from *Descent of Man*, that 'strictly social animals' might bear characteristics that are 'beneficial only to the community'. And we noted that he saw sterility as being really problematic only when it occurred in non-social organisms. How exactly he saw natural selection working in social groups, we don't know. But it was clearly the insects' well developed social structure that enabled him to face their altruism with relative equanimity.

There is a final point to bear in mind about Darwin's views on altruistic characteristics. As we saw earlier, Darwin's interest in social adaptations is strongly influenced by his interest in human descent. Because of this, he examines sociality under the heading of 'moral sense', as one of the mental faculties that is common to humans and other animals. This leads him to think of altruism in terms of kindliness rather than costs, in terms of well-intentioned acts rather than behaviour that is a disadvantage to its bearer but an advantage to others. In this context it is not surprising that the ways of the ant are passed over as an unpromising source of moral sensibility in favour of those of the higher social animals (Darwin 1871, i, p. 74). Indeed, ironically, Darwin uses the social insects to illustrate the very opposite of niceness. He is defending his claim that our moral sense arises from our social instincts combined with intelligence. He points out that different social instincts would give rise to different moralities. Our moral code would be utterly transformed if, for example, we lived like hive-bees. But, far from being admirable, our practices would, by our present standards, be contemptible: 'unmarried females would ... think it a sacred duty to kill their brothers, and mothers would strive to kill their fertile daughters; and no one would think of interfering' (Darwin 1871, i, p. 73). And these same social insects are also uniquely nasty in their 'unusual ... feeling of hatred between the nearest relations, as with the worker-bees which kill their brother-drones, and with the queen-bees which kill their daughter-queens' (Darwin 1871, i, p. 81).

We have seen Darwin's views on the social insects and we have seen what Darwinians know now. Why was so little progress made in the intervening period, even after the lead given by Fisher and Haldane? The answer is that on this issue, above all, Darwinian thinking was dominated by greater-goodism. So entrenched was this view that the social insects' altruism, far from being seen as problematic, was seen as blending seamlessly into the social organisation of the whole community – and that organisation, of course, was seen as good.

Consider, for example, *The Social Insects*, a standard text of the 'greater-good' decades by O. W. Richards, an entomologist whom we have already come across (Richards 1953). This book has the added interest that according to Wynne-Edwards it is a precursor of his own group selectionism (Wynne-Edwards 1962, p. 21). Richards certainly assumes that selection acts on the community as a whole. But it turns out that this is not because he sees altruistic sterility as a problem. On the contrary, his stress on the good of the community obscures the problem. For Richards, sterility is just one among several adaptations that social insects have developed in the cause of cooperation. He groups them all together even though many of the other characteristics bear no trace of apparent altruism and certainly do not involve what is apparently the ultimate Darwinian sacrifice – forgoing reproduction. He suggests that sterility has evolved as a check on population for the good of the group: 'The problem of too rapid multiplication in the social insects [was met by establishing] ... a sterile caste which either did not breed at all or did so only to a limited extent and under certain conditions' (Richards 1953, p. 194). No mention of the disadvantages to the sterile! Indeed, he seems to see the group as a buffer against individual disadvantage: 'In a solitary species, individuals with reduced fertility will not often survive in competition with others which are more fertile. But in social species any change which benefits the group as a whole is likely to be preserved' (Richards 1953, p. 202). And here is that sentiment writ large:

There is ... [a] process which operates, not only in the ant colony but in any social animal. The unit whose efficiency determines whether the species shall survive or become extinct is the colony rather than the individual. An individual which is useful to the colony may survive though it would be quickly eliminated in a solitary species. This happens to a considerable extent in man. In civilised societies, many members are supported who contribute only very indirectly to the provision of the food and shelter necessary for life. Others who contribute nothing are enabled to survive because our social behaviour benefits the whole species and not merely the bread-winners. In an ant-colony there is an analogous situation: the worker which is sterile, or capable only of producing male offspring, is a good example of a type which could not survive apart from the colony. Some of the more fantastic types of soldier ants seem to be an even more extreme example of the same thing. They might be described as freaks for whom, during the evolutionary process, a use has been found, just as the circus has found a use for dwarfs.

(Richards 1953, pp. 145–6)

That passage makes little sense. Many social adaptations, unlike sterility, are not in the least problematic. Neither is sterility similar to bearing only males or being 'fantastic' or 'freakish' (whatever that may be). Socially 'parasitic' humans are not analogous to sterile workers; if anything, the analogue of worker castes would be the 'breadwinners' that support the parasites. Even when Richards draws parallels between the behaviour of the social insects and human morality (Richards 1953, pp. 205–6) he shows no awareness that altruism in either group poses a problem.

Greater-goodism went so far that it became common to regard a colony of social insects as a single organism – not figuratively but literally. William Morton Wheeler, Professor of Entomology at Harvard and author of several well-known books on the social insects, held that 'the personal organism ... is the prototype ... [but colonies are also] real organisms and not merely conceptual constructions or analogies' (Wheeler 1911, p. 309; see also Wheeler 1928, pp. 23–4). On this view, altruism, far from being problematic, is naturally expected. After all, if the community is really a single individual, 'altruism' is nothing more than specialisation of function. It becomes no more reasonable to ask why the sterile workers nurture others than to ask why the heart pumps for the good of the rest of the body. (This, remember, was before the days when organisms were viewed as vehicles of selfish genes; nowadays one might ask that question even of the heart.) An insect colony is not a mob of potentially conflicting interests but a well-integrated whole, with 'correlation and coöperation of parts ... and the resulting physiological division of labor' (Wheeler 1911, pp. 324–5). This single-organism model was particularly attractive to critics of red-in-tooth-and-claw theories. According to Wheeler, it was a mistake of 'aggressive, individualistic' Darwinism to see cooperative behaviour as problematic at all (Wheeler 1928, p. 5); on his view of social insects, 'our attention is arrested not so much by the struggle for existence, which used to be painted in such lurid colors'

(Wheeler 1911, p. 325) as by the colony functioning biologically as a single entity. In the same spirit, Emerson said: 'Like the organism, the group unit exhibits analogous division of labor, integration, development, growth, reproduction, homeostasis, ecological orientation, and adjustment. The term supraorganism seems amply justified for the insect society' (Emerson 1958, p. 330). The views of Wheeler and Emerson were typical of the period during which higher-level explanations were given free rein. For Darwinians of this stripe, altruistic sterility presented no problem.

In the light of modern knowledge it may seem that classical Darwinism could not have got further than Darwin in recognising the problem of altruism in social insects or in solving it because heredity in general, and the relationships in the insect communities in particular, were not adequately understood. But this is unnecessarily generous. One doesn't need sophisticated insights to see that, for example, suicidal stinging presented some kind of problem. Neither does one need Crick and Watson to solve the problem; Mendel is sufficient. The greater-goodists even went backwards from Darwin's own analysis. For Darwin, the problem of 'levels of selection' just wasn't much of an issue; but it didn't need to be – he was on the right track. Greater-goodists, however, were only too much aware of higher levels – and it took them straight up the garden path. Hamilton's observation on group selection applies equally to this case: 'Until the advent of Mendelism uncritical [failure to see the problem] ... could be understood partly on grounds of vagueness about the hereditary process ... But in the event neither the rediscovery of Mendel's work nor the fairly brisk incorporation of Mendelism into evolutionary theory had much effect' (Hamilton 1975, p. 135). Our debt is all the greater to Hamilton himself.

14

Make dove, not war: Conventional forces

It was all very well for Darwin's contemporaries to bridle at his explanation of peacocks' tails. But why didn't they do the same for the other half of his theory of sexual selection? Darwin claimed that male rivalry, not female choice this time but direct competition among males, could also explain the evolution of horns, claws and muscles, of spurs, combs and ruffs, of fighting, roaring and staring. His claim was regarded as uncontroversial, the acceptable (albeit less beautiful) face of sexual selection. After all, aren't mighty battles, ferocious clashes, just what one would expect between rival males? And aren't weapons and armour needed anyway for other purposes, for cornering a food cache or staking out territory, for pinning down prey or fending off predators?

Well, yes, they are. And that is precisely the problem. Because some of the 'fights' that Darwin described look more like posturing than trials of strength, and some of the 'weapons' more ornamental than deadly. It transpires that combat between males of the same species can sometimes be rather kidgloved:

Although wild boars fight desperately together, they seldom ... receive fatal blows, as these fall on each other's tusks, or on the layer of gristly skin covering the shoulder, which the German hunters call the shield ...

The male baboon of the Cape of Good Hope ... has a much longer mane ... than the female ... [It] probably serves as a protection, for on asking the keepers in the Zoological Gardens, without giving them any clue to my object, whether any of the monkeys especially attacked each other by the nape of the neck, I was answered that this was not the case, excepting with the above baboon.

(Darwin 1871, ii, pp. 263, 267)

In some species of fish, the males fight by seizing the jaws of their opponents – the least effective place, for this is the one part of the body that is protected by a leathery skin. In many species of snakes, males wrestle with each other rather than unleash their deadly fangs. Indeed, these encounters can be even more gentlemanly, the struggle concluded and the victor declared without any physical contact, the whole thing resolved just by a bristling of the fur, an imperturbable gaze, an insistent growl.

Let me hastily say that such politeness by no means always prevails among rival suitors. Indeed, conflicts between males of the same species can be more nasty, more brutish than encounters with members of other species. Darwin clearly documents that for the males of many species – he singles out mammals – 'the season of love' heralds serious injury and fights to the death (Darwin 1871, ii, pp. 239–68).

> Two male hares have been seen to fight together until one was killed; male moles often fight, and sometimes with fatal results; male squirrels 'engage in frequent contests, and often wound each other severely'; as do male beavers, so that 'hardly a skin is without scars' ... The courage and the desperate conflicts of stags have often been described; their skeletons have been found ... with the horns inextricably locked together, shewing how miserably the victor and vanquished had perished. No animal in the world is so dangerous as an elephant in must.
>
> (Darwin 1871, ii, pp. 239–40)

When John Maynard Smith and George Price first published their explanation of ritualised combat (which we're about to come to) it drew a sharp response from Valerius Geist, who had devoted many hours to observing male mammals butt and pummel one another: 'The article ... perpetuates an old ethological myth that animals fight so as not to injure each other, or refuse to strike "foul blows" and, presumably, kill each other ... [But] field studies primarily of large mammals ... have shown ... how dangerous combat is' (Geist 1974, p. 354).

Certainly, combat can be dangerous. But conventional aggression is no myth. And, however much or little it occurs, it raises a serious problem. Why on earth are contestants so restrained? Why do they sing or strut when they could maim or kill? Why do they hold back when they could slaughter? If everyone else is fool enough to obey such rules, why don't individuals break them, bluffing and cheating or going all-out for a quick victory? And why such restraint between members of the same species, where surely rivalry is most intense?

It was the theory of games, the theory of evolutionarily stable strategies, that held the answer. This is what John Maynard Smith and George Price demonstrated in the pioneering paper that provoked Geist (Maynard Smith and Price 1973; see also Maynard Smith 1972, pp. 8–28, 1974, 1976b, 1982; Maynard Smith and Parker 1976; Parker 1974). ESS theory reminds us that it is not enough to snatch a quick victory in a single encounter. What matters to natural selection is whether a strategy is evolutionarily stable. And that involves a very special condition. Any strategy that is successful will end up,

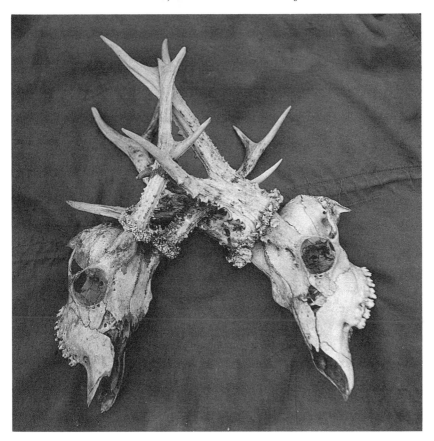

Victims of the rut

"The courage and the desperate conflicts of stags have often been described; their skeletons have been found in various parts of the world, with the horns inextricably locked together, shewing how miserably the victor and vanquished had perished." (Darwin: The Descent of Man)

When deer fight, the antlers occasionally get so entangled that the opponents cannot free themselves. These roe deer stags were found dead in Invernesshire. Such events are rare. The naturalist and former deer-stalker who took this photograph told me that he had heard of only three instances in red deer in his 45 years of experience.

over evolutionary time, encountering itself more than it encounters any other strategy. So if it is to be evolutionarily stable against invasion, it must be able to do better against itself than any other strategy does against it.

We must think, then, not just about a single encounter, nor even about all of a male's encounters over his lifetime, but about the career of a strategy over evolutionary time. From that perspective, things begin to look different.

Imagine a pugnacious bully, throwing his weight about, always ready for a fight, always ready to pursue it to the bitter end; his rival is a coward, sloping off at the first sign of trouble, avoiding a punch-up at all costs. The bully will clearly do better in any particular encounter. But is bullying likely to be evolutionarily stable? Remember that we aren't talking about a particular bullying individual. We are talking about a strategy acting out its bullying role in many different individuals over many generations. Successful strategies will come to be represented in the population in proportion to their success. So eventually any bully will encounter other bullies more often than he encounters cowards. And when the bullying strategy encounters itself, costs will be greater and victory less assured. Bullying may no longer pay. We can see, then, that a strategy of all-out fighting for instant gains may well not be evolutionarily stable. And we can begin to see why, under a range of conditions, conventional combat may well be. To go further, we need to look more closely at the notion of an ESS.

An evolutionarily stable strategy may not be a straightforward single best strategy, a so-called pure strategy, but a mixture of different strategies. A pure strategy can be thought of as a rule of the form: In situation A, always do X – say, bully. In a mixed strategy, the rule is probabilistic: In situation A, do X (bully) with probability p and do Y (be a coward) with probability q. A mixed ESS can be realised in two ways. Everyone in the population could follow the same probabilistic rule, varying their behaviour (during an encounter or from one encounter to another) according to the rule; so everyone would be sometimes a bully and sometimes a coward. Or each individual's behaviour could be fixed, with the frequencies of the different kinds of individuals corresponding to the probabilities in the rule; the population would consist of a proportion p of bullies and a proportion q of cowards. So a mixed ESS amounts to an evolutionarily stable state of critical proportions of different strategies. The proportions will come to be such that, on average, followers of each strategy do equally well. If the proportions aren't fair in that sense then natural selection will balance things up until they are. If there are too many bullies, cowardice is favoured; if too many cowards, bullies prosper.

Evolutionary games can involve any number of players. In some applications of ESS theory, we often think in terms of several players. When we're applying the theory to fighting in particular, we are generally thinking in terms of two-person games.

An ESS, whether pure or mixed, can be conditional. A conditional strategy can be thought of as a rule with an 'if' statement in it: If hungry then bully, if well-fed be a coward. There are good theoretical reasons for thinking that most strategies are likely to be conditional. To see why, we need to make a final distinction.

It is useful to divide games into symmetrical and asymmetrical. Again, this is particularly relevant for games that are to do with combat. The asymmetry could lie in the fighting ability of the contestants (so-called RHP – resource-holding power) or in the value of the resource to them. This could give rise to conditional rules like 'If the larger of the two, bully; if the smaller, be a coward' or 'If it's the last chance to get a mate, bully; if there are plenty of other opportunities, be a coward'. Alternatively, the asymmetry could lie in a purely conventional difference, owing nothing to RHP or differential payoffs – a so-called uncorrelated asymmetry. It could be, say, the asymmetry between owner and latecomer, between a contestant that already just happens to have the mate or food or territory and a contestant that now wants it. In general, the ESS in an asymmetric contest is to let the asymmetry settle the contest with a minimum of escalation. For correlated asymmetries this is intuitively obvious, as long as the contestants can assess what the asymmetry is. If, for example, they can judge their relative strengths without actually fighting, then they could 'agree' on the winner without coming to blows. But for an arbitrary asymmetry it is less obvious. And yet, in theory, the contestants could use even an absurdly arbitrary asymmetry, like 'If you are the northernmost of the contestants, bully, if the southernmost be cowardly'. Why would natural selection favour such an odd rule? Remember that an ESS is defined as a strategy that is uninvadable once it is in a majority. Suppose that, for whatever reason, a majority happens to have formed following the 'North bullies South' strategy. Then most contests are quickly settled, because all individuals 'agree' about who is nearer the north. Anybody that departs from the majority convention has a serious and damaging fight with everybody he meets. Conversely, if the opposite convention, 'South bullies North', had happened to find itself in the majority then it, too, would have been stable. Admittedly, 'North' and 'South' are not very plausible. 'First there', on the other hand, is. The strategy of conventionally allowing mere ownership to settle contests can be an ESS against any strategy that ignores ownership. We can see now why most strategies are likely to be conditional. Natural selection seizes on asymmetries, and nature offers them in abundance.

Now to flesh out those abstract categories. Imagine a summer wood, dappled sunlight spattering the woodland floor. On each patch of sunlight, moving as the sun moves, there is a male speckled wood butterfly (*Pararge aegeria*). Above, in the canopy of leaves, other males are patrolling. The males below are defending a useful resource, which those in the canopy aspire to usurp: a male in a patch courts more females than a male in the canopy. When a rival flies past a patch, the territorial male flies up in defence to meet it. The two of them spiral skywards briefly, then the rival flies off and the owner settles back on his territory. N. B. Davies followed the encounters of male speckled wood butterflies and found, remarkably, that time after time this story was the same: the resident always wins (Davies 1978). What is going on? The males are clearly engaged in an asymmetric game, in which ownership settles contests without escalation. But why do owners win? Obviously strength or some other 'real' asymmetry could be settling the contest. But the butterflies could be using ownership alone as a conventional cue. Davies discovered that they do indeed seem to be observing such a convention. When he netted an owner, then let another male establish himself on that patch, and then released the original owner, the new owner always retained the territory and the spiral flight took no longer than usual; even a few seconds priority was enough to establish ownership. If Davies then removed the new owner from the patch and let the first occupant repossess it, again it was the current occupant that kept it, however short the occupation. So what would happen if both butterflies 'thought' that they were the true owner? ESS theory predicts that their normally brief encounters should escalate dramatically; an asymmetry cannot settle a contest if it is ambiguous. Davies managed to trick both members of pairs of butterflies into simultaneously perceiving themselves to be the owner. The prediction was triumphantly confirmed. In the absence of the unequivocal cue, the spiral flights lasted on average ten times longer – 40 seconds, instead of the usual three or four.

Red deer (*Cervus elaphus*) favour less arbitrary ways of settling things. On the Scottish island of Rhum, during the rut, competition among the deer becomes intense, when harem-holding stags are challenged by other mature males. The encounter begins with a challenger approaching to within about 200 to 300 yards and the two rivals roaring at each other for several minutes; at this point the challenger usually withdraws. If he doesn't, the pair proceed to walk up and down, tensely, in parallel. If the challenger still persists, the two stags lock antlers and push vigorously until one of them is thrown rapidly backwards and runs off; if he should have the misfortune to fall, his opponent attacks him viciously. T. H. Clutton-Brock and his colleagues found that the stags were using this series of conventions to prevent escalation and that the cues that they were employing were non-arbitrary (Clutton-Brock and Albon 1979; Clutton-Brock *et al.* 1979, 1982, pp. 128–39). Fighting is exhausting and dangerous; serious, occasionally even fatal, injury is likely. What's more, a harem-holder is liable to have his harem infiltrated whilst he is fighting. So it is better for both sides if escalation can be kept to a minimum. The cues that the stags use to size each other up are direct indicators of resource-holding power – size, strength and so on. Roaring rate, for example, is a sensitive test because it is heavily dependent on the stag's condition. Each stage of the joint ritual conveys more information than the one before. But each also has higher potential costs. So the stags move from one stage to the next only when assessment fails them and they need to probe further. Significantly, the rare occasions on which fights were not preceded by roaring or walks were when an asymmetry was too obvious to need careful assessment or when an intruder had taken over a harem in the holder's absence.

By the time that two stags have obeyed the conventions and reached the point of locking antlers to pit their strength against each other, they are reasonably well matched. Now they are largely engaged in a war of attrition, a war in which the winner will be the one that is prepared to continue for longer than his opponent. The more the contest goes on, the more the costs mount (and there's the added danger that a false move could end in serious injury). The choice of strategy open to each opponent amounts to the staying time that he is prepared to endure. Natural selection will see to it that no contestant continues for longer than the resource is worth to him. And his strategy must, of course, be unpredictable – otherwise his opponent would adopt the strategy of going just that bit longer. In a war of attrition, then, no pure strategy (no fixed time) can be an ESS; the evolutionarily stable strategy will always be mixed.

It may seem odd for an animal that is not fighting-fit to advertise the fact. The reason is that a strategy of honest advertisement is stable whereas the conditional strategy 'Don't advertise if resource-holding power is below a certain level' is not. Consider a male whose RHP is just that bit lower than any male would want to admit. Suppose that he decides not to advertise at all. In the absence of further information, his opponents will reasonably assume that he is near the average of the below-the-critical-RHP-level group. So, because he is in fact above average for that group, he will do better by advertising and being assessed at his true worth. But now the critical advertising level is lower. Selection will favour males that advertise when they are only slightly below *that* – and so on downwards until all males, however feeble, are compelled to advertise. Honest advertisement is the ESS.

Modern Darwinism has been able to explain how encounters can be conventional: ESSs are refereeing. How did Darwinism deal with the problem for its first hundred years? In short, it didn't. Classical Darwinism largely failed to acknowledge the 'altruism' that is involved, the apparent anomaly of even strong, healthy males at the height of the mating season refusing to act the bully.

We left Darwin describing ferocious fights among all kinds of mammals. Such encounters, of course, require no special explanation. They are to be expected in the struggle for reproduction. And, as Darwin points out, the specially developed weapons can also be turned against enemies of other

Conventional but not arbitrary

Three stages of escalation between red deer

First, a roaring contest ...

... then a parallel walk ...

... and, as a last resort, a fight: antlers locked in a trial of strength.

species (Darwin 1871, ii, p. 243): 'The elephant uses his tusks in attacking the tiger ... The common bull defends the herd with his horns; and the elk in Sweden has been known ... to strike a wolf dead with a single blow of his great horns' (Darwin, 1871, ii, pp. 248–9). What's more, horns can be turned into ploughshares – and into various other tools:

> The elephant ... scores the trunks of trees until they can be easily thrown down, and he likewise thus extracts the farinaceous cores of palms; in Africa he often uses one tusk, this being always the same, to probe the ground and thus to ascertain whether it will bear his weight ... One of the most curious secondary uses to which the horns of any animal are occasionally put, is [by the wild-goat of the Himalayas and the ibex] ... namely, that when the male accidentally falls from a height he bends inward his head, and, by alighting on his massive horns, breaks the shock.
>
> (Darwin, 1871, ii, pp. 248–9)

Weapons that are so useful present no problems; on the contrary, 'it is a surprising fact that they are so poorly developed or quite absent in the females' (Darwin, 1871, ii, p. 243).

But are there not horns and tusks and antlers that are too elaborate, too baroque to be efficient weapons? Darwin admits that there are. 'With stags of many kinds the branching of the horns offers a curious case of difficulty; for certainly a single straight point would inflict a much more serious wound than several diverging points ... [One observer] actually came to the conclusion that their horns were more injurious than useful to them!' (Darwin 1871, ii, pp. 252–3). Darwin won't go that far: 'this author overlooks the pitched battles between rival males' (Darwin 1871, ii, p. 253); he points out that they use the upper antlers for pushing and fencing and, in some species, for attack. Nevertheless, he does agree that 'Although the horns of stags are efficient weapons, there can ... be no doubt that a single point would have been much more dangerous than a branched antler ... Nor do the branching horns ... appear perfectly well adapted for [fighting rival stags] ... as they are liable to become interlocked' (Darwin 1871, ii, p. 254). No, says Darwin, these magnificent structures cannot be entirely utilitarian; surely they are also out to impress. They are doing double duty for sexual selection:

> The suspicion has ... crossed my mind that they may serve in part as ornaments. That the branched antlers of stags, as well as the elegant lyrated horns of certain antelopes, with their graceful double curvature, are ornamental in our eyes, no one will dispute. If, then, the horns, like the splendid accoutrements of the knights of old, add to the noble appearance of stags and antelopes, they may have been modified partly for this purpose, though mainly for actual service in battle ...
>
> (Darwin 1871, ii, pp. 254–5)

Let's concede Darwin a measure of ornament. Even so, there is still some apparently puzzling forbearance to be explained. We have seen Darwin describing how wild boars 'fight desperately together' and yet 'seldom receive fatal blows' because they confine their attacks to specially protected areas; and how the only monkeys that attack one another on the nape of the neck are also the only ones that have a protective mane. Why do these males go for the armour when they could go for the jugular? Why do they seem to obey the Queensberry rules rather than the law of the jungle? Darwin presumes that contests were less gentlemanly in the past, and that is why the boar evolved his shield and the baboon his mane. For him, the problem ends there. But to a modern Darwinian, that is where the problem really begins. The convention of attacking only the shield or mane must have evolved along with those defences. How did it come about and how is it maintained? On such questions, Darwin is silent. The reason is familiar. Darwin, it seems, did not appreciate the apparent costs of holding back, of letting rivals go first, of failing to hit below the belt.

Nevertheless, because he envisaged at least some weapons as partly ornamental, Darwin did not consider male combat to be unremittingly red-in-tooth-and-claw. Most Darwinians, however, did. They were anxious to drive a wedge between Darwin's two kinds of sexual selection, female choice and male combat. The former they wanted to deny; the latter they wanted to assimilate smoothly into the struggle for existence. Wallace, for example, stressed that the outcome of male rivalry was exactly what natural selection would anyway favour: 'there necessarily results a form of natural selection which increases the vigour and fighting power of the male animal, since, in every case, the weaker are either killed, wounded, or driven away ... It is evidently a real power in nature; and to it we must impute the development of the exceptional strength, size, and activity of the male, together with the possession of special offensive and defensive weapons' (Wallace 1889, pp. 282–3). From this perspective, conventions, conciliations, concessions pass into the shadows, known but unseen. Those luxuriant antlers, for example, that set Darwin musing about the knights of old are, for Wallace, clear evidence of natural selection's down-to-earth preference for 'stronger or better armed males' and 'vigour and offensive weapons' (Wallace 1889, p. 282).

Wallace thought of his view as no-nonsense natural selection, with males rearing for all-out victory (and, as a bonus, Darwin's fanciful ornaments cut down to size). But even without a sophisticated ESS analysis it is not difficult for a modern Darwinian to see that a male might well do better by showing some restraint. After all, a policy of let-rip could be very costly. Even a strong male in his prime could have a lot to lose. Opportunity costs, for example: time and energy that he devotes to vanquishing rivals can't be devoted to catching prey or attracting mates. And then there's the fact that, however useful it is to have a rival out of the way, it's equally useful for his other rivals, and it's he that has paid the removal costs. What's more, if the animal that he is fighting already possesses the mate or territory that he wants, the possessor was presumably once a victor, so he is challenging a former champion. In short, as always, the advantages must be set against the costs. Wallace's failure to see all this is perhaps not surprising. He and his contemporaries failed to see the costs of conventions. Failing to see the costs of combat is just the other side of that same coin.

Gradually, as 'good-for-the-species' thinking began to permeate Darwinism, conventional combat shed its invisibility. 'Ritualization ... has been very important' said Julian Huxley 'in reducing intra-specific damage, by ensuring that threat can ensure victory without actual fighting, or by ritualizing combat itself into what Lorenz calls a tournament ... [T]ournament fights provide maximum damage-reduction' (Huxley 1966, pp. 251–2). Indeed, ritualised combat came to play a starring role in greater-goodism. What better evidence that natural selection works for the good of the species than that two hefty rivals, capable of tearing one another limb from limb, choose to settle matters peaceably, with a nod and a grunt?

This line of thinking culminated in the 1960s with Konrad Lorenz's book *On Aggression* (1966). 'Though occasionally, in territorial or rival fights, by some mishap a horn may penetrate an eye or a tooth an artery, we have never found that the aim of aggression was the extermination of fellow-members of the species concerned' (Lorenz 1966, p. 38). By contrast, aggression towards other species is no-holds-barred. Or so, at least, Lorenz seems at times to be telling us. And he has certainly been widely criticised for taking a group- or species-level view (e.g. Ghiselin 1974, p. 139; Kummer 1978, pp. 33–5; Maynard Smith 1972, pp. 10–11, 26–7; Ruse 1979, pp. 22–3). But, if his critics discern so clear a message in his murky pronouncements, they are too kind. Although he constantly talks of natural selection acting 'for the good of the species', it is difficult to know what he is really saying. Sometimes he seems to mean no more than straightforward individual advantage: '"What for" ... simply asks what function the organ or character under discussion performs in the interests of the survival of the species. If we ask "What does a cat have sharp, curved claws for?" ... [we can] answer simply by saying, "To catch mice with"' (Lorenz 1966, p. 9). So here 'the survival of the species' refers to nothing but individual selection. But does Lorenz mean the same thing when he asks about aggression within a species or has he switched to species-level advantage?: 'What is the significance of all this fighting? In nature, fighting is such an ever-present process, its behaviour mechanisms and weapons are so highly developed and have so obviously arisen under the selection pressure of a species-preserving function that it is our duty to ask this Darwinian question' (Lorenz 1966, p. 17). His answer is no clearer than his question: when 'animals of different species fight against each other ... every one of the fighters gains an obvious advantage by its behaviour or, at least, in the interests of preserving the species it "ought to" gain one. But intra-specific aggression ... also fulfills a species-preserving function' (Lorenz 1966, p. 22). Is he contrasting, on the one hand, individual-level selection in fights between species with, on the other hand, species-level selection in fights within species (though he refers to the preservation of the *species* in both cases)? Lorenz's Darwinism is so confused that it is impossible to tell what exactly he did have in mind – and one begins to suspect that, if challenged, he himself would not have been able to say. We have seen how greater-goodism failed to grasp the problems posed by altruism and failed to recognise any unorthodoxy in appealing to a higher level. With Lorenz, we are looking at one of greater-goodism's blithest practitioners.

It is to Wynne-Edwards that one must turn both for an explicit recognition that conventional combat poses a problem and for an explicit attempt to explain it by group selection:

the wholesale wounding and killing of members by one another is generally damaging to the group and has consequently been suppressed by natural selection ... [A]ny immediate advantage accruing to the individual by killing and thus disposing of his rivals for ever must in the long run be overridden by the prejudicial effect of continuous bloodshed on the survival of the group as a whole ... [C]onventions ... have evolved to safeguard the general welfare and survival of the society, especially against the antisocial, subversive self-advancement of the individual.

(Wynne-Edwards 1962, pp. 130–1)

At least one knows where he stands, even if it is resolutely in the wrong place.

A mere twenty miles or so separates the islands of Bali and Lombok, just beyond the eastern tip of Java. Wallace discovered to his astonishment that to traverse those few miles was to step from Asia to Australia, to cross from one creation to another. When I read Lorenz's *On Aggression* in the late 1960s, I put down the book disappointed, bewildered and confused. If this was Darwinism's understanding of conventional conflict, the theory was sadly lacking. Just a few years later, I read *The Selfish Gene*, where I came across Maynard Smith and Price's ESS analysis of the same problem. Here was a different world. Darwinism had entered a new era. By the time I put down that book, I had crossed my own Wallace line.

15

Human altruism: A natural kind?

Man's inhumanity to man may indeed make countless thousand mourn. But it is man's *humanity* that gives Darwinians pause. Darwinians were slow to detect altruism in industrious ants and ritualised aggression. Human morality, however, presents an obvious challenge to Darwinian theory. And, from the very first, Darwinians tried to meet it. We'll look at the diverse responses of four nineteenth-century evolutionists: three leading Darwinians – Darwin himself, Wallace and T. H. Huxley – and Herbert Spencer, only part-Darwinian but an enormously influential thinker. This small group covers a wide spectrum of Darwinian stances on human nature. We'll examine modern parallels and contrasts along the way.

Darwin: Morality as natural history

Let us begin with Darwin. For him, natural selection was not only part of the problem; it was also the solution. Human morality, he urged, should be explained in just the same way as the hand or the eye, as an adaptation, a product of natural selection: 'morals and politics would be very interesting if discussed like any branch of natural history' (Darwin, F. 1887, iii, p. 99). 'This great question' – the origin of our moral sense – 'has been discussed by many writers of consummate ability; [but] ... no one has approached it exclusively from the side of natural history' (Darwin 1871, i, p. 71). No one, that is, until Darwin himself, in *Descent of Man* (Darwin 1871, i, pp. 70–106, 161–7).

Darwin set about his task, as he did throughout *Descent of Man*, by looking for continuities between us and other animals. He wanted to find in them some incipient moral sense, some feeling for others, that would form a link with what we know as morality, a link with our highly developed conscience, our sense of duty, our willingness to die for a cause. It was to what he called 'the social instincts' that he turned. Social behaviour, he says, brings with it the first stirrings of morality because it demands a concern for others as well as oneself: 'the so-called moral sense is aboriginally derived from the social instincts, for both relate at first exclusively to the community' (Darwin 1871, i, p. 97). To this add intelligence and the result is full-fledged morality: 'any animal whatever, endowed with well-marked social instincts, would inevitably acquire a moral sense or conscience, as soon as its intellectual powers had become as well developed, or nearly as well developed, as in man' (Darwin 1871, i, pp. 71–2). Other animals go so far as to act as sentinels, groom one another, hunt communally. The roots of our morality lie in social acts such as these. It is to our intellect that we owe what more we have, our codes of ethics and justice, our finely-tuned sense of principle.

So here, at last, we have Darwin explicitly acknowledging a problematic case of altruism and, what's more, systematically documenting evidence of 'altruistic' behaviour in other animals (behaviour that would be regarded as altruistic if performed by humans). And yet even this does not bring him to generalise beyond human morality to the wider problem of altruism in the Darwinian sense. How can Darwin record instance after instance of apparent self-sacrifice and yet miss its significance for his own theory? Why did he not go on to ask how natural selection tolerates such self-abnegation? We shall examine one explanation in a moment. For now, it is enough to remind ourselves of the familiar reason why altruism went unappreciated for so long: Darwin's analysis of social living in other animals is predictably rich in suggestions about selective benefits but predictably poor in identifying costs. Take this passage, for example:

> The most common service which the higher animals perform for each other, is the warning each other of danger by means of the united senses of all ... Many birds and some mammals post sentinels, which in the case of seals are said generally to be the females. The leader of a troop of monkeys acts as the sentinel, and utters cries expressive both of danger and of safety.   (Darwin 1871, i, p. 74)

Darwin describes those poor, lone sentinels as enjoying mutual aid. But, on the contrary, they seem to be bearing the burden for the whole group.

Moving on to humans, however, Darwin does recognise that here there may be real self-sacrifice; moral considerations are likely to clash with our selfish interests, even overriding our bid for self-preservation. And he fully acknowledges that this poses a problem for his theory. How, he asks, could altruism possibly arise through natural selection? How did we evolve from being merely social to being moral (Darwin 1871, i, pp. 161–7)? His analysis (albeit confined to humans) begins promisingly.

Darwin starts by considering competition between groups. If a group that has a high proportion of unselfishly devoted members comes into conflict with a group that has a high proportion of selfish members, it is easy to see that the group of altruists will triumph. Their discipline, fidelity, courage and other such qualities will soon ensure victory (Darwin 1871, i, pp. 162–3). But the nub of the problem is to explain how altruistic groups got that way and how they stayed that way. How did altruism ever get off the ground in the first place and how did it grow and prosper?: 'how within the limits of the same tribe did a large number of members first become endowed with these social and moral qualities, and how was the standard of excellence raised?' (Darwin 1871, i, p. 163). Unselfish members would not have the most offspring – quite the contrary:

It is extremely doubtful whether the offspring of the more sympathetic and benevolent parents, or of those which were the most faithful to their comrades, would be reared in greater number than the children of selfish and treacherous parents of the same tribe. He who was ready to sacrifice his life ... rather than betray his comrades, would often leave no offspring to inherit his noble nature. The bravest men, who were always willing to come to the front in war, and who freely risked their lives for others, would on an average perish in larger number than other men.

(Darwin 1871, i, p. 163)

He concedes that the problem looks almost intractable: 'Therefore it seems scarcely possible ... that the number of men gifted with such virtues, or that the standard of their excellence, could be increased through natural selection, that is, by the survival of the fittest' (Darwin 1871, i, p. 163).

Darwin sees two ways out of the difficulty. One is reciprocal altruism: 'each man would soon learn that if he aided his fellow-men, he would commonly receive aid in return' (Darwin 1871, i, p. 163). But when Darwin turns to his other solution, he lets us down. He seems to suggest that individual sacrifice for the sake of the group can evolve because it pays off in competition between groups:

It must not be forgotten that although a high standard of morality gives but a slight or no advantage to each individual man and his children over the other men of the same tribe, yet that an advancement in the standard of morality and an increase in the number of well-endowed men will certainly give an immense advantage to one tribe over another. There can be no doubt that a tribe including many members who, from possessing in a high degree the spirit of patriotism, fidelity, obedience, courage, and sympathy, were always ready to give aid to each other and to sacrifice themselves for the common good, would be victorious over most other tribes; and this would be natural selection.   (Darwin 1871, i, p. 166)

This passage is puzzling. Darwin has specifically said that he is now tackling the problem of how altruism gets established *within* the group; he takes care to remind us 'that we are not here speaking of one tribe being victorious over another' (Darwin 1871, i, p. 163). And yet he seems to be speaking of just that. What's more, uncharacteristically, he seems quite explicitly to be offering a higher-level solution. And yet he fails to suggest any mechanism to deal with the problems that he himself has rightly posed: how self-sacrificial behaviour gets established within the group, how it is developed and how maintained. It is hard to guess what he had in mind here. Reluctantly, I feel one must conclude (as Hamilton does (1972, p. 193, 1975, p. 134)) that when Darwin dealt with human altruism, he saw the problem, he discussed it, but he nevertheless left it unsolved.

All the same, Darwin's analysis raises several points that are worth exploring. Consider first how he actually undertook his task of tracing the legacy of natural selection on our moral nature. He didn't do it in the style in which today's Darwinians typically go about it. But his approach could turn out to be a fruitful way for us to study ourselves. It is what we might nowadays characterise (well, caricature really) as 'psychological' rather than 'ethological' or 'sociological'. Darwin is interested in our emotions rather than our actions. Whereas the majority of today's Darwinian investigations of human nature might look at the incidence of homosexual behaviour, comparative divorce rates, social hierarchies, aggressive encounters, family relationships, Darwin was more interested in feelings, in feelings of love and hate, of jealousy and generosity, of pride and shame, of resentment and gratitude, of sympathy and spite. Let's look at why he adopted this way of doing things and what it can offer.

Darwin stumbled upon his method inadvertently. We saw when we examined classical Darwinism that Darwin's preoccupation with human descent and his search for continuities deflected him from looking at behaviour and led him to concentrate instead on the mental states that accompanied it. In searching for a precursor of human morality, he was less interested in the social behaviour of other animals than in their social instincts, less interested in the selective costs and benefits of what they did than in how they felt about it. We'll see in a moment that this didn't do much for his understanding of other animals, particularly when he was dealing with altruism. But in the case of humans it could prove to be a good way of going about things.

This is because there is a problem in treating human behaviour as just another adaptation. The problem arises from our unnatural environment. Most of the time, most of our genes express themselves phenotypically pretty much as natural selection intended. Although we're no longer on the savannah where selection shaped our upright gait, our visual acuity, our manual dexterity, nevertheless our genes for plantigrade feet, colour vision, opposable thumbs, express themselves phenotypically as they were designed to do. Such genes are not much perturbed by the vast difference between where we began and where we are now. But for some of our genes our modern environment is likely to metamorphose their phenotypic expression from that which natural selection originally smiled upon. And genes for behaviour are the most prominent among these. An animal that is adapted to dwell nomadically in smallish bands, to sleep when night falls, to gather and hunt, may find its bodily structure largely undisturbed by a world of crowded cities, electric lights and food that is ready-foraged (and even partly predigested!). Much of that animal's behaviour, however, is likely to change beyond recognition.

There is, of course, nothing surprising about phenotypes venturing outside natural selection's expectations. There is no such thing as 'the' phenotypic effect of any gene. Phenotypes are always the result of an interaction between gene and environment. We have learnt, at tragic cost, that even genes for plantigrade feet and opposable thumbs will not express themselves as expected in the environment of a womb that has been exposed to some of the inventions of the drug industry. Sadly, too, it is possible that genes for conserving energy efficiently in conditions of scarcity may express themselves as diabetes when their bearers take to a modern, Western diet. Or consider the evolutionarily puzzling behaviour of homosexuality. It could be an adaptation, as some have suggested (e.g. Trivers 1974, p. 261; Wilson 1975, p. 555, 1978, pp. 142–7), or pathology, as most of the medical profession long maintained. But (Ridley and Dawkins 1981, pp. 32–3) if there are 'genes for homosexuality' they could be genes that, in our Pleistocene environment, which differed from our modern world in some crucial respect (say, sleeping always with parents rather than alone), would have expressed themselves as something quite different, perhaps a useful ability to pick up the scent of prey or to shin fast up high trees. The details of that fanciful example should not, of course, be taken seriously. But the model of how we need to think of phenotypic expression should.

So the problem of phenotypes that are far from natural selection's intentions is not peculiar to humans nor to behaviour. But it is in that conjunction that the problem is most acute. And the reason is obvious. Humans are not, like the poor digger wasp, condemned to a behavioural straitjacket: if *Sphex ichneumoneus* has to repair one completed step in her routine of provisioning her burrow, she then proceeds to perform the next step all over again – forty times in one experiment (it was the experimenter who could bear it no longer) – unable to appreciate that she need only have taken up where she left off (Hofstadter 1982, pp. 529–32). Natural selection has not cast us in a sphexish mould. It has endowed us with enormous behavioural flexibility. This flexibility was an excellent tactic for evolution. But it makes things hard for the evolutionist, hard to find out what our evolutionary legacy is. When we look at someone praying or cheating or helping their neighbour or getting into a fight, are we seeing something close to a repertoire hallowed by ancestral use? Or is the behaviour, although generated by those same ancient rules, transformed into something quite alien by the environment in which the genes for those rules now find their phenotypic expression? We're familiar with stepping far outside natural selection's range when we intervene consciously – contraception, bottle-feeding, high-speed travel, wearing clothes, using spectacles. It's obvious in those cases that we're not doing what natural selection envisaged for us; indeed, at least with contraception, we're doing what it specifically meant us not to do. But how can we discover natural selection's designs in the less obvious cases?

Worse, our unnatural environment can also raise almost the opposite problem: that we behave too much as nature intended. Suppose that praying or helping looks irrelevant to Darwinian needs or even downright maladaptive. Should we dismiss Darwinian explanations? Or should we remember that behaviour that natural selection honed assiduously on the savannah could well look non-adaptive on city streets?

How, then, can we judge how we were evolved to behave? One standard way is as follows. Suppose that we wanted to know whether we are naturally monogamous or naturally polygamous, and whether men and women differ in their predispositions. We would do a comparative survey of an entire group – the primates, for instance – tracing which characteristics correlate with which mating systems. So, for example, among primate species generally the degree to which males are larger than females and the degree to which they mature later than females both correlate with intensity of polygyny (one male, more than one female) in that species. As we are mildly dimorphic in size and men mature somewhat later than women, our natural mating system, on this reasoning, is slightly towards polygyny (e.g. Daly and Wilson 1978, pp. 297–310, particularly pp. 297–8).

There is an obvious difficulty with this method, one we have already met in connection with testing theories of sexual selection. It is the familiar problem of the inflation of 'n', the problem of non-independence of data. What is to count as a unit? If a majority of sexually dimorphic primates are polygynous, is this a genuinely revealing correlation or have both attributes simply been inherited from a sexually dimorphic and polygynous common ancestor? Fortunately, as we have noted, this problem is – at least in principle – solvable.

A second standard method may be called the method of robust invariance. Are there some patterns of human behaviour that persevere in expressing themselves almost regardless of varying conditions? And by 'varying' I mean conditions that are different, perhaps vastly different, both from the

Pleistocene plain and between one culture and another today. Darwin thought that this was true of smiling, for example: 'With all the races of man the expression of good spirits [by smiling] appears to be the same' (Darwin 1872, p. 211). A century later, the Austrian ethologist Eibl-Eibesfeldt tested this assertion (Eibl-Eibesfeldt 1970, pp. 408–20). He clandestinely filmed people from a very wide variety of cultures. And he concluded that he could detect little difference in the pattern or circumstances of smiling; much of this similarity extended even to children born blind, who had never seen a smile to copy (Eibl-Eibesfeldt 1970, pp. 403–8). Possibly there were misinterpretations – cultural and very likely sexist biases – but he certainly found some common ground:

> To give just one example, we found agreement in the smallest detail in the flirting behavior of girls from Samoa, Papua, France, Japan, Africa (Turcana and other Nilotohamite tribes) and South American Indians (Waika, Orinoko).
> The flirting girl at first smiles at the person to whom it is directed and lifts her eyebrows with a quick, jerky movement upward so that the eye slit is briefly enlarged ... After this initial, obvious turning toward the person, in the flirt there follows a turning away. The head is turned to the side, sometimes bent toward the ground, the gaze is lowered and the eyelids are dropped. Frequently, but not always, the girl may cover her face with a hand and she may laugh or smile in embarrassment. She continues to look at the partner out of the corners of her eyes and sometimes vacillates between looking at him and an embarrassed looking away. We were able to elicit this behavior when girls observed us during our filming. While one of us operated the camera the other would nod toward the girl and smile.
> (Eibl-Eibesfeldt 1970, pp. 416–20)

'Wide agreement is also found in many other expressions. Thus arrogance and disdain are expressed by an upright posture, raising of the head, moving back, looking down, closed lips, exhaling through the nose – in other words through ritualized movements of turning away and rejection' (Eibl-Eibesfeldt 1970, p. 420).

More remarkably, when we look at murders, at who murders whom and why, we find an astonishingly steady pattern repeating itself down the centuries and across cultures (Daly and Wilson 1988, pp. 123–86, 1990). Murder rates differ over time: an Englishman today is only about one-twentieth as likely to die at the hands of an assassin as an Englishman seven centuries ago. And they differ from place to place: rates nowadays in Iceland are 0.5 homicides per million persons per annum, whereas in most of Europe they are 10, and in the United States over 100 (Daly and Wilson 1988, pp. 125, 275); murders in which victim and killer are of the same sex and unrelated vary from as few as 3.7 per million persons per annum in England and Wales (1977–86) to a hefty 216.3 in Detroit (1972) (Daly and Wilson

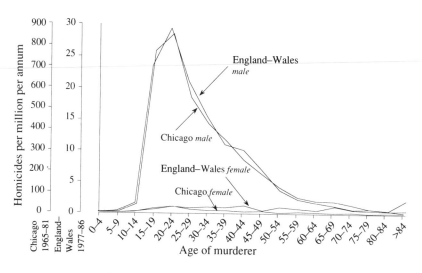

Who murders whom? A robust invariance

These are the age- and sex-specific rates of killing nonrelatives of one's own sex in England–Wales and Chicago over roughly the same period. Although the *absolute numbers* differ enormously, the *shapes* of the curves are astonishingly similar: these murders are overwhelmingly committed by men and overwhelmingly by young men.

1990). And yet go from thirteenth-century Oxford, to Miami in 1980, via Iceland (1946–70), the !Kung San in Botswana (1920–55), the Tzeltal Mayans in Mexico (1938–65), and many another society, from Australia to Germany to India to Africa, and the same pattern of homicide emerges (Daly and Wilson 1988, pp. 147–8). Murders are overwhelmingly committed by men – '*The difference between the sexes is immense, and it is universal.* There is no known human society in which the level of lethal violence among women even begins to approach that among men' (Daly and Wilson 1988, p. 146). And men, but not women, are triggered by what one sociologist has labelled an 'Altercation of relatively trivial origin; insult, curse, jostling, etc.' (Daly and Wilson 1988, p. 125); where homicide rates are high these altercations account for a high proportion of murders, so much so that they 'surely constitute a very large proportion of all the world's killings' (Daly and Wilson 1988, p. 126). What is more, these male murderers are overwhelmingly young – in their mid-20s; so, for example, in spite of the enormous difference that we noted between murder rates in England–Wales and Detroit, the median ages of males who killed unrelated males were 25 and 27 respectively; in Canada (1974–83) and Chicago (1965–81), where rates fell between those two extremes, the median ages were 26 and 24 (Daly and Wilson 1990, p. 93). Individually, these statistics might look like mere demographic contingencies. But put them together and we are faced with patterns that are invariant across huge cultural magnitude shifts: 'Overall homicide rates vary tremendously and can be conceived of as cultural, but the fact of a sex difference transcends cultural variation' (Daly and Wilson 1990, p. 88) – as does the age of the killers and their motive. Now, this doesn't tell us that natural selection's design is for young men to kill for apparently trivial motives, nor even to kill at all. But it does suggest that we are on natural selection's trail.

By the way, this conclusion is not undermined by the fact that most men are not murderers. Certainly they are not. But most murderers are men. And it is the robust cultural invariance of that sex difference that requires explanation. Equally, this line of reasoning is not affected by the fact that women in the most violent societies are more likely to commit murder than are men in the least violent. For even in the most violent societies, men murder more than women do. And, again, this is what we need to explain.

Clearly, the method of 'robust invariance' is also beset by the problem of the inflation of '*n*'. Consider again the question of discovering our own ancestral mating pattern. Of 849 human societies tabulated in Murdock's *Ethnographic Atlas*, 708 are polygynous, 137 are monogamous and 4 polyandrous (see Daly and Wilson 1978, p. 282). A point, it seems, in favour of polygyny as the primitive human mating system. But if 700 of those 708 polygynous societies took their mores from the Koran, we would have there one datum not 700. The robust invariance would come from a book not from genes. We'll return later to the example of murder to illustrate one way out of this problem.

A third standard method is to look not for invariance in human behaviour, nor for patterns of resemblance between humans and other animals, but for adaptively revealing differences within the human species. Take, for example, the widespread social system known as the 'avunculate' or 'mother's brother effect', in which the 'father role' is taken on not by the mother's husband but by her brother. This seems at first sight to challenge our ideas about kin selection. Indeed, Richard Alexander, the influential American zoologist, says that this was one of 'the two most prominently used arguments against a biological explanation of kinship systems' (Alexander 1979, p. 152). Alexander conjectured that, on the contrary, it is a kin-selected adaptation. In societies where promiscuity makes biological fatherhood uncertain, males can be more confident of their genetic relatedness to their sisters' offspring than to their 'own' children. Alexander tested his idea by comparing promiscuous societies with monogamous ones, predicting that the maternal uncle effect would be more prevalent in promiscuous societies. And he found some evidence in favour of the prediction (Alexander 1979, pp. 152, 168–75).

Darwin's 'psychological' method offers a different solution. If we want to know what natural selection intended us to do, we are just as likely to find the answer in how it equipped us for making behavioural responses as in the responses themselves. Following Darwin's inadvertent lead, we could study our emotions as well as our acts, study our brains as well as our behaviour. Our behavioural repertoire, built for flexibility, is very likely to be distorted by our highly unnatural environment; our emotional, motivational and cognitive repertoire, built for generating appropriate behaviour, perhaps less so. Maybe, then, in the case of human evolution, we could bypass the distortions that unnatural environments wreak on our behaviour by going directly to a study of the psychological mechanisms that bring it about.

It is certainly perfectly plausible to assume, as this approach does, that natural selection endowed us with a specific psychological makeup in order to promote non-specific adaptive behaviour. Natural selection shaped our brains just as it shaped our hands and eyes and other organs. And, even more than with the hand or the eye, the brain could incorporate highly specialised capacities in order to respond appropriately to a wide variety of situations. We can imagine how this might work from what we know of human language. The philosopher Jerry Fodor quotes a striking remark of one of his colleagues: 'What you have to remember about parsing ... is that basically it's a reflex' (Fodor 1983, p. vi). But a flexible reflex. Although we are not born speaking English or Chinese, we are born with a capacity that is both sufficiently highly structured and at the same time sufficiently open-ended to learn not only those languages but a host of others. The story is much the same with our ability to recognise human faces (which natural selection seems to have valued, judging by the large area of our brain that it dedicated to the task). This is a highly specific adaptation. But it enables us to recognise an astonishing number of people (far more in our modern environment than natural selection could have bargained for); and more reliably than with most other cues ('I remember your face but not your name'); and to do so even on very minimal information (a fuzzy photograph depicting a blob in a crowd). The point about minimal information is important. We act on our specialised rules in the light of accumulated information from the past and the very latest update. But this information will often be incomplete and one task of the rules is to help us to act adaptively in the face of uncertainties. In short, then, we are familiar with the idea that natural selection's legacy includes specific, specialised, psychological machinery that is designed to generate plastic, flexible – and thereby adaptive – behavioural responses, even on the basis of incomplete knowledge. Natural selection gives us the rules, and we finish the job.

Having been led to this way of thinking by reading Darwin, I was gratified to find that several modern Darwinians who are actively working in the field had converged on Darwin's approach. Among the names to look out for are Martin Daly and Margo Wilson, Leda Cosmides and John Tooby, and Donald Symons (see e.g. Barkow 1984; Cosmides 1989; Cosmides and Tooby 1987, 1989; Daly 1989; Daly and Wilson 1984, 1988, 1988a, 1989, 1990; Rozin 1976; Shepard 1987; Symons 1979, 1980, 1987, 1989, 1992; Tooby and Cosmides 1989, 1989a, 1989b; Trivers 1971, pp. 47–54, 1983, pp. 1196–8). I am not suggesting that this is necessarily always the best way of understanding ourselves from a Darwinian point of view. But it undoubtedly promises to be a fertile method, well worth exploring. Let's now get a more concrete idea of what this approach can offer by looking at two recent attempts to apply it – two very different attempts but both in the spirit of Darwin's 'psychological' method.

Darwin mentioned the kind of psychological responses that we might examine if we are interested in ourselves as social, and specifically moral, beings: it is not 'probable that the primitive conscience would reproach a man for injuring his enemy: rather it would reproach him, if he had not revenged himself' (Darwin 1871, 2nd edn., pp. 172–3, n27); 'the praise and blame of our fellow-men' and the 'love of approbation and the dread of infamy' is a 'powerful stimulus to the development of the social virtues' (Darwin 1871, i, p. 164). Nowadays, at least when it comes to altruism, we have more precise ideas about the responses that we might look for. This is an area in which we can now draw on quite detailed models. We know, for example, that we are likely, certainly to some extent, to have evolved as reciprocators. And we know that reciprocal altruism is not evolutionarily stable unless most cheating doesn't pay. So we should expect to find sensitive mechanisms for detecting cheats, for revealing non-reciprocation if (as is likely) the information is incomplete; and we should expect these mechanisms to operate without our having to apply them consciously.

Such propensities have indeed been searched for – and perhaps found. This was the work of Leda Cosmides (Cosmides 1989; Cosmides and Tooby 1989). The story is slightly complicated but worth telling. People tend to make certain systematic logical errors, and Cosmides suspected that the direction of those errors could well be revealing. Just as psychologists have used visual illusions to uncover the rules of the normal workings of the brain, or errors in grammar acquisition to decipher natural selection's linguistic signature, her idea was to exploit logical errors to uncover deeply built-in social propensities. Experimental psychologists have long known that our reasoning powers are affected by the content and not merely the logical structure of arguments. This shows up in people's responses to the so-called Wason selection task, a test of logical reasoning in which people are asked to determine whether a conditional rule has been violated (see e.g. Wason 1983). With some rules, a high proportion of people respond illogically, picking out irrelevant conditions and failing to pin down the relevant ones. Some rules, but not all. A change in the content of the rules can transform the results dramatically. Some subject matters elicit a high percentage of logical answers. This is known as the 'content effect'.

Consider, for example, the following problem:

Part of your new clerical job at the local school is to make sure that student documents have been processed correctly. Your job is to ensure that the documents conform to the following alphanumeric rule:

If a person has a 'D' rating, then his document must be marked code '3'

You suspect that the secretary whom you replaced did not categorise the documents correctly. The cards below have information about the documents of four students at the school, each card representing one person. One side of a card states a person's letter rating and the other side of the card states that person's number code.

Indicate only those card(s) that you definitely need to turn over to see if the documents of any of these people violate this rule.

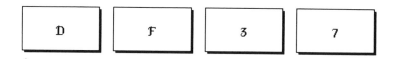

What you are being asked to do, then, is to decide, in the absence of complete information, whether a conditional rule has been violated in each of the four cases. What you should do – the logically correct answer – is to turn over only two cards: *D* and *7*. The reasoning behind this is as follows. The conditional rule can be expressed as 'If P (*D* rating) then Q (code number *3*)'. The only condition that violates that rule is 'P and not-Q' (*D* rating but not-*3*-code). So the only situations that you need to follow up are 'P' (to check that it is Q) and 'not-Q' (to check that it is not-P). That amounts to following up any *D* rating (to check that it is *3*) and any not-*3*-code (to check that it is not-*D*). You can ignore 'not-P' (not-*D*-rating) and 'Q' (*3* code). There is no potential violation of the rule in those cases so they need not concern you. The logic of the problem, then, looks like this:

If a person has a 'D' rating, then his document must be marked code '3'
[If P then Q]

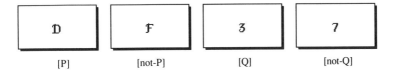

Well, if you would have turned over both *D* and *7* and nothing but *D* and *7*, then you are unusual. People generally perform poorly on a test like this. Typically, only between 4% and 10% see that 'P and not-Q', and only that, violates the rule. Most overlook the relevance of *7* (not-Q) and choose *D* (P) and *3* (Q) or *D* (P) alone (e.g. Wason 1983, pp. 46, 53).

Now consider another problem involving a conditional rule:

In its crackdown against drunken drivers, Massachusetts law enforcement officials are revoking liquor licences left and right. You are a bouncer in a Boston bar, and you'll lose your job unless you enforce the following law:

If a person is drinking beer, then he must be over 20 years old

The cards below have information about four people sitting at a table in your bar. Each card represents one person. One side of a card states what a person is drinking and the other side of the card tells that person's age.

Indicate only those card(s) that you definitely need to turn over to see if any of these people is breaking this law.

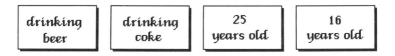

The logic of enforcing the rule is, of course, exactly the same. Deductive logic doesn't change when the content of an argument changes. It looks like this:

If a person is drinking beer, then he must be over 20 years old
[If P then Q]

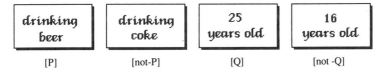

So, again, the only cards that you need to check are 'P' and 'not-Q' – in this case, 'drinking beer' and 'being under 20 years old'. It turns out that when people are charged with enforcing this rule, they perform markedly better.

They seem to be far more logical. The proportion who see that only 'P and not-Q' violates the rule typically shoots up to 75%.

Why this difference? Why are people's reasoning powers apparently so far superior on the 'under-age drinking' kind of test? Psychologists have looked for some systematic bias in this content effect, some property that the subject matter of the rules has in common. And they have generally assumed that it must have something to do with people's previous experience.

But Leda Cosmides suspected that the solution to the puzzle might lie not in individual experience but in our ancestral experience, our Darwinian propensities. She examined the rules that evoked a content effect and concluded that they were almost always to do with social exchanges. According to her analysis, they had the structure:

> If you take the benefit, then you pay the cost
> [If P then Q]

This is the structure of a social contract, a contract that relates perceived benefits (rationed goods that are valued by the recipient) to perceived costs. Cosmides conjectured that there is a good adaptive reason why we perform relatively well when enforcing conditional rules of this kind. We are drawing on responses built into us by natural selection. Selection has given us the means to behave as reciprocal altruists in just the same way as it has given us the means to run or breathe or reproduce. If reciprocal altruism is to evolve, to be established and maintained, then we need certain specific skills, skills for regulating social contracts. For one thing, we must have a way of assessing costs and benefits so that, for any individual, the costs do not, on average, exceed the benefits. We must also be capable of remembering who has cheated so that we can retaliate; this is perhaps one reason why natural selection has taken such trouble to ensure that we recognise human faces – ensuring that 'the defecting individual not be lost in an anonymous sea of others' (Axelrod and Hamilton 1981, p. 1395). Moreover, we must be able to detect cheats. And this, Cosmides hypothesised, explains why people are so much better at applying a conditional rule when it involves a social contract than when it has nothing to do with social exchange. People are operating a search-for-cheats procedure. This is why they seize on both the 'P' and 'not-Q' conditions. Potentially, either of them could involve taking the benefit and not paying the cost – cheating! It is as if people are primed to be alert to cases in which others take benefits but don't reciprocate. They are apparently all set to pounce on anyone who has taken the benefit, P (to see whether they have shelled out the cost) and anyone who has not paid the cost, not-Q (to see whether they have absconded with the benefit). So although people look as if they are being more logical, the impression is mistaken. What they are actually doing is policing social contracts. They are using the adaptive rules of mutual cooperation, not the logician's rules of the propositional calculus. It just happens that in situations like the bar bouncer's job the policing rule coincides with the logical rule. In both cases 'P and not-Q' is the situation to be alert to. This convergence is, however, merely accidental.

So what would happen if the two did not coincide? If Cosmides' conjecture is right, then people should come up with a 'look for cheats' response when enforcing social contracts even if that response is not sanctioned by formal logic. And that, Cosmides concluded from her experiments, is what they indeed tend to do. She made up conditional rules that had the structure of a 'switched social contract':

> If you pay the cost, then you take the benefit

(Switching the position of the contractual terms in the 'if-then' structure of a standard social contract transforms it into a switched one and vice versa.) Imagine, for example, a society in which cassava root is a rationed benefit that must be earned and having a tattoo is the cost or requirement that earns it. A standard social contract would be:

> If a man eats cassava root, then he must have a tattoo on his face
> (If a man takes the benefit, then he pays the cost)

A switched social contract would be:

> If a man has a tattoo on his face, then he may eat cassava root
> (If a man pays the cost, then he takes the benefit)

The cost–benefit structure of the rules, or lack of it for non-social-contract rules, was supplied by the story in which the rule was embedded. So, for example, in the social contract version of the cassava–tattoo rule, the tale was that the scarce cassava root was a powerful aphrodisiac in a society in which only married men were tattooed and sexual relations between unmarried people met with deep disapproval. In the non-social-contract version, the tale was that cassava roots just happened to grow exclusively in the area where the tattooed men just happened to live. In some experiments the social contract rules were expressed explicitly in ethical terms (such as 'If a man eats cassava root, then he must have a tattoo on his face'), in other experiments they were not ('If a man eats cassava root, then he has a tattoo on his face'). But it turned out that people's responses were apparently not influenced by whether or not the appropriate 'musts' and 'mays' were made explicit. If the rule embodied a social contract, supplied by the story, then people appeared to supply 'musts' and 'mays' implicitly for themselves. Conversely, it transpired that people apparently did not treat a rule as a social contract just because it included the word 'must'; it also had to have the appropriate cost–benefit structure.

Although a switched rule has been reversed in its social aspects, the logical structure (If P then Q) is, of course, unchanged. The only condition under which the rule is violated is, as before, 'P and not-Q'. Suppose that you are charged with detecting violations of the switched rule. If you do what is logical, you pick out 'paying cost and not taking benefit' (P and not-Q); you ignore 'not paying cost and taking benefit' (not-P and Q). Well, if that sounds odd and counter-intuitive to you, that perhaps makes the point! For pure logic will lead you to ignore potential cheats. But if you are following a 'look for cheats' procedure, you will pounce instead on 'not-P and Q' (not paying costs and taking benefit). You will be using the same reasoning as for the standard social contract problem, picking out the same condition (not paying cost and taking benefit); but that condition has now changed its place in the logical structure. The look-for-cheats response on the switched contract (not-P and Q), unlike the standard contract, diverges from the logical response (P and not-Q).

Cosmides found that 'look for cheats' is overwhelmingly what people appeared to do. In experiments on the standard social contract, over 70% of subjects chose 'P and not-Q' (the same result as for the 'beer-drinking' standard social contract). The switched social contract produced dramatically different results. Only a tiny proportion – 4% in one experiment, none at all in another – got the logically correct answer: that 'P and not-Q' (paying costs and not taking benefit) was the sole condition that potentially violated the rule. If they were following a look-for-cheats procedure, this is the kind of response that one would expect. Obviously natural selection wouldn't tune our vigilance on behalf of others; a reciprocal altruist has no special need to ensure that others who pay costs receive their benefits. What is more, a high proportion – 67% in one experiment, 75% in another – responded illogically to the switched social contract problem, chasing after costs unpaid and benefit taken ('not-P and Q') even though that condition was utterly irrelevant to enforcing the rule. By contrast, this overwhelmingly popular 'not-P and Q' response was extremely rare when the problem was not a switched social contract. Out of several experiments – and they included standard social contracts and abstract problems like the *D* ratings and *3* code case – only one person ever chose 'not-P and Q' in response to a problem that was not a switched social contract.

So, according to Leda Cosmides, we can find a 'logic' behind these errors of reasoning, just as we can with persistent visual illusions. People are 'erring' systematically in the direction of detecting cheats. (Note that, unlike the case of visual illusions, part of the problem itself is to identify when people are 'erring' at all; after all, the laws of mutual cooperation prompt them to come up more reliably with logically correct answers than when they try to employ the laws of logic alone.) In the case of standard social contracts, logic and adaptive responses happen to coincide. In the case of switched social contracts, they do not. And in the case of conditional rules that are not to do with social contracts at all, we have to rely on our powers of reasoning alone. It is these differences that show up the rules behind people's errors and present us with a window into their minds. Natural selection, it seems, has endowed us with a propensity to pursue a search-for-cheats procedure because it is likely to be adaptively useful. Normally this gives the appearance of improving our logical prowess. Occasionally it diverges from what is logically justifiable; when people have to deal with switched social contracts, they don't perform very impressively as pure logicians although they are apparently reasoning very efficiently as reciprocal altruists who are evolved to detect and punish cheats in standard social contracts. In both cases, standard and switched, people are not thinking logically but they are, it seems, thinking adaptively – a triumph of morals over mind. If this conclusion is correct, it appears that the mind does have its reasons that reason does not know. And, what's more, that those reasons are adaptive.

If we have evolved machinery for running a system of reciprocal altruism, we might also expect to see the cultural (or even biological) emergence of means for keeping that machinery well-oiled. Robert Axelrod has investigated this possibility – not an empirical investigation of what we actually do but a computer simulation of how moral rules might develop in human societies (Axelrod 1986). His findings suggest that, if we are playing games involving cooperation, sanctions against defection and so on, then we should expect the emergence not only of norms that regulate our behaviour but also of 'metanorms'. Metanorms reinforce norms by making people willing to punish anyone that doesn't enforce them. He quotes a memorable example:

A little-lamented norm of once great strength was the practice of lynching to enforce white rule in the South. A particularly illuminating episode took place in Texas in 1930 after a black man was arrested for attacking a white woman. The mob was impatient, so they burned down the courthouse to kill the prisoner within. A witness said 'I heard a man right behind me remark of the fire, "Now ain't that a shame?". No sooner had the words left his mouth than someone knocked him down with a pop bottle. He was hit in the mouth and had several teeth broken.' This is one way to enforce a norm: punish those who do not support it. In other words, be vengeful, not only against the violators of the norm, but also against anyone who refuses to punish the defectors. This amounts to establishing a norm that one must punish those who do not punish a defection.   (Axelrod 1986, pp. 1100–1)

'Going meta-' is, indeed, a potent means of reinforcement: 'to iterate one's powers in this way, to apply whatever tricks one has to one's existing tricks, is a well-recognized breakthrough in many domains' (Dennett 1984, p. 29).

All this suggests that if we want to know whether we are evolved for reciprocal altruism, we could examine not only practices like the exchange of gifts but also propensities like detecting and punishing cheats. If we want to know the same for kin selection, we could study not only social relationships within families but also our unconscious skills at recognising close relatives. And if we want to know about monogamy and polygamy, we could compare not only mating patterns across cultures and species, but also precisely what triggers jealousy in men and in women. Like Darwin, we could focus not only on what we do, but on what our psychology suggests we are designed to do.

I don't want to go into whether Leda Cosmides' conclusions are right in detail. It would not be surprising if such pioneering work got some things wrong (see e.g. Cheng and Holyoak 1989) – although it is remarkable how far she has managed to anticipate criticisms and to show, by crucial experiments, how well her theory fits the facts compared with apparently plausible alternatives (such as the theory that the content effect reflects differences in familiarity with the subject matter). For my purpose, her work serves as an example of one approach to the problem of testing conjectures generated by Darwinian psychology. Her solution was to probe for adaptively revealing errors, using carefully contrived experiments. Let's now look at a very different way of tackling the same problem.

Consider again an example that we took earlier: the overwhelming preponderance of men among murderers, young men above all, and, in particular, the persistent thread of apparently trivial altercations that escalate into the ultimate conflict. Such robust invariance across cultures and across time suggests that perhaps something more than mere cultural conditioning is going on. But what? Martin Daly and Margo Wilson set out to answer this and a host of similar questions in their book *Homicide* (Daly and Wilson 1988; see also Daly and Wilson 1990). Their analysis is a model of Darwinian 'psychological' reasoning about human behaviour (and is, by the way, highly readable – more so, I should imagine, than most murder mysteries). Daly and Wilson decided to look at patterns of homicide because murder springs from the very stuff of Darwinian adaptations: conflicts of interests. They didn't assume that the act of murder is an adaptation, that it is of Darwinian advantage to the killer. What they did assume was that the human mind is adapted in such a way that, under certain circumstances, murder is a likely outcome. It is not the behaviour itself, then, either in any specific case or on average over our evolution, that they attempt to explain adaptively, but the psychological propensities that bring it about.

So what can be said about those consistent patterns of sex and age and motive among murderers? Some unDarwinian analyses indulge in wide-eyed wonderment that a man could risk his life 'over a 10 cent record on a juke box, or over a one dollar gambling debt from a dice game' (quoted in Daly and Wilson 1988, p. 127). Against this, several social scientists have stressed that, contrary to first appearances, something important is at stake: 'A seemingly minor affront ... must be understood within a larger social context of reputations, face, relative social status, and enduring relationships ... In most social milieus, a man's reputation depends in part upon the maintenance of a credible threat of violence' (Daly and Wilson 1988, p. 128). But why is reputation so important? Why do men so value these intangible resources that they will pursue them even unto death?

To answer this, Daly and Wilson turn to Darwinian theory and to the impact of sexual rivalry (Daly and Wilson 1988, pp. 123–86; Wilson and Daly 1985). 'If selection has shaped this aspect of the human psyche, it would appear that the answer must somehow take the following form: Such social resources are (or formerly were) means to the end of fitness' (Daly and Wilson 1988, p. 131). And they sift through the evidence in order to demonstrate that this is indeed so:

> *Homo sapiens* is very clearly a creature for whom differential social status has consistently been associated with variations in reproductive success. Men of high social rank have more wives, more concubines, more access to *other* men's wives than men of low social rank. They have more children and their children survive better. These things have consistently been the case in foraging societies, in pastoral societies, in horticultural societies, in state societies.
>
> (Daly and Wilson 1988, pp. 132–3)

Why, though, the difference between men and women, and why young men in particular? The answer, of course, lies in sexual selection. Several lines of evidence point to a human history of polygynous competition (albeit mild polygyny). The differences in reproductive success are greater among men than among women, and are more strongly correlated with social status. Men, but not women – young men above all – have powerful incentives to fight for that status. And, whether or not natural selection intended them to go so far, they will fight even literally, and sometimes even fatally.

Status-conscious males are one thing. But it is commonly said that, when it comes to violence and homicide, the family is one of the most dangerous places to be. The apparent implication that murderers kill their kin seems embarrassing for kin-selection theory. But when Daly and Wilson looked more carefully at the American data, it turned out that most 'family' victims were the murderer's spouse! If the FBI were of a more Darwinian turn of mind, they would analyse their statistics in a crucially different way – as would many a social scientist who has tried to account for murder. Trawling carefully through figures from many sources, Daly and Wilson concluded that, far from undermining kin-selection theory, patterns of murder fit neatly with its expectations. Not only was violence more likely to escalate the more distantly were people related, but also people were more likely to find common cause in murderous disputes the more closely they were related; so co-offenders are more closely related on average than victim and offender (Daly and Wilson 1988, pp. 17–35).

And yet, infanticide within families does happen, although to kill one's own child is surely to commit Darwinian suicide. But, once again, Daly's and Wilson's detailed analysis finds that, on the contrary, infanticide fits well with the evolved inclinations that we would expect in the allocation of scarce parental resources (Daly and Wilson 1988, pp. 37–93). Perhaps most tellingly, stepchildren turn out to be enormously more at risk than natural children (Daly and Wilson 1988, pp. 83–93). So, for example, in 1967 an American child living with one or more substitute parents was 100 times as likely to be fatally abused as a child living with natural parents; Canadian figures are similar; and, in North America as a whole, stepparents are more over-represented among homicides than among non-fatal abuse cases. Incidentally, revealing as these figures are, they are not readily revealed in the official statistics. As with other 'family' matters, data on children are gathered under doggedly unbiological categories: 'Astonishingly, census bureaus in the United States, Canada, and elsewhere have never attempted to distinguish natural parents from substitutes, with the result that there are no official statistics on the numbers of children of each age who live in each household type' (Daly and Wilson 1988, p. 88). It is indeed astonishing. And wasteful. If social scientists refuse to admit that a parent's genetic relationship to its offspring is a wellspring of human action, perhaps they should let zoologists gather the statistics. Daly and Wilson use the same Darwinian method to illuminate parricide, the killing of spouses and many other patterns of murder. Between large-scale demographic data on the one hand and general Darwinian principles such as kin selection, parental care and sexual rivalry on the other, they succeed in placing an evolved human psychology.

To think of the human mind in the structured way that this method favours may look less like a step forward than a leap backwards, into the nineteenth century and even beyond. In those dark recesses of scientific history lurk 'faculty psychologies' that divided the mind into sealed compartments with fixed capabilities; and in the murkiest corner moulders the cult of phrenology (Fodor 1983, particularly pp. 1–38). Such associations may understandably in the past have deterred Darwinians from thinking about our behaviour in terms of specific psychological faculties. But the recent revolution in our Darwinian understanding of behaviour takes us far from all that. It has given us powerful insights into what we might have been built to do and what psychological makeup we might have needed to do it. We can, then, start to construct a respectable faculty psychology, a Darwinian psychology, which bears no resemblance to those long-forgotten, strangely-mapped skulls.

This view of the mind, by the way, makes no assumptions about the architecture of the brain. It does not imply, for example, that our capacities are neurologically localisable (though, like the ability to recognise faces, they may be). Since Kant, most philosophers have routinely assumed that our minds are packed with synthetic a priori ideas but they have rightly felt no need to show us exactly where these ideas are sitting in the brain. Until we know more about our neurology and physiology, we can respectably think of the psychological endowments of natural selection simply as Darwinised synthetic a priori ideas – as Darwin himself did, on the evidence of this memorable jotting from one of his Notebooks: 'Plato ... says in Phaedo that our "imaginary ideas" arise from the preexistence of the soul, are not derivable from experience. – read monkeys for preexistence' (Gruber 1974, p. 324). Neither need we assume that every one of our faculties is highly specific; some of them (memory, for example) are more plausibly very general. All we are assuming is that, rather than specifying our design, sphex-like, down to the last behavioural detail, natural selection gave us the means, in the form of computation rules, to act adaptively in the light of information about our environment.

John Maynard Smith has suggested that 'often, we understand biological phenomena only when we have invented machines with similar properties' (Maynard Smith 1986, p. 99). We find it relatively easy to fathom the adaptive significance of hearts and lenses and wings. By contrast, we have made painfully slow progress in embryology: 'understanding how structures develop is one of the major problem areas in biology. One reason why we find it so hard to understand the development of form may be that we do not make machines that develop' (Maynard Smith 1986, p. 99). Perhaps a Darwinian understanding of our minds has been hampered by the fact that we do not make machines that 'think'. Until recently, novelists and biographers were probably our major purveyors of models of the mind; maybe that is part of their work's fascination. Now we have analytic machinery like distributed networks (see e.g. McClelland *et al.* 1986), Turing machines and modern logic. Perhaps at last we have something that will concentrate our minds for us.

Until quite recently, psychologists did not see it as their task to investigate the mind at all in the way that Darwin did: 'In the *Descent of Man* Darwin wrote of the combination of intellectual faculties forming "the higher mental powers": curiosity, imitation, attention, memory, reasoning and imagination. The list of topics Darwin covered reads almost like an inventory of subjects chronically neglected by twentieth-century psychologists until the upsurge of cognitive psychology beginning in the 1950s' (Gruber 1974, p. 236). Behaviourism turned its back on all such studies; the belief was that if we were ultimately to understand our minds, it would only be through understanding our behaviour. A Darwinian psychological approach goes in exactly the opposite direction; the adaptive significance of our behaviour may be obscure but we have some hope of understanding it by understanding our minds.

Darwinian attempts to explain human behaviour have often been condemned for latching on to the wrong things to be explained. Stephen Gould, for example, says of E. O. Wilson that he 'has made a fundamental error in identifying the wrong level of biological input. He looks to specific behaviors and their genetic advantages, and invokes natural selection for each item. He tries to explain each manifestation, rather than the underlying ground that permits their manifestation as one mode of behavior among many' (Gould 1987a, p. 290). Darwin's psychological method provides one means of winkling out the right units to be explained. As we saw in an earlier chapter, there is no easy way to decide on candidates for adaptive explanation. We may pity the poor moth, compelled to immolate itself on the candle-flame. But we must pity, too, the poor Darwinian, compelled to explain the moth's apparently non-adaptive genetic imperative. The answer in the case of this favourite example is well known: we should be explaining not an attempt at suicide but an attempt to steer a straight course. In the environment in which natural selection wired up the moth's navigation rules. the only light source was the moon; because celestial bodies are at optical infinity, their rays are parallel when they strike a moth, so the moon could safely be used as a compass to navigate a straight line. In the moth's normal environment, then, its inbuilt rules generate behaviour that is adaptive. Only in the unusual environment of candles and electric lights do those rules let it down. So the adaptation that Darwinians need to explain is the rule, not the behaviour. It is the same with humans. We need to find the right descriptive categories, the right candidates for adaptive explanation. Darwin's approach points us in the direction of the rules. We should not be surprised if our behaviour looks non-adaptive in the candle-flame of modern life. Indeed, some distortions may be such that we never discover their evolutionary roots: the connection between what we do and what we were meant to be doing may be so tortuous that to prise it out would be like asking for the moon – and, unlike the explanation of the moth, not getting it (Dawkins 1986a, pp. 66–72).

But if we do try for a Darwinian account, then the rules of our psychology might help us to find the steady course that natural selection intended us to steer.

Now let's look at the other side of Darwin's coin. How did Darwin's concentration on feelings rather than behaviour influence his views about altruism not in humans but in other animals? As we saw with the social insects, one important effect is that it makes him less ready to appreciate that there is any problem at all. The Darwinian problem of altruism is to do with costs to the altruist. But Darwin pays more attention to the sentiments that accompany altruism than to its apparent disadvantages; he cares less about whether behaviour is costly than whether it is caring. This is one reason why he is able to catalogue what is apparently unselfish behaviour in other animals without seeing it as problematic. Darwin's interest is in the source of the milk of human kindness rather than the bitter fruit of self-sacrifice.

Ironically, Darwin's very own approach gave him the means to do exactly the opposite: to get straight to the general Darwinian problem of altruism, whether in humans or any other living things. To arrive at this general problem, the trick is to refuse to get bogged down with questions of moral conscience, concentrating instead on the selective advantages and disadvantages of an animal's (or plant's) behaviour (or structure), particularly if it involves apparent self-sacrifice. Darwin failed to take this path. But his own approach made it available to him. Let's see how.

We need to begin with a point from ethical theory. Moral philosophers make much of the distinction – most famously insisted upon by Kant – between merely acting in accordance with a rule and acting on a rule, between actions that just happen to conform to duty and actions done for the sake of duty. It's the difference between not stealing the money merely because you didn't realise it was there and not stealing it because you believe theft to be wrong, the difference between making somebody else happy inadvertently (even unknowingly) and making them happy because you believe in doing good. Only if an agent is acting on a maxim can the action be moral (or immoral); only agents that are capable of adopting maxims can be moral (or immoral) beings. We wouldn't call a dog moral because it left its master's money undisturbed, nor immoral if it dragged the money off to its basket (though perhaps we would be tempted to think in moral terms if the dog furtively snatched its master's steak from his table or looked longingly at it but resisted temptation). Darwin uses the terms 'material' and 'formal' morality for the same idea (Darwin 1871, 2nd edn., p. 169, n25); material morality is about the practice of morality (behaving in accordance with moral rules) whereas formal morality is about moral consciousness (the knowledge of those rules). One can see why, for ethics, the division is crucial. It marks off moral acts and agents from the realm in which moral considerations do not apply.

Darwin, however, rejects any sharp distinction between an act that just happens to have good effects although undertaken without conscious design and the full flourish of a moral act that is consciously performed out of a deep sense of duty (Darwin 1871, i, pp. 87–9). Important as the difference may be to moral philosophers, Darwin insists that it is unworkable: 'it appears scarcely possible to draw any clear line of distinction of this kind' (Darwin 1871, 2nd edn., p. 169). He points to cases in which it seems to him that the philosophers' elevated criterion for what is moral gives us the wrong answer, excluding from the moral sphere acts that we would surely want to put within it. Many instances, for example, 'have been recorded of barbarians, destitute of any feeling of general benevolence towards mankind, and not guided by any religious motive, who have deliberately as prisoners sacrificed their lives, rather than betray their comrades' (Darwin 1871, i, p. 88). If it is true that these 'barbarians' are not compelled by 'exalted motives' (Darwin 1871, 2nd edn., p. 169), by general ethical maxims (although it is not clear why Darwin assumes this), then they do not satisfy Kantian standards for acting morally; and yet surely we would rightly want to call their actions moral. Surely, too, we are witnessing noble heroism 'when a Newfoundland dog drags a child out of the water, or a monkey faces danger to rescue its comrade, or takes charge of an orphan monkey' (Darwin 1871, 2nd edn., pp. 170–1); but for philosophers these deeds would also fail the test of morality because in their view dogs and monkeys lack the ability to grasp abstract moral principles, an ability that is essential for an agent to be a moral agent. So Darwin rejects any hard-and-fast demarcation, pointing instead to grey areas, to overlaps, to proto-morality, to continuities between mere sociality and a high moral sense.

Now, this offers him a freedom denied to ethical philosophers, denied to those who cling to the notion of moral conscience. It offers him the freedom to characterise altruism as it presents a problem for Darwinian theory, as the biologist's problem of altruism rather than the moralist's – to characterise both human and non-human altruism not as behaviour that is 'moral' but as behaviour that is costly, apparently too costly to have been favoured by natural selection. As a result of his own approach, Darwin had it within his grasp to look at animal altruism in the amoral, non-anthropomorphic way that modern Darwinians do – purely from the point of view of the practice of 'morality' and the selective effects of such behaviour, rather than from the point of view of its mental accompaniments. But the irony is, as we have seen, that Darwin used this freedom to take exactly the opposite path. He wanted to see the brave dog and monkey as embryonic moralists, as taking the first, faltering steps to a Kantian consciousness. He wanted to reveal elemental, inchoate signs of human morality in other animals, to pull their actions into, or at least closer to, the ambit of the moral.

There is a further irony. Darwin's critics complained that, by ignoring the philosophers' distinction, he failed to appreciate that it was our possession of a moral sense that was distinctive of human morality. Mivart, for example, protested: 'Mr Darwin is continually mistaking a merely beneficial action for a moral one; but ... it is one thing to *act well* and quite another to be a moral agent. A dog or even a fruit-tree may act well, but neither is a moral agent' ([Mivart] 1871, p. 83). In much the same vein, some critics today (e.g. Midgley 1979a, pp. 444–6) wax indignant about the Darwinian idea of altruism because, so they claim, it neglects the motives and emotions that must enter into altruistic acts. But a dog's fidelity or a fruit-tree's generosity (presuming the dog or tree incurs some cost) is precisely what the problem of altruism is about. The irony is that, from the standpoint of that problem, Darwin is far too *much* concerned with what goes on in our hearts and heads.

The Descent of Man (the first half, on human evolution) is one long argument for continuities between us and other species. What better way to establish our pedigree than through connections, comparisons, affinities, homologies, rudiments? It is a standard Darwinian method and an immensely powerful one. But I cannot help feeling that it served Darwin less well for human morality than for our bones and muscles, our use of tools, our feats of memory. Darwin and many others saw in our moral attributes the greatest gap between us and other living forms: 'of all the differences between man and the lower animals, the moral sense or conscience is by far the most important' (Darwin 1871, i, p. 70). All the more difficult, then, for natural selection to explain. And all the more need, one might think, to establish continuities. But perhaps where the gap is greatest, it would be more fruitful to concentrate on the adaptive reasons as to why it is so wide rather than on trying to narrow it, more helpful to study what is adaptively different and special than what is similar and common. It seems likely that Darwin expected the divide to be considerable. This was for much the same reasons as he expected sexual ornaments to be exaggerated. We noted that he thought of sexual selection, unless natural selection clamped down on it, as capable of escalating indefinitely, pushing itself ever onwards under its own steam. He regarded this as unusual. Unusual but not unique. Mental development in humans also, he believed, had 'no definite limit':

In many cases the continued development of a part, for instance of the beak of a bird, or of the teeth of a mammal, would not be advantageous to the species for gaining its food, or for any other object; but with man we can see no definite limit, as far as advantage is concerned, to the continued development of the brain and mental faculties.   (Darwin 1871, i, p. 189)

For Darwin, our 'mental faculties' included our moral sense; he discusses morality under the topic of 'mental powers' (Darwin 1871, i, pp. 70–106). Perhaps, then, he saw our moral qualities as one of the peacocks' tails that flourish in our mental world, the result of selective pressures to which there is no natural end. If so, Darwinians should not be alarmed at the vast gulf that morality puts between us and the 'lower animals'. We might even expect it, expect an evolution so rapid and dramatic that it would carry us far even from our closest living relatives. But then maybe Darwin should not have devoted himself so assiduously to establishing continuities. Maybe he should have taken a feather or two from the peacock's tail, should have explored instead the adaptive nature of this explosive growth and of the gaps that it can leave in its expansive train.

Perhaps we tend to take it too much for granted that Darwinians should be concerned with continuities. If Darwin did think that our morality has a peacock's-tail-ish quality about it, then to look for affinities with other animals might not be helpful. Admittedly, continuities are essential for establishing history – and history was, of course, Darwin's prime concern in *Descent of Man*. But when he discussed the burgeoning of peacocks' tails and the human mind, his concern was not with phylogeny but with the ways in which natural selection works, the ways in which adaptations are wrought. And on issues of principle continuities may have little to offer. Anyway, Darwin unfortunately failed to elaborate on why 'selection unlimited' would be likely in either of these cases. Presumably he regarded sexual ornament and mental qualities as peculiarly self-reinforcing, peculiarly liable to generate positive feedback. As we noted with sexual selection, it is probably no coincidence that these were two of those rare cases in which he recognised that the most salient selection pressures were social forces.

The question of continuities brings us to a common criticism of Darwinian studies of human behaviour: that they are founded on 'the conviction that since humans are animals who have evolved in much the same ways as other animals they must be explicable in much the same way' (Montagu 1980a, p. 5). I'm not sure what that 'much the same' covers; it could span a multitude of methodological (and political) sins. But it is worth noting that an impeccably Darwinian approach could lead to quite the opposite conclusion. Darwin's concentration on psychology rather than behaviour could make the study of humans markedly different from that of other animals. Admittedly Darwin himself applied his method to them as well as to us. But modern Darwinism has got further by concentrating on the behaviour of other animals than on their less accessible minds.

'Much the same' anyway requires no defence if it means trying to apply the same general Darwinian principles to any animal or plant. We don't assume that ants believe sisterhood to be powerful; but we do consider that their behaviour can be explained by the principle of kin selection. We don't assume that chromosomes have a moral conscience; but we can reasonably speculate about whether the lottery of cell division sets up a game of Prisoner's Dilemma and whether chromosomes have evolved a Tit-for-Tat response. Conversely, we can apply the theory of kin selection to humans without having to assume that our staple food is wood, or that we recognise members of our family by smell, or that our sibs are more genetically valuable to us than our offspring. So we can assume 'much the sameness' of principles without making the absurd assumption that humans and termites and chromosomes implement their strategies in the same way.

Indeed, to complain about attempts to explain humans in much the same way as 'animals' is to assume implicitly that all non-human animals can be explained in much the same way as each other – that tortoises, leopards, ants, ostriches (and, presumably, primroses and bacteria) all fall into one single explanatory category whereas we alone stand apart, an entirely different explanatory realm. Now, that assumption really is mistaken – and speciesist to boot. There are many, many ways of being a Darwinian strategist. And they don't divide neatly into 'human ways' and 'all the rest'. The reason that we are justified in assuming sameness of strategic principles is that, although behaviour is manifested in organisms, strategies belong ultimately to genes. And genes are not speciesist.

What is more, to erect a biological apartheid of 'us' and 'them' is to cut ourselves off from a potentially useful source of explanatory principles. Once we have understood ourselves as naturally selected tacticians, we might have a suggestive heuristic guide to the tactics that natural selection has employed with other living things. If, following Darwin, we look at how natural selection has shaped our minds, we are studying an area to which we have privileged access, an area that is sadly so profoundly hidden from us in all other species that, by comparison, the tricky problem of how we know human minds other than our own looks trivial. This is a rich source of information, surely too rich to be kept under intellectual house-arrest for fear of anthropomorphism. We needn't assume that the outcome will tell us about the workings of the minds of other animals – though it may do. Nor need the guidance be more than heuristic – though it may be. All we need to imagine is that, in pursuit of the same strategies as ours, other living organisms might have converged on the same tactics. There's nothing unduly anthropomorphic about that. We're not assuming that other organisms think as we do. We're not even assuming that they think at all. After all, chromosomes and plants manage to implement Darwinian principles even without brains. It is natural selection that has done their 'thinking'. Nevertheless, their strategic choices and ours could run parallel, the structure of their behaviour could be the same, because natural selection has implemented its strategies in similar style. Admittedly, we are unique. But there's nothing unique about being unique. Every species is in its own way. Understanding how we as strategists think could help us to anticipate how other strategists might behave. Our minds could provide a working model of one possible way of going about things. We could serve for other species as their guinea pigs, their rats-in-mazes.

In a note to himself, Darwin declared: 'He who understand baboon would do more toward metaphysics than Locke' (Gruber 1974, p. 281, [M] 84). His public statement was more temperate. Ethical philosophers, he said, should acknowledge that our moral feelings are part of our evolutionary endowment:

> Mr J. S. Mill speaks, in his celebrated work, 'Utilitarianism', of the social feelings as a 'powerful natural sentiment', and as 'the natural basis of sentiment for utilitarian morality' ... But ... he also remarks, '... the moral feelings are not innate, but acquired'. It is with hesitation that I venture to differ from so profound a thinker, but it can hardly be disputed that the social feelings are instinctive or innate in the lower animals; and why should they not be so in man? ... [Several thinkers] believe that the moral sense is acquired by each individual during its lifetime. On the general theory of evolution this is at least extremely improbable. The ignoring of all transmitted mental qualities will, as it seems to me, be hereafter judged as a most serious blemish in the works of Mr Mill.   (Darwin 1871, 2nd edn., pp. 149–50, n5)

It was not only philosophers who felt that monkeys and metaphysics did not mix. Darwinian scientists, too, were in those ranks. We'll come soon to some of their stated reasons for rejecting Darwin's programme. Here we'll glance at some of the extra-scientific motives.

Darwin's nineteenth-century opponents laid great stress on our moral superiority – even, as Kropotkin regretfully noted, to the point of refusing 'to admit well-proven scientific facts tending to reduce the distance between man and his animal brothers' (Kropotkin 1902, p. 236). This need to keep a distance suggests that a Darwinian account of ethics was seen as threatening our elevated position; morality would be denigrated if it was shared (albeit in minute quantities) with 'the lower animals'. But a Darwinian account of the origins of morality need not, of course, threaten our moral pre-eminence – any more than a cheetah's claim to being the best of sprinters is undermined by natural selection sharing its glory. Darwinians could hold, as Darwin did, that our moral sense has evolved but is nevertheless unique and highly sophisticated. Darwin's critics very likely also feared the encroachment of relativism – the denial that there is any absolute, single moral standard that holds for all moral agents at all times. For if our practice of morality depends on our evolutionary development then perhaps moral principles also change over evolutionary time. Didn't Darwin himself say that our morals just happen to be as they are because of our social system (a social system that is contingent on our biology)?

I do not wish to maintain that any strictly social animal, if its intellectual faculties were to become as active and as highly developed as in man, would acquire exactly the same moral sense as ours. In the same manner as various animals have some sense of beauty, though they admire widely different objects, so they might have a sense of right and wrong, though led by it to follow widely different lines of conduct.   (Darwin 1871, i, p. 73)

And he goes on to cite an example that we have already noted:

If ... men were reared under precisely the same conditions as hive-bees, there can hardly be a doubt that our unmarried females would, like the worker-bees, think it a sacred duty to kill their brothers, and mothers would strive to kill their fertile daughters; and no one would think of interfering.   (Darwin 1871, i, p. 73)

If what we believe to be right depends so heavily on our being humans rather than intelligent bees or baboons, then how do we know that we are correct about what we believe to be right? Indeed, perhaps the very notion that there is an objective moral code at all is only an illusion, a belief built into us by natural selection. And unlike, say, our propensity to experience the world as three-dimensional or to internalise a twenty-four hour clock, it could be a belief that corresponds to nothing 'out there'. It could be no more than a reinforcer, just another of natural selection's tricks for oiling the machinery of altruism. For those of Darwin's contemporaries who feared slippery slopes, his line of thinking might well have felt perilously like the beginning of a dizzy incline.

Wallace: Wise before the event

The distinction between stated reasons and background motives for rejecting Darwin's programme brings us to the strange case of Wallace. We are by now accustomed to meeting him as the ever-vigilant defender of natural selection, the ultra-adaptationist, the most Darwinian of Darwinians. And yet, when it came to humans, particularly to our moral sense ... Well, here are Wallace's own words: 'It will ... probably excite some surprise among my readers to find that I do not consider that all nature can be explained on the principles of which I am so ardent an advocate; and that I am now myself going to state objections, and to place limits, to the power of natural selection' (Wallace

1891, p. 186). Although Wallace remained a staunch Darwinian to the end of his life, he also gradually became increasingly convinced of the reality and power of supernatural forces (Durant 1979; Kottler 1974, 1985 pp. 420–4; Schwartz 1984, pp. 280–8; Smith R. 1972; Turner 1974, pp. 68–103). At an early age he had taken up phrenology and mesmerism; in the mid-1860s, he turned to spiritualism. As these convictions grew, he came to believe that natural selection could not account for several of our distinctly human qualities, above all our advanced mental attributes (1864 revised version, 1869, 1870, pp. 332–71, 1870a, 1877, 1889, pp. 445–78):

> The Origin of Man as an Intellectual and Moral Being: On this great problem the belief and teaching of Darwin was, that man's whole nature – physical, mental, intellectual, and moral – was developed from the lower animals by means of the same laws of variation and survival; and, as a consequence of this belief, that there was no difference in *kind* between man's nature and animal nature, but only one of degree. My view, on the other hand, was, and is, that there is a difference in kind, intellectually and morally, between man and other animals; and that while his body was undoubtedly developed by the continuous modification of some ancestral animal form, some different agency, analogous to that which first produced organic *life*, and then originated *consciousness*, came into play in order to develop the higher intellectual and spiritual nature of man.   (Wallace 1905, ii, pp. 16–17)

This 'different agency' was a spiritual one: 'man's body may have been developed from that of a lower animal form under the law of natural selection; but ... we possess intellectual and moral faculties which could not have been so developed but must have had another origin; and for this origin we can only find an adequate cause in the unseen universe of Spirit' (Wallace 1889, p. 478). Putting it roughly (but not unfairly), nature gave us our bodies and our lower mental capacities but our souls are a gift of the supernatural. This is a familiar position. It is the standing argument that religion is still having with Darwinism. Darwinian theory, so the argument goes, furnishes an excellent explanation of the organic world but it cannot explain the spiritual aspect of our being (although Wallace, unlike most religious commentators, held that spiritual forces were amenable to scientific investigation).

Our interest here is in Darwinism, not in whatever other ideas Darwinians happen to hold. So we shall not follow Wallace into realms ethereal. Fortunately, we can examine his position without having to do so. Whatever his extra-Darwinian motives, Wallace, being the true Darwinian that he was, provided a pert Darwinian defence for his non-Darwinian account of human morality.

The problem with humans, said Wallace, is that we are more advanced, more sophisticated, better-prepared for modern living than Darwinian forces could have made us. Natural selection can never do more than solve the problems it is presented with. It has no foresight, makes no provision for the future. It cannot give rise to characteristics that are useless or harmful, even if it turns out that they would have been useful at some later date. Natural selection has 'no power to advance any being much beyond his fellow beings, but only just so much beyond them as to enable it to survive them in the struggle for existence. Still less has it any power to produce modifications which are in any degree injurious to its possessor' (Wallace 1891, p. 187). We should remember this when we study human beings:

> If ... we find in man any characters, which all the evidence we can obtain goes to show would have been actually injurious to him on their first appearance, they could not possibly have been produced by natural selection. Neither could any specially developed organ have been so produced if it had been merely useless to him, or if its use were not proportionate to its degree of development.   (Wallace 1891, p. 187)

Now look at us. Look in particular at our brains. They were clearly built surplus to requirements, surplus to adaptive needs. On the one hand, the human brain is large in proportion to our body size as compared with 'lower' apes; its size is constant across races today and has not changed since prehistoric times; and brain size is the major determinant of mental ability. On the other hand, the demands that prehistoric peoples and 'savages' make of the brain fall far below its capabilities: 'The higher feelings of pure morality and refined emotion, and the power of abstract reasoning and ideal conception, are useless to them, and rarely if ever manifested, and have no important relations to their habits, wants, desires, or well-being. They possess a mental organ beyond their needs' (Wallace 1891, p. 202). The brain could not be the product of natural selection, for selection can work only on faculties that are exercised, not on potentialities: 'Natural selection could only have endowed savage man with a brain a few degrees superior to that of an ape, whereas he actually possesses one very little inferior to that of a philosopher' (Wallace 1891, p. 202). So natural selection could not have been responsible for 'the higher feelings of pure morality' – 'the constancy of the martyr, the unselfishness of the philanthropist, the devotion of the patriot, ... the passion for justice, and the thrill of exultation with which we hear of any act of courageous self-sacrifice' (Wallace 1889, p. 474). Neither can natural selection be credited with 'the present gigantic development of the mathematical faculty'; it is absent or unexercised in primitive societies and yet has flourished 'during the last three centuries ... [in] the civilised world' (Wallace 1889, pp. 465, 467). Our musical faculty tells the same story – hardly exercised at all in the 'rude musical sounds ... [and] monotonous chants' of 'lower savages' but suddenly, since the fifteenth century, advancing 'with marvellous rapidity' (Wallace 1889, pp. 467–8). Our philosophical faculties, too, 'spring suddenly into existence' as we shed our primitive ways (Wallace 1889, p. 472). And 'the peculiar faculty of wit and humour, ... almost unknown among savages, ... appears more or less frequently as civilisation advances' (Wallace 1889, p. 472).

Not only do these higher feelings and refined capacities go unexercised in 'uncivilised' societies but, worse, some would even be a downright nuisance, possibly a danger:

> in his moral and aesthetic faculties, the savage has none of those wide sympathies with all nature, those conceptions of the infinite, of the good, of the sublime and beautiful, which are so largely developed in civilised man. Any considerable development of these would, in fact, be useless or even hurtful to him, since they would to some extent interfere with the supremacy of those perceptive and animal faculties on which his very existence often depends, in the severe struggle he has to carry on against nature and his fellow-man. (Wallace 1891, pp. 191–2)

The brain and our mental powers pose the most serious problem. But we also come ready fitted-out with other features for which we cannot thank natural selection – some of the very features that we need for sophisticated, cultured modern living. Our superb manual dexterity, for example, seems to go far beyond the demands of a primitive society: 'the hand of man contains latent capacities and powers which are unused by savages, and must have been even less used by palæolithic man and his still ruder predecessors. It has all the appearance of an organ prepared for the use of civilized man, and one which was required to render civilization possible' (Wallace 1870, pp. 349–50). Our loss of hair on the back would surely have been more harmful than helpful when it occurred. And how could the utilitarian force of natural selection account for the exquisite musicality of the voice, its 'wonderful power, range, flexibility, and sweetness' (particularly, Wallace says wistfully, in the female sex), when 'savages' manage no more than 'a more or less monotonous howling' (Wallace 1870, p. 350)? But, although none of these things would have been adaptive when they first arose, they are just what we need in civilised society. They are, in fact, exactly what a far-seeing designer would have specified.

And Wallace points to what he saw – perhaps wrongly – as another oddity about some of our higher faculties. They vary far more in any population than one would expect for utilitarian characteristics. Any fox is pretty well as good as the next one at catching rabbits; any rabbit is pretty well as good as the next one at running from foxes. But we can't say the same of artists and musicians and writers. If we really needed to be witty and philosophical and musical, why are there just a few geniuses, with most of us trailing far behind, and some of us even resoundingly bad?

In the light of all this, Wallace insists, he is no apostate when it comes to his Darwinian principles. Far from reneging on them, he is sticking to them resolutely. But is he? Is he the hardline, ultra-respectable natural selectionist that he would have us believe?

Any Darwinian has to admit that we humans present some awkward cases for natural selection. We should not take it as unproblematic that evolution has equipped us with hands that can type or play the violin (even though we have shaped these activities to our endowments). Still less is it obvious why we possess the faculty of enjoying a Schubert quartet (not to mention that rare, precious faculty of composing one). Wallace was not alone among his contemporaries in feeling uneasy about such endowments. He quotes a remark that Huxley was said to have made about his enjoyment of music and scenery: 'I do not see how they can have helped in the struggle for existence. They are gratuitous gifts' (Wallace 1889, p. 478). Darwin made a similar point: 'As neither the enjoyment nor the capacity of producing musical notes are faculties of the least direct use to man in reference to his ordinary habits of life, they must be ranked amongst the most mysterious with which he is endowed' (Darwin 1871, ii, p. 333). Wallace quotes Weismann as saying that talents such as mathematical or artistic ability 'cannot have arisen through natural selection, because life is in no way dependent on their presence' (Wallace 1889, p. 473). And Romanes commented: 'why it is that beauty attaches to architecture, music, poetry, and many other things – these are questions which do not especially concern the biologist. If they are ever to receive any satisfactory explanation in terms of natural causation, this must be furnished at the hands of the psychologist ... As biologists we have simply to accept this feeling as a fact' (Romanes 1892–7, i, p. 404).

The answers favoured by many of his fellow-Darwinians were often uncongenial to Wallace. Non-adaptive explanations offended his strict adaptationism; and sexual selection, as we have seen, didn't satisfy him even for peacocks' tails – all the less so for human attributes.

Take the loss of body hair. Darwin considered several suggestions as to how natural selection could have favoured it but found them wanting and finally settled for sexual selection (Darwin 1871, i, pp. 148–50, ii, pp. 318–23, 375–81). He agreed with Wallace that 'The loss of hair is an inconvenience and probably an injury to man ... No one supposes that the nakedness of the skin is any direct advantage to man, so that his body cannot have been divested of hair through natural selection' (Darwin 1871, ii, pp. 375–6); 'man, or rather primarily woman,' he concluded 'became divested of hair for ornamental purposes' (Darwin 1871, i, p. 149). Other critics (e.g. Bonavia 1870; Wright 1870, pp. 291–2) suggested that hair loss was merely a non-adaptive side effect of selection; hairlessness was an inevitable accompaniment to selection for some useful characteristic – it did, after all, correlate in particular with increase in brain size. Chauncey Wright (a civil servant in Massachussets who was a keen Darwinian) added to this argument by turning one of Wallace's own arguments against him. Wallace had argued that, at a certain point in our evolution, human ingenuity had the effect of shielding our bodies from natural selection (Wallace 1864: 1891 reprint, pp. 173–6). Perhaps, said Wright, the loss of hair was originally a non-adaptive side effect. But natural selection would have had no incentive to restore our protective hairy coat once we were coping with the problem: 'Every savage protects his back by artificial coverings. Mr Wallace cites the fact as a proof that the loss of hair is a defect which Natural Selection ought to remedy. But why should Natural Selection remedy what art has already cared for?' (Wright 1870, p. 292). It was a similar story with our musical development. Darwin attributed it to sexual selection (Darwin 1871, i, p. 56, ii, pp. 330–1, 336–7); but – his usual objections apart – Wallace claimed, as we have noted, that the use of the human singing voice 'only comes into play among civilized people' and sexual selection 'could not therefore have developed this wonderful power' (Wallace 1870, p. 350). Weismann, with no justification that Wallace could see, concluded that all talents like musicality, ability to paint and mathematical aptitude are merely by-products of the human mind (Wallace 1889, pp. 472–3, n1).

So we shouldn't simply dismiss Wallace's arguments as meretricious special pleading. They do tackle some serious problems for Darwinism. And the answers are not obvious. Nevertheless, a Darwinian judgement on Wallace must be 'Could try harder'.

For a start, Darwinism need not be embarrassed by apparent foresight on the part of natural selection. There is an orthodox way of dealing with it, well-known to nineteenth-century Darwinians. Wallace should have given this standard argument proper consideration, even if eventually rejecting its application to this case. The reasoning is as follows. Any adaptation has 'unintended' features. These may serve no useful purpose when they first appear. But natural selection can press them into service later if there is a suitable job for them to do. So 'preadaptations', as they are called, need not violate Wallace's principle of utility. The lung of primitive fish was subsequently recycled as an excellent swim-bladder. Birds' feathers turned out to be good both for insulation and for flying, although natural selection originally favoured only one of those functions (experts aren't agreed as to which). Now, these unintended characteristics may be manifested from the beginning, like the buoyancy of the fishes' lung. But they may be only potentialities, latent, untapped, not showing themselves until called upon.

And, as some of Wallace's critics were quick to point out, this is surely how we can think, above all, of the astonishing capacities of the human brain.

Chauncey Wright, for example, suggested that the use of language demands an enormously powerful brain: 'even the smallest proficiency in it might require more brain power than the greatest in any other direction' (Wright 1870, pp. 294–8). So perhaps those unused mental abilities of 'savages' that so worried Wallace are emergent properties. Darwin agreed with Wright (Darwin 1871, i, p. 105, ii, pp. 335, 391; 2nd edn., p. 72). And he explained some aspects of our musical ability in much the same way: 'Many ... cases could be advanced of organs and instincts originally adapted for one purpose, having been utilised for some quite distinct purpose. Hence the capacity for high musical development, which the savage races of man possess, may be due ... simply to their having acquired for some distinct purposes the proper vocal organs' (Darwin 1871, ii, p. 335). According to some of Wallace's critics this 'distinct purpose' was just communication; the fact that Europeans, even trained singers, cannot reproduce many of the sounds of 'savages' shows that 'the accurate cultivation of the throat and windpipe ... is necessary, not merely for those highest requirements of art, but also for the commonest sounds and cries of savages little elevated above the beasts' (Dohrn 1871, p. 160; see also Wright 1870, p. 293).

Most Darwinians nowadays would agree with the general principles of these arguments, if not the details. I can't resist quoting the following example of the kind of process that Wallace's critics had in mind; it is not about humans but about some captivating (and unfortunately captive) behaviour in cetacea:

> Dolphins and whales have evolved large brains relative to their bodies, so that they are relatively brainier than most other mammals except monkeys and apes. As one would expect, these large brains are associated with sophisticated learning abilities. This includes the ability to achieve what is called second-order learning. For example, roughtoothed dolphins were taught by standard conditioning methods to perform novel behaviour in order to gain a reward. They soon made the intuitive leap that new behaviour was required and began to pour out large numbers of invented patterns never before seen in captivity or sea, such as corkscrew swimming and gliding upside-down with the tail out of the water.   (Trivers 1983, pp. 1205–6)

Nineteenth-century Darwinians must have been familiar with many a mechanical device that was built for one purpose and turned out to have unexpected powers for other tasks. Nowadays computers provide us with an even better model of the evolution of the brain as envisaged by Wallace's critics. Although computers were built for calculation they automatically possess latent skills, a potential that can be put to other uses. It would be quite hard to design a machine for programmable calculation that could *not* be easily reprogrammed for word-processing or for holding a library's reference system. We have startling evidence of emergent properties in our own on-board computers every time we read or write. These powerful skills depend on natural selection's gift but far transcend its intentions. They were not purpose-built; presumably they spilt over from the cornucopia of our linguistic capacities. It's really rather surprising, by the way, that dysfunctions of reading and writing, like dyslexia, are not more common. Natural selection has no means of eliminating them directly. Perhaps, insofar as they are dealt with biologically (rather than culturally), they are corrected automatically as a by-product of improvements in our linguistic skills.

Wallace's critics also pointed out that he failed almost entirely to apply his rigorous criterion of utility to any living beings other than humans. Had he done so, he might not have managed to stake out the unique place in nature that he claimed for us. Other species, too, show 'preadaptation' of an apparently emergent kind. Darwin, for example, maintained that 'there is nothing anomalous in ... [human musicality lying dormant]; some species of birds which never naturally sing, can without much difficulty be taught to perform; thus the house-sparrow has learnt the song of a linnet' (Darwin 1871, ii, p. 334). Chauncey Wright (1870, p. 293) also cited unused singing powers in birds, quoting Wallace himself (in an essay reprinted in the same volume as the essay on man that Wright was criticising): some species 'which have naturally little variety of song, are ready in confinement, to learn from other species, and become much better songsters' (Wallace 1870, p. 221). (Wallace might have felt that this was not convincing; answering a similar point from another critic (1870a), he argued that some non-singing birds have a redundantly complex larynx because their ancestors did sing.) Huxley (1871, pp. 471–2), too, cited cases of what he claimed to be development beyond needs in 'lower' animals. 'The brain of a porpoise', for example, 'is quite wonderful for its mass, and for the development of the cerebral convolutions. And yet ... it is hard to believe that porpoises are much troubled with intellect' (Huxley 1871, pp. 471–2) (though how could he be so sure?).

In fairness to Wallace it should be said that, as weapons for the adaptationist, these 'preadaptation' arguments about potentialities can be double-edged. They rely on the idea that some side effects of adaptations, which become positively useful when conditions change, are until then just lying around dormant. Such arguments, unless they are applied with discrimination, could end up peppering the world with a multitude of characteristics that have no Darwinian purpose (even though they eventually get put to good use). Wallace might have had adaptationist compunctions about allowing a proliferation of these functionally idle entities. Nevertheless, if a case is to be made anywhere for emergent properties (as surely it should be), the brain must be a prime candidate.

Finally, Wallace didn't make much attempt to deal with the adaptive explanations that were available – including his own! He himself at one time seemed to think that natural selection provided pressures enough for our moral advancement: 'It is the struggle for existence, the "battle for life", which exercises the moral faculties and calls forth the latent sparks of genius. The hope of gain, the love of power, the desire of fame and approbation, excite to noble deeds, and call into action all those faculties which are the distinctive attributes of man' (Wallace 1853, p. 83). That was a passing remark in his diatribe against slavery, from an early work, *Travels on the Amazon and Rio Negro*. It was not until about fifteen years later that he came to argue that natural selection could not account for the human brain and advanced mental qualities and for certain physical characteristics (Wallace 1869, 1870, pp. 332–71). And in the case of some of those physical characteristics (such as our hairlessness, loss of the prehensile foot and development of an opposable thumb) he eventually returned to his original adaptive explanation (compare e.g. Wallace 1870, pp. 348–50 and 1889, pp. 454–5).

On the specific question of our mental and moral capacities, several of Wallace's contemporaries disagreed with him that they must have been superfluous in the early stages of our development. According to Darwin, for example, they were crucial to our evolution (along with our bodily structure):

Man in the rudest state in which he now exists is the most dominant animal that has ever appeared on the earth. He has spread more widely than any other highly organised form; and all others have yielded before him. He manifestly owes this immense superiority to his intellectual faculties, his social habits ... and to his corporeal structure ... Through his powers of intellect, articulate language has been evolved; and on this his wonderful advancement has mainly depended. He has invented ... various weapons, tools, traps, &c ... He has made rafts or canoes ... He has discovered the art of making fire ... This last discovery, probably the greatest, excepting language, ever made by man, dates from before the dawn of history ... I cannot, therefore, understand how it is that Mr Wallace maintains, that 'natural selection could only have endowed the savage with a brain a little superior to that of an ape'.   (Darwin 1871, i, pp. 136–8)

Huxley (1871, pp. 470–1) also insisted that the 'primitive' life was mentally demanding, quoting Wallace's own essay 'On instinct in man and animals' (Wallace 1870, pp. 201–10) (which, again, appeared in the same volume as the work on man that Huxley was criticising). Wallace's essay takes off from the fact that many people have thought 'savages' possess some 'mysterious power', so astonishing is their proficiency at finding their way unerringly through unfamiliar countryside. Wallace, denying that 'savages' have some special instinct, argues that these impressive feats of navigation rely on intricate knowledge – the pooling of meticulously detailed information, acute observation and excellent memory. So on Wallace's own admission, says Huxley, the primitive's world is hardly undemanding. Nor, in the light of Wallace's own evidence, does Wallace go far enough: 'The Civil Service Examiners are held in great terror by young Englishmen; but even their ferocity never tempted them to require a candidate to possess such a knowledge of a parish, as Mr Wallace justly points out savages may possess of an area a hundred miles, or more, in diameter' (Huxley 1871, p. 471). Huxley suggested that social living in particular made heavy demands – indeed, that social pressures could have been one of the major selective forces that pushed us into developing advanced mental faculties (Huxley 1871, pp. 472–3):

> the conditions of our present social existence exercise the most extraordinarily powerful selective influence in favour of novelists, artists, and strong intellects of all kinds; and it seems unquestionable that all forms of social existence must have had the same tendency ... [T]he conditions of social life tend, powerfully, to give an advantage to those individuals who vary in the direction of intellectual or aesthetic excellence.   (Huxley 1871, pp. 472–3)

'The savage who can amuse his fellows by telling a good story over the nightly fire', for example, 'is held by them in esteem and reward, in one way or another, for doing so' (Huxley 1871, p. 472).

Most modern Darwinians would go further on the potential importance of social pressures as selective forces. As we have seen, today's Darwinism is acutely aware of the immense selective power that can be generated by 'others-like-oneself'. In the case of our mental qualities, the psychologist Nicholas Humphrey, for example, has argued that it is to the complexities of social living that we owe the evolution of our self-awareness (Humphrey 1976, 1986). People constitute especially difficult and complicated bits of our environment, requiring skilled and sensitive handling. In order to understand and manage others, we make a picture in our mind of the human being to whom we have privileged access – ourselves – and this serves as a model of what it is like to be someone else. Natural selection, then, has made us into 'natural psychologists' and, in so doing, has endowed us with consciousness. This kind of argument is notably different from the long-standing and widely-held view that one of the principal driving forces in the evolution of our intelligence was the need for practical invention.

So, in appreciating the general principle of the importance of our social environment, we have come a long way since Wallace and his contemporaries. But on the empirical question of exactly what it is that our mental attributes contribute to our Darwinian success and how they do so, we haven't got much further. What did our ancestors use their brains for? Huxley thought that 'savages' find it really useful, for example, to be able to crack a good joke round the camp fire. By contrast, Wallace thought that the 'peculiar faculty of wit and humour ... is almost unknown among savages [and] ... is altogether removed from utility in the struggle for life' – so far removed that most people are 'totally unable to say a witty thing or make a pun even to save their lives' (Wallace 1889, p. 472). What, then, is the Darwinian advantage of spinning a good yarn? Trying to gather anthropological data on that would be no joke. (The difference between Huxley's and Wallace's views probably reveals more about their personalities than about human history.) But we might be better equipped to answer Wallace's problem of overdesign if we knew, for example, whether intelligence correlated with quality or quantity of mates, number of offspring, tubers dug, animals caught or whatever.

The arguments of both Wallace and his critics rest on the idea that 'savages' have the same average intelligence as 'us' and that they reason in much the same way. If we are to develop a Darwinian understanding of our minds, of the evolution of our mental and moral faculties, it is essential that we think of human beings in this way, as unified by natural selection. Unfortunately, the majority of anthropologists, the experts on 'others' (if not on 'us'), have long insisted that, on the contrary, different societies have fundamentally different modes of thought. One extreme version of this attitude, for example, reached a peak around the turn of the century, when the idea of the 'savage' having a 'pre-logical mind' got a grip on some anthropological circles, under the influence of the French anthropologist Lucien Lévy-Bruhl. In his *La Morale et la Science des Moeurs* (1903; published in 1905 as *Ethics and Moral Science*) Lévy-Bruhl developed the idea that 'primitive peoples' possess a 'primitive mentality', with reasoning processes that are vastly different from those that we employ in civilised societies; their thinking, he claimed, is not governed by the laws of logic, and violates in particular the law of contradiction. This was admittedly a flagrant case. Nevertheless, the tyranny of 'different cultures, different systems of thought' permeated much of anthropological thinking then and for decades after. According to the anthropologist Maurice Bloch, it was one of the founding fathers of modern sociology, Émile Durkheim, that was largely to blame (Bloch 1977; Symons 1979, pp. 44–5). Durkheim held that our knowledge is socially constructed, that culture, not nature, determines our categories of understanding and that different cultures have fundamentally different modes of classification.

Now, it may seem odd to lump together, in one breath, Lévy-Bruhl's idea of a 'primitive mentality' and Durkheim's cultural relativism. After all, the idea of a 'primitive mind' is born of cultural imperialism, whereas cultural relativism has often been the standard liberal response to such imperialism in the social sciences. But from a Darwinian point of view they have a common failing. We find in both stances the same fragmentation of humanity, the same stress on cultural differences, differences so profound that our Darwinian unity is overlooked. Perhaps this was one of the many factors that hindered Darwinian progress in understanding the evolution of the human mind. Our theories must be premissed on a fundamental affinity among human beings across a multitude of cultures and down through time.

But back to Wallace. Some critics have agreed with him that his views on human evolution are entirely consistent with his Darwinian principles, an inevitable outcome of applying utilitarian criteria to the bitter end (e.g. Gould 1980, pp. 53–4; Kottler 1985, p. 422; Lankester 1889; Smith R. 1972). Unlike Wallace, however, they have of course taken this not as justifying his views on human evolution, but as exposing a weakness in the principles. Gould sees the whole sorry episode as a cautionary warning against the excesses of hyper-adaptationism. E. Ray Lankester lamented Wallace's insistence that Darwinism was the only scientific explanation; when natural selection failed him, he had nowhere to go but outside science: 'Mr Wallace seems so much convinced of the importance and capability of the principle of natural selection, that when it breaks down as an explanation he loses faith in all natural cause, and has recourse to metaphysical assumption' (Lankester 1889, p. 570). Similarly, David Hull has said: 'When Wallace became convinced that natural selection was inadequate to account for the superabundant powers of the human brain, he had no auxiliary naturalistic hypotheses to fall back on and was forced to posit a supernatural agency' (Hull 1984, p. 799). But, as we have seen, all this takes Wallace's own protestations a bit too much at face value. Things weren't that bad for adaptationism on the human front.

How damaging to Darwinism was Wallace's position that natural selection cannot explain human morality? According to Wallace, not damaging at all. Natural selection, he insisted, is not undermined by the fact that we humans 'evolve' new plants and animals when we practise domestic selection; why, then, should it matter if natural selection has not had much of a hand in our mental evolution? His views, he said,

> do not in the least affect the general doctrine of natural selection. It might be as well urged that because man has produced the pouter-pigeon, the bull-dog, and the dray-horse, none of which could have been produced by natural selection alone, therefore the agency of natural selection is weakened or disproved. Neither, I urge, is it weakened or disproved if my theory of the origin of man is the true one.
>
> (Wallace 1905, ii, p. 17)

Some commentators have disagreed with Wallace. Joel Schwartz, for example, has claimed that Darwin took the opposite view and (or so Schwartz seems to think) that he was right to do so (Schwartz 1984). Darwin, he says, was 'aware that his whole concept of "evolution by natural selection" was endangered by Wallace's insistence that natural selection was not the only factor in the evolution of man. If one essential part of the theory was denied, the entire theory was called into question' (Schwartz 1984, p. 288). But if Darwin had come to this conclusion (Schwartz offers no evidence that he did and I know of none) his reaction would surely have been unduly alarmist. Darwin was certainly preoccupied with our moral sense. The subject absorbs nearly a quarter of his entire argument about human evolution in *Descent of Man*. He discussed several other candidates for human uniqueness (or, at least, several other apparently major discontinuities) – among them language use, introspective thought, brain–body ratio, upright posture, digital dexterity and tool-making; but these all received relatively short shrift. Certainly, by the 1870s, morality had come to be a last, lone redoubt of the uniqueness argument (whereas earlier critics tended to concentrate as much on rationality (Herbert 1977, p. 197; Richards 1979, 1982)). So if Darwin could storm this little fortress he would certainly add plausibility to the Darwinian story of human descent. But such a victory wasn't crucial to the story's acceptance. By the time that *Descent* was published, it was very widely agreed – even by non-scientists (Ellegård 1958, pp. 293–331) – that evolution (either wholly or partly by natural selection) could make claim both to our bodies and to some of our mental attributes, even if not to our moral sense. So the Darwinian case for human descent was – rightly – not seen as hingeing upon the issue of morality. Even less was this one point seen as a test case for the 'entire theory' of natural selection – again, rightly.

Before we leave Wallace, let's glance at his views on the very beginnings of morality. Although he did not think that natural selection could explain our highly developed moral sense, nevertheless he did think that it was responsible for morality at some crude level, for its origins albeit not its development (1864 original and revised versions, 1864a, 1869, 1870, pp. 332–71, 1870a, 1889, pp. 445–78). As we have noted, he proposed the theory that, at a certain point in our evolution, selection for our mental qualities became more important than selection on our bodies. Among these mental qualities is our capacity for being 'social and sympathetic':

By his superior sympathetic and moral feelings he [man] becomes fitted for the social state; he ceases to plunder the weak and helpless of his tribe; he shares the game which he has caught with less active or less fortunate hunters, or exchanges it for weapons which even the weak or the deformed can fashion; he saves the sick and wounded from death; and thus the power which leads to the rigid destruction of all

animals who cannot in every respect help themselves, is prevented from acting on him.
                                                    (Wallace 1864: 1891 reprint, p. 184)

Wallace recognises that such altruism may appear to run counter to natural selection:

> we meet with many difficulties in attempting to understand how those mental faculties, which are especially human, could have been acquired by the preservation of useful variations. At first sight, it would seem that such feelings as those of abstract justice and benevolence could never have been so acquired, because they are incompatible with the law of the strongest, which is the essence of natural selection.   (Wallace 1891, pp. 198–9)

Like Darwin, he points to the advantage that altruistic groups will experience in competition with other groups:

> we must look, not to individuals, but to societies; and justice and benevolence exercised towards members of the same tribe would certainly tend to strengthen that tribe and give it a superiority over another in which the right of the strongest prevailed, and where, consequently, the weak and the sickly were left to perish, and the few strong ruthlessly destroyed the many who were weaker.
>   (Wallace 1891, p. 199)

But this is no explanation. What is natural selection favouring? One could equally argue that a group with a relatively high proportion of weak and sickly members would be at a considerable *dis*advantage.

What is more, Wallace tends to overlook the possibility of a conflict between what is good for the race or tribe and what is good for the individual. Uncharacteristically, he seems to assume in some vague way that natural selection will have both interests at heart. Even though some of the qualities that he talks about are obviously self-sacrificial, he doesn't seem to appreciate the costs:

> mental and moral qualities will have increasing influence on the well-being of the race. Capacity for acting in concert for protection, and for the acquisition of food and shelter; sympathy, which leads all in turn to assist each other; the sense of right, which checks depredations upon our fellows; the smaller development of the combative and destructive propensities; self-restraint in present appetites, and that intelligent foresight which prepares for the future, are all qualities that from their earliest appearance must have been for the benefit of each community, and would, therefore, have become the subjects of natural selection. For it is evident that such qualities would be for the well-being of man, would guard him against external enemies, against internal dissensions, and against the effects of inclement seasons and impending famine, more surely than could any merely physical modification.
>   (Wallace 1864: 1891 reprint, pp. 173–4)

As with Darwin's talk of selection between groups, one can only wonder what he had in mind.

Huxley: Morality at enmity with nature

If it seems odd to find Wallace the arch-adaptationist stopping short at a Darwinian explanation of morality, it is no less odd to find T. H. Huxley, Darwinism's self-appointed public relations officer, doing the same (albeit for different reasons). Huxley eventually came to believe that our morality must be the result of cultural evolution alone, a battle against the dictates of natural selection, a conscious and arduous intervention in nature's course. The good in us cannot have arisen from evolutionary forces; the struggle for existence is so profoundly red-in-tooth-and-claw that it would strangle a developing morality at birth (Huxley 1888, 1893, 1894; see also Paradis 1978, pp. 141–63).

Well, almost at birth. Like Darwin and Wallace, Huxley did see natural selection as cosseting the first glimmerings of goodness. There can, after all, be adaptive advantages in worthy behaviour like cooperation. Think of the beehive, Huxley urges us, and one can immediately see how natural selection could favour what is ethically right. Unfortunately, the way that Huxley sees it seems to be group-selectionist, although he appears to be unaware of the fact – or, if he is aware of it, unaware that group selection is not orthodox natural selection. According to Huxley, bees and many other social species prosper in the struggle for existence because some individuals selflessly sacrifice themselves for the good of the group, and groups that practice such enlightened social behaviour are at an advantage in competition with those that don't:

Social organization is not peculiar to men. Other societies, such as those constituted by bees and ants, have also arisen out of the advantage of cooperation in the struggle for existence ... Now this [bee] society is the direct product of organic necessity, impelling every member of it to a course of action which tends to the good of the whole ... [T]he devotion of the workers to a life of ceaseless toil for a mere subsistence wage, cannot be accounted for either by enlightened selfishness, or by any other sort of utilitarian motives ...   (Huxley 1894, pp. 24–5)

And, as with bees, so – in the beginning – with us:

at its origin, human society was as much a product of organic necessity as that of the bees. The human family, to begin with, rested upon exactly the same conditions as those which gave rise to similar associations among animals lower in the scale ... And, as in the hive, the progressive limitation of the struggle for existence between members of the family would involve increasing efficiency as regards outside competition.   (Huxley 1894, p. 26)

So at least the origins of morality emerge from and are part of the struggle for existence. What Huxley calls 'the ethical process' (the development of morality) has firm roots in what he calls 'the cosmic process' (evolution by natural selection):

> strictly speaking, social life, and the ethical process in virtue of which it advances towards perfection, are part and parcel of the general process of evolution ... Even in ... rudimentary forms of society [such as the beehive], love and fear come into play, and enforce a greater or less renunciation of self-will. To this extent the general cosmic process begins to be checked by a rudimentary ethical process, which is, strictly speaking, part of the former ...   (Huxley 1893, pp. 114–15)

But humans are not bees. '[R]ivalries and competition are absent from the bee polity' because 'the members of the society are each organically predestined to the performance of one particular class of functions only' (Huxley 1894, p. 26). With humans, however, the struggle for existence brings with it conflicts of interest. Humans have an inbuilt selfish desire 'to do nothing but that which it pleases them to do, without the least reference to the welfare of the society into which they are born ... That is their inheritance ... from the long series of ancestors, human and semi-human and brutal, in whom the strength of this innate tendency to self-assertion was the condition of victory in the struggle for existence' (Huxley 1894, p. 27). Among prehistoric and primitive peoples,

> the weakest and stupidest went to the wall, while the toughest and shrewdest, those who were best fitted to cope with their circumstances, but not the best in any other sense, survived. Life was a continual free fight, and beyond the limited and temporary relations of the family, the Hobbesian war of each against all was the normal state of existence.   (Huxley 1888, p. 204)

'The history of civilization ... is the record of the attempts which the human race has made to escape from this position' (Huxley 1888, p. 204). If we are to be moral beings, we must rise above our biological heritage, we must struggle against it. Our weapons must be culture and education. The development of morality cannot be a Darwinian development, for morality must work against nature: 'since law and morals are restraints upon the struggle for existence between men in society, the ethical process is in opposition to the principle of the cosmic process, and tends to the suppression of the qualities best fitted for success in that struggle' (Huxley 1894, pp. 30–1).

the practice of that which is ethically best ... involves a course of conduct which, in all respects, is opposed to that which leads to success in the cosmic struggle for existence. In place of ruthless self-assertion it demands self-restraint; in place of thrusting aside, or treading down, all competitors, it requires that the individual shall not merely respect, but shall help his fellows; its influence is directed not so much to the survival of the fittest, as to the fitting of as many as possible to survive. It repudiates the gladitorial theory of existence ... the ethical progress of society depends, not on imitating the cosmic process, still less in running away from it, but in combating it.
(Huxley 1893, pp. 81–3)

How is this achieved? Again, Huxley seems to assume that what is best at some higher level will prevail over individual selfishness, that members of society will practise voluntary self-sacrifice for the greater good: 'Morality commenced with society. Society is possible only upon the condition that the members of it shall surrender more or less of their individual freedom of action ... Thus the progressive evolution of society means increasing restriction of individual freedom in certain directions' (Huxley 1892, pp. 52–3).

So human beings are the product of natural selection, but to be humane we must civilise our natural legacy: 'ethical nature, while born of cosmic nature, is necessarily at enmity with its parent' (Huxley 1894a, p. viii). 'We cannot do without our inheritance from the forefathers who were the puppets of the cosmic process; the society which renounces it must be destroyed from without. Still less can we do with too much of it; the society in which it dominates must be destroyed from within' (Huxley 1894a, p. viii). But once we have reached a high level of moral development, Darwinian forces can no longer shape us. By ensuring that all members of society have the means of existence, human beings deprive natural selection of its power.

For Huxley, then, culture necessarily contravenes natural selection's preferences. Cultural evolution had to steam ahead in opposition to genetic evolution in order to make us what we are. And for a Darwinism as red-in-tooth-and-claw as Huxley's, it would surely have to. Today's Darwinism, however, knows better. Our most admirable qualities may indeed be the legacy of culture rather than natural selection. But there is no need to assume that they must be, that we have to depend on cultural evolution if we are to rise above the selfishness of our genes. Natural selection does not preclude self-sacrifice, good deeds, kindness, concern for others. Darwinian paths can lead to altruism. And they can do so by several routes, most obviously by mutual cooperation and kin selection. So Huxley was wrong to think that if we meet a moral act then we must necessarily attribute it entirely to culture and learning. Natural selection could have been the instructor. (For a varied selection of modern attempts to relate culture to our Darwinian inheritance see e.g. Alexander 1979, 1987; Boyd and Richerson 1985; Cavalli-Sforza and Feldman 1981; Lumsden and Wilson 1981, 1983 – but on Lumsden and Wilson's 1981, note that, although some commentators have taken it seriously (e.g. Ruse 1986), it also has powerful detractors (e.g. Maynard Smith and Warren 1982).)

The idea that cultural norms somehow manage on the whole to incorporate a higher-level good probably reached its peak in the first half of the twentieth century with functionalist theories of sociology and anthropology. These theories often explicitly purported to be Darwinian. But they were notoriously vague about the mechanisms by which the higher level prevailed over individual selfishness (for criticisms see e.g. Elster 1983, pp. 49–68; Jarvie 1964, pp. 182–98). Recently, some social scientists have attempted to develop more sophisticated functional analyses. Nevertheless, they have not always managed to avoid the group-selectionist trap. Consider, for example, the commendably readable book *Cows, Pigs, Wars and Witches* (1974) by the American anthropologist Marvin Harris. (This is not to suggest that Harris's work is egregiously groupish; I have used it in illustration in part because he himself draws attention to what he perceives as its distinctively Darwinian character, and in part because its influence extends well beyond the world of academic anthropology.) Harris explains a wide variety of cultural practices, from sacred cows to unclean pigs, as biologically functional. Unfortunately, his idea of biological optimality often seems to be covertly group-selectionist. He assumes that what is 'good' biologically will evolve culturally but he doesn't always ask at which level natural selection or cultural selection acts. Take his discussion of pig-love among the Tsembaga, a tribe in New Guinea. He describes a regular cycle, recurring about every twelve years, which involves, among other things, clan warfare, ancestor appeasement, and a year-long pig festival that wipes out the herd. According to Harris: 'Every part of the cycle is integrated within a complex, self-regulating ecosystem, that effectively adjusts the size and distribution of the Tsembaga's human and animal population to conform to available resources and production opportunities' (Harris 1974, p. 41). That sounds more like the vague greater-goodism of old-fashioned ecology than a respectable Darwinian model. And his explanation of the Jewish and Islamic view that pigs are unclean is also suspect: 'the Bible and the Koran condemned the pig because pig farming was a threat to the integrity of the basic cultural and natural ecosystems of the Middle East' (Harris 1974, p. 35). I should mention that Harris's more recent book, *Good to Eat* (1986), which argues that many apparently arbitrary cultural food preferences are actually of biological advantage, tries to be more careful about levels of selection and less groupish: 'bad foods, like ill winds, often bring someone some good. Food preferences and aversions arise out of favorable balances of practical costs and benefits, but I do not say that the favorable balance is shared equally by all members of society' (Harris 1986, pp. 16–17). It would, however, be more reassuring if he didn't try to look for 'balances' at a group level at all. And even more reassuring if his notion of 'favorable' didn't smuggle in Darwinian overtones, ranging loose, as it does, over financial profit, ecological benefit, and so on – 'favorable' perhaps, but surely not Darwinian adaptations.

Spencer: Darwinian bodies, Lamarckian minds

Our last nineteenth-century evolutionist is the social philosopher Herbert Spencer. Spencer thought that Huxley's position was, to use his term, 'ridiculous'. Here is his neat summary of Huxley's view. Negate every statement, and you will have an equally neat summary of Spencer's own position:

> his view is a surrender of the general doctrine of evolution in so far as its higher applications are concerned, and is pervaded by the ridiculous assumption that, in its application to the organic world, it is limited to the struggle for existence among individuals under its ferocious aspects, and has nothing to do with the development of social organization, or the modifications of the human mind that take place in the course of that organization ... The position he takes, that we have to struggle against or correct the cosmic process, involves the assumption that there exists something in us which is not a product of the cosmic process ...   (Duncan 1908, p. 336)

Spencer, then, favoured a thoroughly biological account of human morality. To put 'Man and Nature in antithesis' was, he thought, profoundly mistaken (Duncan 1908, p. 336). Morality, he argued, is peculiar to humans but it is nevertheless the result of biological evolution. This far he was with Darwin rather than Wallace or Huxley. But when it came to the question of which evolutionary force was responsible, natural selection was firmly ruled out. According to Spencer, only the inheritance of acquired characteristics could have been the agent.

In some species, Spencer argues, it is the most intelligent members that win the struggle for existence because their intelligence enables them to respond the most creatively to selective pressures (Peel 1972, pp. 125, 127). This is more true of humans than of any other species, and more true of 'cultured' humans than of 'primitives'. Humans can, in particular, increase their efficiency. This will involve division of labour. This in turn will involve interdependence and a vast network of social relationships. And altruism is likely to form a strand in that network (Peel 1971, pp. 138–9, 1972, pp. 25–6,

36–7, 160–1). Altruism, then, is very much a human attribute, far removed from the 'lower animals'.

Why is the development of altruism Lamarckian rather than Darwinian? It is because, according to Spencer, Darwinism is merely an exterminating force whereas Lamarckism is creative. Spencer views natural selection (what he calls 'indirect equilibration') as purely passive, whereas Lamarckism ('direct equilibration') (Peel 1971, pp. 142–3, p. 295, n42) involves an intelligent adaptive response on the part of the organism (although, unlike most Lamarckians, he believes the response to be in some way mechanical rather than willed (Bowler 1983, pp. 69–71)). So, as organisms increase in intelligence, the importance of natural selection declines and that of Lamarckian forces rises. Among 'civilised races' this has gone so far that the work of natural selection is restricted to the destruction of the feeble (Spencer 1863–7, i, pp. 468–9). By contrast, Lamarckian evolution is busiest at this stage. Altruistic behaviour is the pinnacle of evolution (Peel 1971, pp. 152–3, 1972, p. xxxiv) and it evolves through creative cooperation; in Spencer's view, it must therefore be Lamarckian (e.g. Peel 1971, p. 147).

For Spencer, Lamarckism had a special appeal. He believed that 'a right answer to the question whether acquired characters are or are not inherited, underlies right beliefs, not only in Biology and Psychology, but also in Education, Ethics and Politics'; 'as influencing men's views about Education, Ethics, Sociology, and Politics, the question whether acquired characters are inherited is the most important question before the scientific world'; 'a grave responsibility rests on biologists ... since wrong answers lead ... to wrong beliefs about social affairs and to disastrous social actions' (Spencer 1863–7, revised edn, i, pp. 650, 672, 690; see also Spencer 1887, pp. iii–iv). His vision was that the inheritance of acquired characteristics would bridge biological and cultural evolution, forging them into one grand seamless process (Peel 1971, p. 143; Young 1971, p. 495).

As we saw when we looked at Lamarckism, a belief in Lamarckian inheritance has often been fuelled by that same vision. The hope is that the best ideas of one generation will automatically be transferred to the next without the grind of education, training and indoctrination. We also saw then that, ironically, such aspirations rely on Lamarckism for the one thing that it is necessarily incapable of delivering. Let us leave aside the questionable assumption that such inheritance would be more progressive than conservative, that people would be liberated rather than imprisoned by genes that are committed to the attitudes of their parents. Even without this problem, Lamarckian forces could never be in the vanguard of social change. Lamarckian evolution is an instructive rather than a selective mechanism. And one thing above all that instructive mechanisms cannot do is to initiate, to take creative steps. For real novelty, they must in the end rely on selective mechanisms. So Lamarckian processes must ultimately be shaped by Darwinian ones. Admittedly, I am not sure what this Lamarckian model would look like when transposed to the world of ideas, which is what we are talking about here. But, if it really was impeccably Lamarckian, then presumably the same problems would arise with innovation in ideas as with novel structures and behaviour. Contrary to Spencer's ardent wishes, then, the inheritance of acquired characteristics could never be the spearhead of social engineering. At best, it could only reinforce changes that are sparked off by other forces.

It has often been remarked that cultural evolution is Lamarckian. When Spencer said it, he meant it literally. Nowadays Darwinians mean it only figuratively: 'Psychosocial evolution ... is an evolution in the Lamarckian style, in the sense that a father's particular knowledge and skills and understanding can indeed be transmitted to his son, though not (as Spencer supposed) through genetic pathways' (Medawar 1963, p. 217). Cultural transmission, then, can be construed as the 'inheritance' of acquired characteristics; what is learnt in one generation is acquired by the next. But we don't need to go to Darwin only for our genes and cross over to Lamarck for our culture. Cultural evolution, too, can be construed in a Darwinian way. It depends on what we take as being fundamentally Darwinian, what we take as being diagnostic of Darwinian processes. Darwinism can be understood in its most general form as a theory of the selection of replicators (as we saw when we analysed higher-level explanations of altruism). On this analysis, genes steered by natural selection need not be the only candidates for Darwinian models. 'Memes' (cultural units of replication) steered by cultural selection could also fit Darwinian specifications (Dawkins 1976, pp. 203–15, 2nd edn., pp. 322–31, 1982, pp. 109–12; for other theories of cultural evolution on Darwinian lines see e.g. Boyd and Richerson 1985; Cavalli-Sforza and Feldman 1981). If we think of Darwinism in this way, then to say that cultural evolution is Darwinian need be no mere analogy. Cultural evolution could make some claims to being as Darwinian as the evolution of life on earth.

Spencer's reason for rejecting natural selection as the force behind human morality turns a familiar line of argument almost completely on its head. The more standard position is Huxley's: Darwinian forces are too cruel, too relentless to have fostered altruism. For Spencer, however, Darwinian struggle, far from being too officiously self-seeking, is too passive to account for the complexity of social relations that morality involves. According to him, altruism requires a biological mechanism that can incorporate active responses, especially cooperation. Spencer dismisses Darwinian struggle as the agent of moral development not because it involves too much striving but because it involves too little.

Incidentally, if you have ever tried to wade through Spencer's writings and wondered why many of his contemporaries thought of him as one of the greatest thinkers of the time, you might be comforted by Darwin's comments on his work. Darwin is very frequently quoted as saying of Spencer (in a letter to E. Ray Lankester in 1870): 'I suspect that hereafter he will be looked at as by far the greatest living philosopher in England; perhaps equal to any that have lived' (Darwin, F. 1887, iii, p. 120). And three other letters in *Life and Letters* and *More Letters* are also laudatory – although less surprisingly, as they were to Spencer himself (Darwin, F. 1887, ii, pp. 141–2, iii, pp. 165–6; Darwin, F. and Seward 1903, ii, p. 442). But elsewhere in those volumes Darwin's praise is more equivocal: 'wonderfully clever ... even in the master art of wriggling ... If he had trained himself to observe more ... he would have been a wonderful man ... [A] prodigality of original thought. But ... each suggestion, to be of real value to science, would require years of work ... With the exception of special points I did not even understand H. Spencer's general doctrine; for his style is too hard work for me' (Darwin, F. 1887, iii, pp. 55–6, 193; Darwin, F. and Seward 1903, ii, p. 235; see also pp. 424–5). And in letters that were not published in those collections he was even more forthright. In 1860 he told Lyell that Spencer's essay on population was 'Such dreadful hypothetical rubbish' and in 1865 he confided 'somehow I never feel any wiser after reading him, but often feel mistified [sic]'; in 1874 he wrote to Romanes: 'I have so poor a metaphysical head that Mr Spencer's terms of equilibration &c. always bother me and make everything less clear' (Freeman 1978, pp. 263, 264). Coming from someone with that view of philosophy, perhaps even the accolade 'greatest living philosopher' is not all that it seems. It was, after all, Darwin who said of another philosopher: 'he is a metaphysician, and such gentlemen are so acute that I think they often misunderstand common folk' (Darwin, F. and Seward 1903, i, p. 271).

Huxley developed his views partly as a criticism of Spencer. He saw Spencer as making a plea for a laissez-faire economy on the grounds that struggle was beneficial. Some commentators have protested that Huxley and the majority of subsequent critics have been mistaken about Spencer's defence of Victorian capitalism. He did not claim, they say, that although development involved struggle, the struggle was productive; on the contrary, he envisaged industrialisation, even as experienced in that period, as requiring cooperation rather than conflict (Carneiro 1967, p. 62; Peel 1971, pp. 125, 146, 151, 1972, p. xxi, pp. 170–1). Now, it may be that Spencer viewed Victorian England in this optimistic way. But his optimism cannot serve as justification of his position. Spencer blithely assumes that individual and societal benefits will tend to coincide (Carneiro 1967, pp. 62–71). Biologically, his conclusion rests on a 'greater good' view of evolution. And we have noted what is wrong with that. Politically, his conclusion rests on a liberal voluntarist view of cooperation, which sees relationships like landlord and tenant, boss and worker as necessarily of mutual benefit. If Spencer is to be exonerated from Huxley's charge, it is the implausible premiss that voluntarism aptly describes Victorian capitalism that requires defence.

Spencer also optimistically assumed that human social and moral progress would continue indefinitely. He thought of evolution in general as inherently progressive, and human altruism, spurred on by the selective pressures of social life, as particularly so. When Wallace read Spencer's *Social Statics* he, too, became convinced that our social qualities would undergo indefinite improvement (this was before he became converted to the view that our more sophisticated mental faculties were overdesigned and required supernatural explanation): 'If my conclusions are just, it must inevitably follow that the higher – the more intellectual and moral – must displace the lower and more degraded races; and the power of "natural selection", still acting on his mental organisation, must ever lead to the more perfect adaptation of man's higher faculties to the ... exigencies of the social state' (Wallace 1864: 1891 reprint, pp. 184–5; in a footnote in the original paper he acknowledges the inspiration of Spencer (p. clxx)). Darwin also, as we have seen, took the view that our moral evolution could continue without limit. So Darwin, Spencer and (the Darwinian) Wallace all believed that good eventually emerges from nature's own course, that at least some of her ways are ways of gentleness and some of her paths are peace. From this point of view, Huxley is the odd man out. To him, the natural state was 'bad' and progress towards 'goodness' could be achieved only by an uphill struggle, an unnatural intervention (e.g. Huxley 1894, pp. 81–3, 1888, p. 203).

From another point of view, however, Huxley is closer to Spencer than either of them would probably have liked. Although Spencer saw morality as a natural outcome of evolution, he also, as a Lamarckian, saw human striving as an essential contribution to that process. What's more, in stressing the role of struggle, Spencer's view finds itself with another unlikely bedfellow: a Marxist interpretation of human development. Marxists have, of course, traditionally been opposed to Spencer, arch-apologist of capitalism, for all of Huxley's political reasons and more.

And that brings us to one of today's views of human development that we haven't yet touched on: a modern Marxist view. (I cautiously say 'a' rather than 'the'; Marxism is not a place to look for consensus.) This has influenced critics of Darwinian explanations of human behaviour far beyond specifically Marxist circles; although I call this a 'Marxist' view, I mean to include those wider circles, too – 'marxist' with a very small 'm'. This position stresses the importance of non-biological forces in shaping human social life, the importance of economic, social and political influences as compared with Darwinian factors. (We have already noted that a similar outlook is endemic in anthropology and it is common in the social sciences generally; claims of this kind are not peculiar to 'marxism'.) According to this school of thought, the very notion of 'human nature' is misguided: 'evolutionary positivism purports to establish "human nature", a concept which is inherently ideological in the sense that it establishes a certain model of humanity as essential and thus "natural" ... we must affirm that humans are social beings and that their socialization cannot be removed like a veneer to reveal the naked human nature underneath' (Miller 1976, p. 278). The fixed component of our makeup is so insignificant and so general, the argument goes, that Darwinism can tell us little of interest about human affairs. It can inform us, for example, that all humans have loves and hates and fears and preferences, that we all want to eat when hungry, to find a mate, to be neither too hot nor too cold. But the greater part of our behaviour and psychology is not universal, not handed to us by evolution. It is culture-specific, specific to particular economies or social organizations: 'human biological universals are to be discovered more in the generalities of eating, excreting and sleeping than in such specific and highly variable habits as warfare, sexual exploitation of women and the use of money as a medium of exchange' (Allen *et al.* 1975, p. 264) – or, indeed, 'anthropological and sociological observations indicate that even the most basic and widespread human functions such as sleeping, eating, and excreting are irrevocably socially conditioned' (Miller 1976, p. 278). So, for example, it is part of our Darwinian endowment that we all have a capacity for love, attachment, fondness. But to understand any of the forms this has taken in human history, such as the notion of romantic love, we have to look at the particular conditions of the society in which it arose – in this case Europe in recent centuries. A Darwinian explanation will necessarily be superficial, lacking in detail, it will necessarily fail to explain one of the most striking aspects of human behaviour, culture and social institutions: their diversity.

I don't want to get involved in romantic love. But I do want to point out that all of this need not be so far from a Darwinian stand as rhetoric would suggest. It is a mistake to assume that if we are equipped with behavioural rules then we are in the steel grip of a sphex-like human nature. Putting that the other way, it is a mistake to assume that behavioural plasticity demands an entirely open-ended, all-purpose mind. We have seen that, on the contrary, natural selection can enable us to act adaptively by equipping us with specially-tailored information-processing machinery, with specific, content-full rules that will generate flexible behaviour. It is obvious that rules for behaviour need not be rules for behavioural rigidity. Equally, it is obvious that an empty-slate mind or brain would be a very unDarwinian apparatus, which, far from rescuing us from rigidity, would leave us unable to behave at all (let alone adaptively); even a lowly induction machine cannot get off the ground without prior guidance, without some rules about what constitute sameness, repetition, pattern. So, if we observe humans behaving in a multitude of different ways, we needn't conclude that natural selection has had no hand in our behaviour. And if we do allow that natural selection has done more than merely shape our bodies whilst leaving our minds vacant, we have not thereby condemned ourselves to be natural selection's slaves.

I have used the idea of the *tabula rasa* as a shorthand for one set of views. But I should stress that this caricatures even the most diehard of Lockeans, including Locke himself. As Donald Symons points out, the dividing line between views of human nature has not been between innate-ists and *tabula-rasa*-ists but between innateness that is specific and hightly structured and innateness that is less so. 'Historically, there have been two basic conceptions of human nature: the Lockean, empiricist conception ... in which the brain/mind is thought to comprise only a few, domain-general, unspecialized mechanisms; and the Kantian, nativist conception, in which the brain/mind is thought to comprise many, domain-specific, specialized mechanisms' (Symons 1992). 'All psychological theories, including the most extreme empiricist/associationist ones, assume that mind has structure. No one imagines that a pile of bricks, a bowl of oatmeal, or a blank slate will ever perceive, think, learn, or act, even if given every advantage' (Symons 1987, p. 126). 'Every theory of human behaviour implies a human psychology. This includes theories that attribute human behaviour to "culture": if human beings have culture, while rocks, tree frogs, and lemurs don't, it must be because human beings have a different psychological makeup from that of rocks, tree frogs, and lemurs' (Symons 1992).

Incidentally, all that we have just seen suggests that we shouldn't look on free will and biological 'constraints' as pulling in opposite directions. 'On the contrary, one might plausibly argue that the proliferation of biological constraints protects man from manipulation by the environment' (Marshall 1980, p. 24) – indeed, that far from constraining us, these 'constraints' are the very instruments of free will. What's more – though this is just a matter of taste – neither need we look on them as impugning our dignity. On the contrary, doesn't it enhance our dignity more if we come into the world not as some kind of lightly grafitti-ed *tabula rasa* but as a complex bundle of capabilities and propensities, with preferences and tastes, with powers of discrimination and with our own ways of doing things?

Rhetorical skirmishes

Rhetoric litters the writings on human altruism. The samples we have just
seen happened to come from the 'marxist' literature. But it is to be found at
every turn. Take, for example, the notorious chapter on aggression in E. O.
Wilson's *Human Nature*. John Maynard Smith, among others, rightly took it
to task:

> It opens 'Are human beings innately aggressive? This is a favourite question of
> college seminars and cocktail party conversations, and one that raises emotion in
> political idealogues of all stripes. The answer to it is yes'. Now the reason that this
> opening raises emotion in political ideologues (including this one) is that it will be
> taken to mean that human beings are aggressive come what may, that war is
> inevitable, and it is therefore a waste of time to work for peace. But it turns out that
> Wilson does not mean anything of the kind. By saying that we are innately
> aggressive, he means only that we have shown aggressive behaviour, including
> warfare, in most, but not all, of the cultural environments in which we have so far
> found ourselves. He emphasises that the word aggression has been used to describe
> many disparate patterns of behaviour. He ends the chapter with a discussion of how
> we might circumvent our tendency to be violent towards one another. Given that
> these are his views, I think that the opening words of the chapter are unfortunate.
> They will certainly provoke controversy, but it is likely to be of that singularly
> useless type which takes place between people who do not understand one another.
>
> (Maynard Smith 1978d, p. 120)

And what about the following for unabashed biological determinism?
'Natural selection dictates that organisms act in their own self-interest ...
They "struggle" continuously to increase the representation of their genes at
the expense of their fellows. And that, for all its baldness, is all there is to it;
we have discovered no higher principle in nature'. 'If we are programmed to
be what we are, then these traits are ineluctable. We may, at best, channel
them, but we cannot change them, either by will, education, or culture'.
There's die-hard intransigence for you! But, actually, those quotes come not
from some ardent proponent of an all-in-our-genes view but from Stephen
Gould, a voluble critic of selfish gene-ery in general and of its application to
humans in particular (Gould 1978, pp. 261, 238). Now, wouldn't one expect
him to be saying something more to do with the supremacy of culture over
the apparently 'natural', perhaps something more like this?: 'The hatred of
indecency, which appears to us so natural as to be thought innate, and which
is so valuable an aid to chastity, is a modern virtue, appertaining exclusively
... to civilised life'. Or like this?:

> We have the power to defy the selfish genes of our birth ... We can even discuss
> ways of deliberately cultivating and nurturing pure, disinterested altruism –

something that has no place in nature, something that has never existed before in the whole history of the world. We are built as gene machines ... but we have the power to turn against our creators. We, alone on earth, can rebel against the tyranny of the selfish replicators.

But the first of those quotes is taken not from an implacable opponent of Darwinian human nature but from one of its unswerving advocates: Darwin himself (Darwin 1871, i, p. 96). And the second is from Richard Dawkins – also no idler when it comes to Darwinising (Dawkins 1976, p. 215). Nevertheless, theirs is a position that is often characterised as relentlessly genes-over-culture.

I should like to have approached this chapter in the same way as the other chapters on altruism, by looking at history in the light of the current consensus. But rhetoric stood in my way. There appears to be no current consensus. I suspect that the various positions differ less than many of their proponents would like to believe. If your suspicions are different from mine, I ask you to carry out this little test. Try to state the alternative views on human altruism without making any of them sound ridiculous, without having hastily to qualify them by 'Of course, it must be admitted that ...' or 'Obviously, nobody would deny that ...'. Admittedly, in my wanderings in the literature, I encountered unreconstructed *tabula rasa*-ists, rabid genetic determinists and other fabled beasts. But, significantly, they generally hang out in only two places: in manifesto statements and in the fevered descriptions of their opponents. I can't take either very seriously as science and it is science that is our concern. The difficulty in attempting a classification of views on human altruism is not that they are so diverse but that they are all so much the same.

16

Breeding between the lines

How does one species split into two? This, for Darwinians who are primarily interested in the diversity of life, has always been the problem of problems: the question of the origin of species itself. One issue in particular vexed Darwinians for many years. How do species come to be reproductively isolated? Why, if they try to interbreed, do their efforts usually peter out in sterility or low fertility? Why, if hybrids are born, are they likely to be sterile?

These have been important questions for Darwinism. But why discuss them under altruism? After all, for modern Darwinians there is no particular connection. The answer is that from before the time of Darwin until as recently as a few decades ago, repeated confusions have led to the idea that speciation, the separation of lineages into species and in particular the development of interspecific sterility, involves altruistic self-sacrifice. And attempts to correct this error have led to further confusions. As a first step to understanding this history, we shall look at how both the problem and its solution are viewed today.

The origin of species

Fundamentally, the problem of speciation is about how a single ancestral species can split into two without the incipient new gene pool being swamped by the old. When the dividing of the ways came for our ancestors and those of chimpanzees, how did they manage to part? After all, once upon a time, they were brothers and sisters. Why didn't they go on interbreeding and stay as one, locked in mutual embrace? Natural selection has no reason to favour speciation, no reason to view it as a good thing. And yet, with hindsight we can see that species have, as a matter of fact, divided, millions and millions of times. If they had not, all animals and plants would be one vast species (not even separated into animals or plants). The problem is how that splitting has come about.

Modern Darwinism does not assume that selection ever actively tries to make two species where there was one. Speciation is viewed as largely incidental, the main work of splitting populations into incipient species occurring merely by accident, usually a caprice of geography. Natural selection is seen as stepping in, if at all, only at a very late stage, applying the finishing touches, polishing up what has largely been completed.

Imagine that we could take any two quite closely-related modern species and view their history spread out before us, stretching back to the time just before they began to divide. What might we typically see? As we were watching our happily homogeneous interbreeding group, the first change to attract our notice would be a geographical barrier setting itself up. A river would swell, marooning some animals or plants on one side. A mountain range would become impassable. A 'narrow isthmus now separates two marine faunas; ... let it formerly have been submerged, and the two faunas ... may formerly have blended' (Darwin 1859, p. 356). A few animals might find themselves carried off on a tangled raft of leaves and branches, drifting along a mangrove swamp, until eventually setting shore far from their companions. Darwin remarked that 'fishes still alive are not very rarely dropped at distant points by whirlwinds; and it is known that the ova retain their vitality for a considerable time after removal from the water' (Peckham 1959, p. 612). Plants might be transported to distant islands by the unwitting agency of birds, the carrier and its cargo blown by gales vast distances across the seas. Darwin successfully germinated seeds retrieved from birds' crops, stomachs and excrement, some of them having been eaten by fish that the birds had then eaten – he examined, too, the amount of earth that could cling to a bird's foot – and he concluded that 'birds can hardly fail to be highly effective agents in the transportation of seeds' (Darwin 1859, p. 361; see also pp. 361–3). He also found that some seeds, especially if they had been dried, could float in seawater for long enough to be carried across wide oceans, and still germinate in spite of their immersion; and some species that would be killed by salt water within a few days could nevertheless journey unharmed in the floating corpse of an animal: 'some taken out of the crop of a pigeon, which had floated on artificial salt-water for 30 days, to my surprise nearly all germinated' (Darwin 1859, p. 361; see also pp. 358–61). Darwin discovered that newly hatched freshwater snails clung tenaciously to the feet of a duck and survived there, out of water, for up to twenty-four hours: 'in this length of time a duck or heron might fly at least six or seven hundred miles, and would be sure to alight on a pool or rivulet, if blown across sea to an oceanic island or to any other distant point' (Darwin 1859, p. 385). But geographical barriers need not involve heroic distances or mighty mountains. For a very small or not very mobile animal, slight physical separation could be enough, an insurmountable ten-yard gap between two trees, perhaps even the alien world on the other side of a leaf. Some of these barriers admittedly owe their existence to improbable, freak events. But that is no obstacle to speciation. For such events need not happen often. Even one occurrence could be enough. So, as Darwin said, 'what are called accidental means ... more properly might be called occasional means of distribution' (Darwin 1859, p. 358). A single storm-tossed bird, a single volcanic eruption, could profoundly affect the course of speciation, putting asunder what were potential mates, splitting the gene pool in arbitrary fashion.

The history has reached the end of the first stage: our initially interbreeding species has been severed into two or more fragments. But it is still one species. What, then, brings about divergent evolution, the gradual development of different forms? Here, again, nature's workshop offers a choice. Chance can make an impact. We noted, when we looked at the role of chance in evolution, that freak colonisation is unlikely to amount to a miniature replication of the parent species. More likely, the genes of the founder fragment will be some biased sample of the original gene pool, those genes that just happened to get lopped off by the intrusion of the barrier. And chance can also play a role after the initial colonisation, as we again noted, through genetic drift – the genes in any generation being chosen not by natural selection's non-random sampling of the previous generation but by sampling error. Then there is the fact that the parent and founder groups might well be in different environments – drier, hotter, windier – and so subject to different selective pressures. The most important of such environmental differences, as Darwin often insisted, are likely to be other organisms: 'Bearing in mind that the mutual relations of organism to organism are of the highest importance ... there would ensue in different regions, independently of their physical conditions, infinitely diversified conditions of life, – there would be an almost endless amount of organic action and reaction' (Darwin 1859, p. 408); there is a 'deeply-seated error ... [in] considering the physical conditions of a country as the most important for its inhabitants; ... it cannot, I think, be disputed that the nature of the other inhabitants, with which each has to compete, is at least as important, and generally a far more important element of success' (Darwin 1859, p. 400). This is why islands, exposed to strange climates, built from a strange geology, above all stocked with strange creatures, are often powerhouses of creation: a 'richness in endemic forms ... few inhabitants, but of these a great number ... endemic or peculiar' (Darwin 1859, pp. 396, 409). The picture, then, at the end of this second stage of the history is of two or more forms slowly diverging, prevented by physical separation from intercrossing, and thereby prevented from obliterating their ever-increasing differences.

But how do the groups become incapable of interbreeding? How do they become two different species? One thing that could happen is that the divided gene pools evolve so far apart that the two sets of genes become incapable of working together to program the development of a viable embryo. Another possibility is that they manage to form an embryo but that it turns out to be sterile – a 'mule'. Why such hybrids should be perfectly viable and yet sterile is still something of a mystery. In some cases the sterility may be caused by the problem of manufacturing gametes from chromosomes derived from very different parents; a mule's body cells contain intact horse chromosomes and intact donkey chromosomes and it has to bring them together to make mule gametes. Any geographically isolated gene pool will, as time goes by, accumulate its own idiosyncratic rearrangements of its chromosomal material; inversions of parts of chromosomes, and translocations from other parts of the genome, will be tolerated or even encouraged by natural selection. The resulting differences between the chromosomes of separated populations could become so wide that, as a by-product, the two kinds of chromosomes would not match up at meiosis. Of these two outcomes – full intersterility, resulting in no offspring at all, or partial intersterility, of the kind that produces 'mules' – partial 'success' is the worse fate. Total 'failure' is more welcome, for out-and-out sterility wastes little parental effort; the greater the 'failure' the better, the most economical failures being unions, like attempts to unite the chromosomes of humans and other animals, that never really get started. Nature is least obliging when it presents a barrier so meagre that it allows parents to produce viable offspring and lavish resources on them, only to see their efforts end in hybrid sterility, the aspiring river of their germ plasm dammed up forever in a dead-end mule.

Suppose now, as an end to the history, that the geographical barriers disappeared and these two more-or-less non-interbreeding, more-or-less non-interfertile groups mingled. At this point, their intersterility could cease to be incidental. Natural selection could suddenly take an interest in it. Now it could become important to prevent breeding between the two groups so that individuals would not squander their reproductive effort on abortive embryos, mules or their like. There could be selective pressure for complete barriers to interbreeding, for reinforcement of isolating mechanisms. Natural selection could work to perfect barriers that had already arisen without its direct aid, and could seize on any new ones that happened to arise. Differences in mate preferences, unsynchronised mating times, divergent choices of habitat, low interfertility, spontaneous abortion – all could be grist to selection's mill. Natural selection could actively strive to prevent the flow of genes between the two groups. And interspecific sterility could be one of its means, the final barrier against interbreeding when all other defences have failed. Natural selection undoubtedly could, in principle, do all this. Modern Darwinians agree that reinforcement by natural selection is theoretically possible.

Recently, however, some have challenged the widely held view that selection generally does act in this way (e.g. Barton and Hewitt 1985, particularly pp. 121, 137).

That is one history we would see. It is speciation in which geographical barriers play a crucial role (with or without subsequent reinforcement by natural selection). Allopatric ('other place') speciation is its name. Modern Darwinians are agreed that it has been the vital process in the speciation of most animal groups. More controversially, some Darwinians also claim that sympatric ('same place') speciation has been significant. For sympatric speciation, the initial barriers to interbreeding are not geographical (although some of the more slender barriers that we have called geographical, like the two sides of a leaf, might be included here). The clearest case occurs in plants, when instant reproductive isolation occasionally springs up by the sudden doubling of the chromosomes – so-called polyploidy – in hybrids. These hybrid offspring would otherwise have been sterile. But with the doubling of their chromosomes (from two (diploid) to four (tetraploid) or, more generally, any number more than two (polyploid)) they become able to breed with one another, each chromosome having a partner with which to pair at meiosis. At the same time, they are sterile with their parent species. When this occurs nature can announce the imminent birth of a new species. Dahlias, plums, flowering chestnuts, loganberries, swedes and many other plant species have come about in this way. The primrose *Primula kewensis* – the polyploid hybrid offspring of *P. verticillata* and *P. floribunda* – is a famous example.

In animals, however, new species rarely, possibly never, arise by polyploidy. With them, sympatric speciation might come about through, say, a change in feeding habits or choice of breeding place. Imagine a species of insect in which, through chance mutation, some individuals became able to cope with a new food plant. If those same individuals also preferred to lay their eggs on those plants and to mate with individuals with the same inclinations then the species would gradually split. Of course, it would, in John Maynard Smith's words, be 'demanding a miracle to suggest the chance origin of a new genotype which simultaneously influences the capacity of the larvae to grow on a new food plant, and the egg-laying habits and mating preferences of the adult' (Maynard Smith 1958, p. 225). But, as Maynard Smith makes clear, miracles are not in fact required. Suppose that, when laying their eggs, adults follow the rule: Choose the food plant on which you were reared. 'In such a case, the egg-laying preferences of females would be transmitted from generation to generation in the same way as languages are transmitted in our own species; it is genetically determined that human beings can learn to talk, but not that they shall learn English rather than French, or vice versa' (Maynard Smith 1958, p. 226). Suppose, too, that members of the species mate soon after they have emerged, so that they are most likely to mate with individuals on the same type of plant. Then the genetically determined food preference and the more environmentally determined egg-laying and mating decisions will reinforce one another. The incipient species will gradually be eased apart. A similar splitting could be imagined in, say, birds, through the influence of individual experience on nest site and mating preferences: 'To give an extreme example, domestic pigeons are descended from rock pigeons ... and London's pigeons are in turn descended from ... domestic pigeons. Yet the London pigeons remain effectively isolated from their wild ancestors by their choice of buildings instead of cliffs as nesting sites. Given time, they might well evolve as a distinct species' (Maynard Smith 1958, p. 227). On the sympatric model, then, natural selection can be powerful enough to increase adaptive differences between two forms even in the face of interbreeding. As a side effect, the groups gradually diverge and eventually start to become intersterile. At this point, as with allopatric speciation, selection acquires an interest in keeping the forms separate; it takes up a direct role, reinforcing failures to interbreed, including intersterility, and thereby turning the two forms into two species.

As well as allopatric and sympatric speciation, there is an intermediate possibility. Separate species could evolve from populations that, geographically, are neither separated nor intermingled but are continuous. This is called parapatric speciation (or semigeographic, semisympatric or stasipatric speciation). The considerations for and against this possibility are much the same as for sympatric speciation.

We see, then, that, according to modern theory, natural selection is indifferent to much of the process of speciation in general and to the development of interspecific sterility in particular. This is because the splitting into species of a successfully interbreeding group offers no benefits to individuals or genes. If selection intervenes at all, it is not until the final stages, when individual advantage is at stake. Natural selection could, of course, play an important role in adaptive divergence. But if it does, it is the adaptation that is its interest; the divergence is only a side effect. For selection, most of speciation is just an unintended by-product of other activities. Modern Darwinism does not assume that natural selection cares about speciation as such, that it ever actively works to drive a wedge through a species.

Speciating for the greater good

Ideas on speciation have not always been so clear. During Darwinism's period of greater-goodism, the failure of species to intercross was often seen as 'good for the species'. On this view, sterility was saintly, an altruistic refusal to reproduce, a sacrifice for the benefit of the species as a whole. To those whose sights were trained on the greater good, it seemed obvious that selection would foster interspecific sterility. Wasn't it better for each species to maintain its integrity rather than merge into an undifferentiated mass? Wasn't there a species-level advantage in keeping groups entirely separate, each with its own adaptations, its own way of exploiting its particular niche? If two varieties were diverging, wouldn't natural selection work for the general good of both sides and take pains to encourage the split? If a new form arose that could make use of an unoccupied ecological niche, wouldn't nature seize on the opportunity for speciation rather than let the aspirant cross back with the parent species? Certainly, some individuals might find themselves investing time and effort in unrequited courtship, wasting eggs or sperm in unsuccessful union, even condemned, as a hybrid, to a lifetime of sterility. But these individual sacrifices would bring rich benefits to the species.

Such was the thinking, whether explicitly or by default, of many a Darwinian during the decades of greater-goodism that pervaded evolutionary ideas for much of this century. I had a vivid glimpse of how this outlook influenced theories of speciation when I asked W. D. Hamilton for his opinion on today's views. The particular point that I asked him about we'll come to later, for thereby hangs another tale. The general issue was to do with the reinforcement of isolating mechanisms – the question of which mechanisms natural selection is able to reinforce when geographical barriers that formerly separated two incipient species break down. It turned out that I had touched on the very issue that had induced Professor Hamilton to write his first paper for publication. When he was an undergraduate in the late 1950s, he had been directed to the standard textbooks on the subject. There he found all kinds of isolating mechanisms lumped together, with no distinction between the adaptive and the incidental, between mechanisms that could have been the work of natural selection and those that would be outside its power. He realised that he was seeing the indiscriminate ecumenicalism of good-for-the-species thinking. No matter that, say, hybrid sterility was of no advantage to the hybrid or to its parents or to any identifiable individual. If it was serving some higher-level good – as, in a vague, often unarticulated, way it was commonly thought to do – then that was advantage enough. Natural selection would promote hybrid sterility because it was a way of keeping species apart, and keeping species apart was a good thing for those species. It was a paper in *New Biology* by the Professor of Botany at Belfast, J. Heslop-Harrison (Heslop-Harrison 1959), that finally moved Hamilton to express his misgivings. Unfortunately, this over-hopeful undergraduate was unaware that *New Biology* commissioned its articles, and his paper was never published. Fortunately, however, he kept the manuscript and was kind enough to search it out when we talked about it. It stands as eloquent witness to the prevalence of greater-goodism at that time.

There is, Hamilton wrote,

> a very fundamental distinction within the term [isolating mechanisms] which is seldom made ... An I.M. (hereafter used as abbreviation for 'Isolating Mechanism') which operates by failure of sexual attraction is likely to be a completely different *kind* of phenomenon from an I.M. that operates by hybrid sterility and the two should not be classified as equivalent unless it is certain that their nature is the same.

Sexual preference could be moulded by selection: it 'depends simply on the selective disadvantage at which the "hybrids" find themselves: if this is sufficiently severe *fission* of the species takes place by the actual *selection* of any character leading to less likelihood of the hybrids being formed'. Hybrid sterility, however, must be accidental – though it could instigate selection for sexual preference:

> this phenomenon [selection for sexual preference] ... is ... different from the phenomenon of hybrid sterility which is sometimes observed when allopatric subspecies re-meet ... [Hybrid sterility] is its antecedent, giving ideal conditions for the evolution of a mechanism to prevent mating. For as Fisher pointed out: 'The grossest blunder in sexual preference, which we can conceive of an animal making, would be to mate with a species different from its own and with which the hybrids are either infertile or, ... at so serious a disadvantage as to leave no descendants'.

To think of hybrid sterility as an adaptation is a mistake of population-level thinking: 'It is true that hybrid sterility does in fact prevent the exchange of genes, but that it does so must be regarded as quite fortuitous unless we are prepared to envisage the sub-populations themselves as reproductive bodies and selection discriminating between them, preserving the most fit'. And yet, although these two kinds of isolating mechanism come about in such different ways, they 'are usually classified as equivalent (e.g. in the classification of Dobzhansky set out in *Genetics and the Origin of Species*, and in the adaptation from it by Stebbins). The last issue of *New Biology* nowhere explicitly distinguishes them'. And Hamilton quotes the following statement from the article that was his catalyst (adding his own emphasis): 'selection will favour the establishment of a barrier to interbreeding as such, *since* this will protect the adaptive gene complexes of the two populations'. This, he objects, makes it look as if selection is interested in the welfare of the species: 'The conjunction "and" would seem more appropriate here. It is just where ideas of race and species fitness are involved that teleological ways of speaking tend to go astray'.

How, then, can we distinguish between those isolating mechanisms that could be the work of natural selection (which he suggests calling Isolating Devices – I.D.s) and those that could not? The answer, Hamilton says, is to keep our minds firmly on reproductive advantage to individuals. Take, for example, the progression in isolating mechanisms from hybrid sterility to mating preferences. The sterility cannot be the result of natural selection whereas the preferences can. And yet the progression is apparently smooth:

> What I.M.s can possibly be used as I.D.s? ... [T]he sterility of hybrid offspring is a character that cannot possibly be of advantage to an individual of a subspecies, although it might be to the subspecies as a whole. And yet sterility of F1, premature death of F1, failure to develop of zygote, failure of pollen tube penetration, form a series of steps which lead eventually to failure of sexual attraction between members of different subspecies in the case of animals or to pollination by different insects in the case of plants.

It is by concentrating on individual reproductive benefit that we can pin down the point at which the series divides, the split between the adaptive and the incidental:

> These latter phenomena [failure of attraction and pollination] are quite reasonably likely to be I.D.s: from the point of view of reproductive value those earlier in the series are catastrophes whereas in these there is no detriment at all – or hardly any, provided mates or pollinators are abundant. Somewhere a disjunction seems to have occurred and in general it will be at the point where an individual so far commits itself in making the inadvisable union that it reduces its expectation of representation by descendants in distant future generations.

This point will depend on the reproductive habits of the species:

> A wind-pollinated plant inevitably spreads its pollen on all sorts of stigmas besides those of its own species, but since the process of doing so is quite random, pollen which lands on them is no more reproductive value wasted than pollen which lands on the soil; all being allowed for in the abundant production. But if one of the plants so pollinated commits any of its limited number of ovules to development upon this stimulation, it might seriously reduce its expectation of fertile offspring. But again, if the incompatibility were such that the embryo were destined to die at an early stage, and if the ovary contained more ovules than it could possibly make into seeds if all were fertilised, then the plant would not lose reproductive value.

Under these circumstances an I.D. could arise which worked by killing the early embryos. Likewise a flower which sets crowded achenes could probably afford to let a few of them degenerate without decreasing its total out-put of seeds. Similar too would be the case of a tree whose seeds germinated in masses in its close neighbourhood: here again the more unfit (and hybrid) seedlings could die without affecting the tree's reproductive value. But if the seeds were so dispersed that competition for light and soil were with seedlings from another tree of the same species, then it would be a real loss to its parent if by its death the hybrid seedling yielded its place to an unrelated one.

Thus the lateness of the stage at which an I.D. can operate is determined by the circumstances of reproduction pertaining to the species in question. To the case of the wind-pollinated plant above, one might compare the case of an orchid where it could be disastrous to give up the pollinia to an insect that was going to fly straight to flowers of the other sub-species: in this case one might expect an I.D. operating at the stage of insect attraction.

It is clear, then, that there is no unique, universal point in the sequence of reproduction at which selection puts a stop to the reinforcement of isolating mechanisms. Natural selection's decisions will depend on the costs to the would-be parents, above all the opportunity costs. These may well differ from species to species, between male and female, even for the same individual at different stages in its reproductive career.

Would that such ideas had been heeded! Darwinians did, indeed, become aware of the need to sort out adaptive from incidental isolating mechanisms. But the attempt to do so ensnared them in further misconceptions. Several of them, bending over backwards to avoid the 'speciation is altruistic' trap, bent too far, and landed in another pitfall. If we are to understand Darwin's and, particularly, Wallace's solutions to the problem of interspecific sterility, and, what's more, the judgements of today's biologists and historians on those solutions, we need first to understand – and to clear up – these mistakes. This is the subject of the next section.

Selection's great divide: Mating or weaning?

An odd idea stalks the literature on speciation, and has done since the early 1960s. It is the idea that natural selection can reinforce premating isolating mechanisms but has no power over postmating ones. Premating mechanisms include, for instance, a choice of breeding at different seasons or in different places; the failure of sexual attraction would come in here. Postmating mechanisms would be, say, the rejection of a non-viable foetus; hybrid sterility would be placed here. On this view, then, selection can foster a disinclination to mate but not a 'disinclination' to retain an embryo whose destiny is sterility; it can foster a tendency to prefer one mating call to another but not a tendency to prefer one offspring to another.

Of course, in practice, as we have noted, natural selection may have no detectable effect. N. H. Barton and G. M. Hewitt have claimed that, 'despite the popularity of the idea that selection may increase ... isolation within hybrid zones, there is remarkably little evidence for such reinforcement. We could find only 3 plausible cases' (Barton and Hewitt 1985, p. 137). One of the instances that they allow concerns two species of narrow-mouthed frog, *Microhyla olivacea* and *M. carolinensis*, in the United States (Blair 1955). The mating calls of the two species are generally very alike. But in the areas where the species meet, there is a 'striking divergence' in the pure forms (Blair 1955, p. 478). In these places the call of *M. olivacea* is at a higher pitch than *carolinensis*, it averages nearly twice the length of its neighbour's effort, and it also starts with a distinctive peep, which *carolinensis* never uses. Indeed, *olivacea* is so transformed in the overlap zones that 'the call of the Arizona *olivacea* resembles the call of *carolinensis* more than it does that of Texas and Oklahoma *olivacea* [which are in or near the overlap zone]' (Blair 1955, p. 474). This is a text-book example. Most Darwinians assume that it is typical. However, whether or not most Darwinians are right, we are not for the present concerned with what selection actually does. Our concern here is with what it could do, in principle: Could it reinforce post-mating, as well as premating, mechanisms? Time after time, the literature tells us that it can't.

Here, for example, is John Mecham in the early 1960s:

there is good reason to believe that postmating mechanisms are rarely if ever established or intensified through the action of natural selection. Considerable disagreement exists in the literature regarding the significance of natural selection with respect to reproductive isolating mechanisms, and this disagreement is due, at least in part, to the fact that the clear-cut differences in the adaptive significance of premating and postmating mechanisms has not been clearly understood. According to the theory propounded by Dobzhansky (1940 and elsewhere) and based on Fisher (1930), if hybridization takes place to any extent between two different forms the hereditary factors which promote intraspecific (as opposed to interspecific) matings are more likely to be perpetuated (will be selected for). This is due to the fact that offspring produced by crossing between different integrated genotypes are probably adaptively inferior in most cases to the pure parent types, and genes expended in the production of hybrids are thus less likely to survive. It therefore follows that under these conditions reproductive isolating mechanisms between species should become intensified through the action of natural selection in zones of overlap. A very important point here which has generally been overlooked (by Dobzhansky as well as others) is that postmating isolating mechanisms cannot be strengthened through such a process because they in no way act to promote homozygous, as opposed to heterozygous, matings. The theory, therefore, is applicable to premating isolating mechanisms only.

... hereditary factors enhancing reproductive isolation will have superior survival value over those not doing so in cases where some natural hybridization takes place. This theory covers any premating isolating mechanism ... but ... cannot apply to postmating isolating mechanisms, none of which favor intraspecific over interspecific matings.   (Mecham 1961, pp. 43–4, 50)

Moving to the 1970s, we find Theodosius Dobzhansky, the eminent evolutionary geneticist that Mecham cited, having perhaps taken those very words to heart: 'Postmating mechanisms, such as hybrid inviability and sterility, are byproducts of genetic divergence: premating ones are contrived by natural selection to mitigate or eliminate the losses of fitness which result from hybridization of genetically divergent and differentially adapted forms' (Dobzhansky 1975, p. 3640). In the 1980s, Murray Littlejohn, in a comprehensive review of reproductive isolation, said that only premating reproductive isolating mechanisms 'are amenable to the direct action of natural selection for their isolating effect per se' (Littlejohn 1981, p. 300). (Nevertheless, this view is not universal (see e.g. Coyne 1974; Grant 1966, p. 100, 1971, p. 180).)

This position is the very opposite of the stance that goaded Hamilton. He found natural selection, in the name of the greater good, being credited with far-fetched abilities to fine-tune fertility at any point, even to sterilise hybrids or their hybrid offspring. But now, in this distinction between premating and postmating, we find natural selection, on the contrary, being credited with too little control, its influence assumed to end at the meeting of egg and sperm. We shall see that this was a backlash against good-for-the-species explanations, an attempt to oust their vague, all-embracing notion of selection's powers. The aim of these critics was to distinguish between reproductive isolating mechanisms that could be the result of natural selection acting at the level of the individual and those that couldn't. But their solution was mistaken.

Natural selection recognises no such distinction between pre- and postmating. Certainly, it is far less efficient to put up postmating barriers than to prevent mating in the first place. Nevertheless, natural selection could well decide that even an inefficient barrier is preferable to allowing the dead weight of eventual sterility to take its course. Given the chance, selection might well encourage the abortion of a low-fertility hybrid embryo if, by doing so, it freed the mother to find a mate that could endow her with grandchildren. Why should she carry a mule to term if she could bear a fertile son or daughter in its stead? Better to abort it, or to let it die unweaned, than to shell out resources that could fall on more fertile ground. Of course, natural selection would have done better to have prevented the unhappy union from occurring at all. But if an unpromising zygote does slip through the net of other barriers, selection could still act rather than stand idly by. Admittedly, it is hard to imagine how natural selection could go so far along this line as to favour hybrid sterility. Such a theory would have to assume a rather unlikely combination of circumstances. If, say, the children of my fertile hybrid offspring would be sterile (what is called hybrid breakdown) but would compete with my non-hybrid grandchildren, and if, by some means or other, I had the power to manipulate the fertility of my offspring, then hybrid sterility could well be an adaptation; genes in manipulative parents would be favoured over parental genes that let all grandchildren, sterile and fertile alike, scramble for the same limited resources. Well, that is a far-fetched case. But it illustrates the moral: If we are seeking selection's cut-off point, the distinction between pre- and postmating is irrelevant and misleading. The correct distinction is pre- and post'weaning', where 'weaning' stands for any parental investment. It is this that is the great divide. On one side, natural selection could act to save costs to the parents; on the other side, it stands powerless.

I was puzzled when I first found this spurious distinction cropping up. What lay behind it? This was the question that I asked of Professor Hamilton. It was when he told me how he had come to write his unpublished paper that I realised that the confusion had arisen from a reaction against greater-goodism. The distinction to be found in the literature nowadays is a remnant of that reaction. The criticisms were mistaken, an over-reaction. But that remnant has nevertheless survived right into the selfish-gene Darwinism of today.

What had happened was this. The idea of isolating mechanisms goes back, as we shall see, to nineteenth-century Darwinism. But it was Dobzhansky, in his classic *Genetics and the Origin of Species*, who set out the idea systematically and secured its place in the modern synthesis, coming up with a now-famous and much-quoted categorisation (Dobzhansky 1937, 1st edn., pp. 228–58). And yet he showed scant interest in which of those mechanisms could be adaptations and which were likely to be merely incidental by-products of divergence during geographical separation; indeed, only once does he explicitly mention selection (Dobzhansky 1937, 1st edn., p. 258). It was in a paper a few years later that Dobzhansky acknowledged a need to consider natural selection's role, a need to distinguish what could be selected from what must be incidental (Dobzhansky 1940). Subsequent editions of his book reflect the change; compare 'The origin of isolation', pp. 254–8 in the first edition, with 'Origin of isolating mechanisms', pp. 280–8 in the second and 'Reproductive isolation and natural selection', pp. 206–11 in the third.

The usual view of speciation, he said in his paper, is that isolating barriers accumulate only as a side effect. But divergence will not generally be enough to bring about reproductive isolation; that will have to be the work of natural selection. 'The basic problem', then, 'is how frequently and to what extent can the isolating mechanisms be regarded as adaptational by-products arising without the intervention of ... special selective processes' (Dobzhansky 1940, p. 320). This raised the question of why natural selection would want to achieve isolation. To this he answered: 'Each species, genus and probably each geographical race is an adaptive complex which fits into an ecological niche somewhat distinct from those occupied by other species, genera and races' (Dobzhansky 1940, p. 316); hybridization jeopardises the integrity of these adaptive complexes. Now, it is not clear whether Dobzhansky was thinking at an individual level or a higher level. But for Darwinians steeped in vaguely group-selectionist ideas, such talk must have been ripe with ambiguity. We have already found 'adaptive complexes' cropping up in the article that Hamilton criticised. Here is John Moore, in the mid-1950s, in his 'embryologist's view of the species concept', explaining that, contrary to what he takes to be Dobzhansky's thesis, natural selection might not always be interested in isolating mechanisms because they are not always good for the species:

> we must not assume that the wastage of gametes is always a disadvantage to the species. There are some special cases where the wastage of gametes might be a real advantage to the population as a whole. As a hypothetical case let us consider a species whose numbers are held in check by predators and available food. If the source of predation is then removed, there will be greater competition for the available food. In some species severe competition may result in 100 per cent mortality. In this unusual situation, some wastage of gametes would be an advantage.   (Moore 1957, p. 336)

It was against this background that Mecham set up the premating–postmating distinction. It was intended to cut through the proliferation of isolating mechanisms that were being attributed to natural selection, to sort out more carefully what individual selection really could and could not achieve. The ultimate goal was apparently to eliminate group-selectionist assumptions. And, in the minds of some Darwinians, this was certainly how it came to be seen. Dobzhansky, for example, stated in 1970:

> Whether postmating isolating mechanisms can be reinforced by natural selection is ... an open problem. If the progeny of hybrids is inferior in fitness, it would seem advantageous to the species concerned to prevent hybridization, either by premating isolation or, failing that, by such postmating mechanisms as inviability or sterility of F1 hybrids. Group selection could, theoretically, bring such a result about. However,

because the efficiency of group selection is low relative to the selection of individual genotypes, it is doubtful that isolating mechanisms frequently arise in this way.

(Dobzhansky 1970, p. 382)

The fallacies that generated the distinction between premating and post-mating as selection's dividing line are by now familiar to us. The first, which we shall meet again with Darwin and Wallace, is a failure to appreciate costs, most notably, in this case, opportunity costs. For natural selection, the game need never be over as long as there is expenditure that can be saved for further reproductive effort. And, in principle, such opportunity costs could present themselves even at the very end of 'weaning'. However much time and food and energy a parent has expended on a sterile son or daughter, that parent would do better to abandon its charge than to persist in caring for it, if time and food and energy were thereby released for fertile offspring. This would, admittedly, be a profligate way to proceed, but less profligate than to plod on, passing up more promising reproductive opportunities.

The second fallacy results from individual- rather than gene-centred thinking. According to one adage of the premating versus postmating school of thought, natural selection must act on the parental generation (e.g. Murray 1972, p. 81). On an individual-centred view, it's an easy slide from there into the idea that natural selection must act on the parents themselves – not on the hybrid offspring, not even on the foetus. Another dictum has it that natural selection is out to avoid 'wastage of gametes' (e.g. Mecham 1961, p. 43). Again, it's easy to drift into the idea that once a gamete is fertilised its fate is sealed, and natural selection can have no further interest. I'm not sure why, out of the whole vast reproductive effort, gametes get so elevated. But it is clear that they would have shrunk into more reasonable perspective if greater-goodism had been criticised from a gene-centred, rather than a parent-centred, point of view. And the same is true of another conceptual transmutation – from the idea that natural selection tries to avoid 'a wastage of reproductive potential' (e.g. Mecham 1961, p. 26) into the idea that, from the crucial moment of fertilisation, that potential is forever fixed and natural selection's influence over it is at an end.

The problem for Darwin and Wallace

Look up 'sterility' in Ernst Mayr's index to the first edition of the *Origin* and then in Darwin's own index to the same edition and you will find a telling difference. Mayr directs us to 'isolating mechanisms' (Darwin 1859, p. 501). Darwin's entry looks more like a breeder's handbook: sterility in hybrids, the laws and causes of sterility, the conditions that induce it (Peckham 1959, p.

813). In Mayr's index, 'sterility' is to do with speciation, with the origin of species as we understand it today. For the Darwin of 1859, however, the aim of his chapter on sterility was to argue against separate creationism, against the anti-evolutionary view that species owe their origin to divine intervention. His interest was not in isolating mechanisms, or in any of the other tools of speciation, but in showing that there is no important difference between species and varieties.

Now look up the same index entry in later editions of the *Origin* and in Wallace's *Darwinism*. A new issue has entered: whether sterility is adaptive. And whereas Darwin's entry reads 'not induced through natural selection', Wallace's reads 'of hybrids produced by natural selection' (Peckham 1959, p. 813; Wallace 1889, p. 492). During the 1860s, this became the focus of Darwin's interest in sterility. And by the late 1860s, it had become an area of profound disagreement between Darwin and Wallace.

To understand Darwin's and Wallace's views on interspecific sterility and on speciation we need to understand what it was that they were opposing. So we'll look now at their opposition to creationism, at the significance of explaining interspecific sterility adaptively – and at a curious convergence between these issues.

Special creationism rested on the anti-evolutionary idea that species were fixed, immutable, irreducible, that they do not evolve but spring complete from an act of creation, each brought separately into the world. It was a central tenet of this view that varieties and species were fundamentally different. Species were intersterile whereas varieties could intercross. It was an essential part of any species' endowment, part of its precious uniqueness, for its members to be sterile outside their own group – or, if that rule was transgressed, for sterility to be visited on the hybrid offspring. Individuals, then, had been made sterile (either with other species or, in the case of hybrids, utterly sterile) for the good of the species. This went beyond the vague Darwinian greater-goodism that we have just looked at. Interspecific sterility was seen as evidence of an organising hand, striving to keep natural forms apart, preventing them from collapsing into disorder. It was seen as nature's – or God's – way of keeping living things taxonomically tidy:

> The view generally entertained by naturalists is that species, when intercrossed, have been specially endowed with the quality of sterility, in order to prevent the confusion of all organic forms. This view certainly seems at first probable, for species within the same country could hardly have kept distinct had they been capable of crossing freely.   (Peckham 1959, p. 424)

Critics of Darwinism took up this supposed difference between varieties and species (see e.g. Ellegård 1958, pp. 206–9). Natural selection, they said, may be able to accumulate varieties. But how does it manage to go that last step and cleave them into separate species? Until Darwinians had explained how selection could give rise to sterility, they would not have accounted for the origin of species, no matter how many of the other differences between species they had been able to explain. What's more, it was not enough just to account for it. Darwinians needed to demonstrate the development of intersterility between two incipient species before everyone's very eyes.

To make matters worse, some Darwinians took that empirical challenge seriously. Most notably, T. H. Huxley, much to Darwin's exasperation, kept on insisting that 'until selective breeding is definitely proved to give rise to varieties infertile with one another, the logical foundation of the theory of natural selection is incomplete' (Huxley 1893–4, ii, p. vi; see also e.g. Huxley 1860, pp. 43–50, 74–5, 1860a, pp. 389–91, 1863, pp. 148–50, 1863a, pp. 108–17, 1887, p. 198; Huxley, L. 1900, i, pp. 194–6, 238–9). Darwin rightly dismissed the idea that any such empirical demonstration was necessary, putting it on a par with the notion that Darwinians had to show a step-by-step complete transformation of one species into another (e.g. Darwin, F. and Seward 1903, i, pp. 137–8, 225–6, 230–2, 274, 277, 287).

But Huxley's Complaint was contagious. It spread itself for years among all kinds of doubters of Darwinism. Early in the twentieth century, it was picked up by Thomas Hunt Morgan and William Bateson, who, as we have noted, were at some period in their careers both antagonistic to Darwinism. According to Morgan:

Within the period of human history we do not know of a single instance of the transformation of one species into another, if we apply the most rigid and extreme tests used to distinguish wild species from each other. It may be claimed that the theory of descent is lacking, therefore, in the most essential feature that it needs to place the theory on a scientific basis. (Morgan 1903, p. 43)

Twenty years on, Bateson claimed that Huxley's objection was all the more serious now that breeding experiments had been tried but had failed:

that particular and essential bit of the theory of evolution which is concerned with the origin and nature of *species* remains utterly mysterious ... The conclusion in which we were brought up, that species are a product of a summation of variations, ignored the chief attribute of species ... that the product of their crosses is frequently sterile in a greater or lesser degree. Huxley, very early in the debate pointed out this grave defect in the evidence, but before breeding researches had been made on a large scale no one felt the objection to be serious. Extended work might be trusted to supply the deficiency. It has not done so, and the significance of the negative evidence can no longer be denied ... The production of an indubitably sterile hybrid from completely fertile parents which have arisen under critical observation from a single common origin is the event for which we wait. Until this event is witnessed, our knowledge of evolution is incomplete in a vital respect.

(Bateson 1922, pp. 58–9)

Bateson's words, in turn, made a profound impression on William Jennings Bryan, Scopes's fundamentalist prosecutor in the ignominious 'monkey trial' (Clark 1985, p. 284). Some years later, Haldane, pointing out that research on plant breeding had by then supplied the evidence that was being demanded, remarked that, nevertheless, 'Catholic apologists, whom I sometimes read, because their arguments are at least coherent, still taunt us poor Darwinians with our failure' (Haldane 1932, p. 55). That same taunt crops up in Charles Singer's widely read *A Short History of Biology*, which was published at that time; one of 'the three weakest points in the Darwinian position', Singer declared, is 'the absence of any experience of the formation of a species ... Until we see a variety pass into a new species the problem [of the mechanism of evolution] cannot be said to be approaching solution' (Singer 1931, pp. 305–6). (Another of those 'weak points', by the way, was 'the absence of any evidence that the unions of different varieties (i.e. incipient species on Darwin's view) are relatively more sterile than unions of the same variety' (Singer 1931, p. 305).) Even today this 'failure' is regularly trotted out in tracts about the death of Darwinism – usually, as was only too predictable, with triumphant reference to Huxley's weighty authority (e.g. Hitching 1982, p. 105).

Empirical demonstrations aside, Darwin and Wallace did feel that there was a genuine point to be explained. In fact, as we shall see Darwin showed, there was no such absolute difference between species and varieties as critics claimed. These advocates were able to maintain their sharp distinction only by judicious definition – they defined groups that interbred successfully as varieties and those that didn't as species (see e.g. Darwin 1859, pp. 246–7, 268, 277; Wallace 1889, pp. 152–3, 167). Nevertheless, the claim was more or less true and Darwinians needed to account for it. Indeed, Wallace went so far as to assert that 'the remarkable difference between varieties and species in respect of fertility when crossed' was 'One of the greatest, or perhaps we may say the greatest, of all the difficulties in the way of accepting the theory of natural selection as a complete explanation of the origin of species' (Wallace 1889, p. 152). If interspecific sterility could be accounted for, with or without empirical demonstration, adaptively or not, Darwinians would have provided the finishing touches that were needed to turn varieties into species.

Closely connected with all this was the problem of the swamping effect of intercrossing. As critics of Darwinism kept stressing, it seemed that intercrossing would prevent any differentiation into specialised groups, whether varieties or species. As a result, Darwin, Wallace and their contemporaries were on the lookout for anything that would act as a barrier to crossing between different groups – sexual preferences, sterility or whatever.

Did Darwin and Wallace see interspecific sterility as having anything to do with altruism? 'Yes' is the answer of the two historians that have written most extensively on this subject, Malcolm Kottler and Michael Ruse. Kottler asserts that cross- and hybrid sterility are 'useful, at least to the species ... if not to the individual' and that group selection is a possible explanation of how they have come about; one of the 'central issues' of the debate between Darwin and Wallace was, he says 'at what levels does the process of selection operate?', Darwin's answer being the individual and Wallace's the group (Kottler 1985, p. 388; see also pp. 406, 408, 414). Ruse states that 'for selection to effect sterility would contradict the basic idea of selection' (Ruse 1979a, p. 217); 'the sterility between members of different species, or if hybrids were formed, the sterility of these hybrids' was a question 'where Darwin might have been tempted towards a group selection mechanism', hybrid sterility seeming 'almost to beckon a group selectionist approach' – a temptation that Darwin resisted but Wallace did not (Ruse 1980, pp. 619–20, 1982, p. 191; see also 1979a, p. 217, 1980, p. 624). Other commentators have gone along with this (e.g. Sober 1985, pp. 896–7). Ghiselin, independently, has suggested that even self-sterility in plants (which we shall be discussing) could look altruistic, a disadvantage to the individual but offering 'long-term advantages to the species' (Ghiselin 1969, p. 149). We shall see later how far these claims are justified.

We are now ready to look at Darwin's and Wallace's solutions to the problem of speciation, particularly interspecific and hybrid sterility. We shall start with Darwin. (For discussions of Darwin's and Wallace's positions on this problem see Ghiselin 1969, pp. 146–53; Kottler 1985, pp. 387–417; Mayr 1959 particularly pp. 226–8, 1976, pp. 117–34; Ruse 1979a, pp. 214–19, 1980, pp. 619–25, 1982, pp. 191–2; Vorzimmer 1972, pp. 159–60, 168–85, 203–9. For further discussion of the nineteenth-century debate over the role of geographic isolation in speciation see Kottler 1978; Lesch 1975; Mayr 1976, pp. 135–43; Sulloway 1979. For modern views on speciation in general and interspecific sterility in particular see e.g. Barton and Hewitt 1985, 1989; Bush 1975; Endler 1977, pp. 12–13, 142–51 (for parapatric speciation); Maynard Smith 1958, pp. 215–58; Mayr 1959 particularly pp. 226–8, 1963, pp. 89–135, 1976, pp. 117–34; Ridley 1985, pp. 89–120. For Darwin's work on reproduction in plants in the light of later knowledge see e.g. Heslop-Harrison 1958, pp. 276–86; Whitehouse 1959. For the state of modern knowledge on plants see Ford 1964, pp. 218–33; Lewis 1979; Meeuse and

Morris 1984, chs. 1–3; Stebbins 1950, pp. 189–250 (which is thorough, if somewhat dated). For a sexual-selection analysis of the reproductive mechanisms that Darwin discusses see Willson and Burley 1983. Ghiselin 1969, pp. 141–6, 151–3, 1974, pp. 120–4, 243–4 deals tangentially with some relevant points.)

Darwin against creation: Incidental, not endowed

Darwin took the creationist challenge over interspecific sterility very seriously, treating it in the *Origin* as one of his four 'difficulties on theory': 'how can we account for species, when crossed, being sterile and producing sterile offspring, whereas, when varieties are crossed, their fertility is unimpaired?' (Darwin 1859, p. 172). He gives the problem a whole chapter to itself: 'Hybridism' (Darwin 1859, pp. 245–78; Peckham 1959, pp. 424–74).

His line of attack is to show that nature displays none of the neat, systematic relationships of species/sterility, varieties/fertility that creationists have maintained. Far from being a straight, sharp dividing line, the ability to interbreed weaves itself in eccentric, often unpredictable, fashion back and forth, through species and varieties, through first cross and hybrid, through the male of one species mated with the female of another and vice versa, through the mongrel offspring of varieties and the hybrid offspring of species. Consider, for example, the 'singular fact ... that there are individual plants ... which can be far more easily fertilised by the pollen of another and distinct species, than by their own pollen; ... all the individuals of nearly all the species of Hippeastrum seem to be in this predicament' (Peckham 1959, p. 430). Or consider this:

> There is often the widest possible difference in the facility of making reciprocal crosses [male from one species, female from another, and vice versa]. Such cases are highly important, for they prove that the capacity in any two species to cross is often completely independent of their systematic affinity, that is of any difference in their structure or constitution, excepting in their reproductive systems ... To give an instance: Mirabilis jalapa can easily be fertilised by the pollen of M. longiflora, and the hybrids thus produced are sufficiently fertile; but Kölreuter tried more than two hundred times, during eight following years, to fertilise reciprocally M. longiflora with the pollen of M. jalapa, and utterly failed.   (Peckham 1959, p. 438)

'Now', asks Darwin, 'do these complex and singular rules indicate that species have been endowed with sterility simply to prevent their becoming confounded in nature? I think not' (Peckham 1959, p. 440). Why, for example, does sterility differ so from species to species if it is equally essential to all of them? Why do some species intercross easily and yet produce sterile hybrids, whilst others cross with difficulty and yet produce fairly fertile hybrids? Why, indeed, are there hybrids at all?: 'To grant to species the special power of producing hybrids, and then to stop their further propagation by different degrees of sterility ... seems a strange arrangement' (Peckham 1959, p. 440). Sterility, then, is not a 'special endowment': 'neither sterility nor fertility affords any clear distinction between species and varieties'; species 'aboriginally existed as varieties' and there is 'no essential distinction' between the two (Peckham 1959, pp. 427, 474, 470).

Darwin's theory is that 'the sterility both of first crosses and of hybrids is simply incidental or dependent on unknown differences in their reproductive systems' (Peckham 1959, p. 440–1); the 'blending' or 'compounding' of disparate structures lowers fertility (e.g. Peckham 1959, p. 450). (In earlier editions of the *Origin* he also mentions other constitutional and structural differences. But by the last edition, it is changes in the reproductive systems or 'sexual elements' that he stresses above all (e.g. Peckham 1959, p. 466).) By the way, his theory that sterility was incidental did apply equally, as the last quote suggests, to first-cross and hybrid sterility. Admittedly, in the first three editions of the *Origin*, he does distinguish between the two cases. But his distinction is not, as we might expect, about what is incidental and what is adaptive. He is distinguishing between different physiological causes of reproductive disturbance: 'Pure species have of course their organs of reproduction in a perfect condition ... Hybrids, on the other hand, have their reproductive organs functionally impotent ... In the first case the two sexual elements which go to form the embryo are perfect; in the second case they are either not at all developed, or are imperfectly developed' (Peckham 1959, p. 425; see also e.g. p. 447). Later, when he came to the view that the physiological disturbances were much the same, he stopped distinguishing between sterility in first crosses and in hybrids (e.g. Peckham 1959, pp. 424–5, 447–9, 472).

Sterility, then, has no purpose; it is just one of nature's accidents. It 'is not a specially acquired or endowed quality, but is incidental on other acquired differences' (Peckham 1959, p. 425). As he says in *Variation under Domestication*: 'we must infer that it has arisen incidentally during their [species'] slow formation in connection with other and unknown changes in their organisation' (Darwin 1868, ii, p. 188). And to bring out the difference between 'specially endowed' and incidental, Darwin draws an analogy with potential for grafting:

As the capacity of one plant to be grafted or budded on another is so entirely unimportant for its welfare in a state of nature, I presume that no one will suppose that this capacity is a *specially* endowed quality, but will admit that it is incidental on differences in the laws of growth of the two plants.

(Peckham 1959, p. 441; see also pp. 441–3, 471)

In *Variation under Domestication* he cites susceptibility to poison as a familiar example of an incidental property:

> By a quality arising incidentally, I refer to such cases as different species of animals and plants being differently affected by poisons to which they are not naturally exposed; and this difference in susceptibility is clearly incidental on other and unknown differences in their organization.   (Darwin 1868, ii, p. 188)

Darwin adds that his account does not go 'to the root of the matter' (Peckham 1959, p. 451), for he cannot explain why reproduction is thrown out of gear; but nor could he have done without far more knowledge of the mechanisms of heredity. (And even today it is still not understood why hybrids are so often perfectly viable and yet sterile.)

By the way, Darwin was so struck by the arbitrariness of the 'complex and singular rules' he had discovered that he came to view sterility as unreliable for distinguishing species: 'the physiological test of lessened fertility ... is no safe criterion of specific distinction'; 'the evidence from this source graduates away, and is doubtful in the same degree as is the evidence derived from other constitutional and structural differences' (Peckham 1959, pp. 456, 427). These 'strange arrangements' also presented difficulties, of a rather different kind, to those who were still determined to see sterility as evidence of a divine hand. Mivart, for example, persisted in the claim that God's intelligence was manifest; but, he now decided, it was certainly 'not such as ours' (Mivart 1871a, pp. 124–5, 238)!

Darwin against natural selection: Incidental, not selected

So much for the Darwin of the first three editions of the *Origin*. His arguments there are with naturalists of a pre-Darwinian era. But from the fourth edition, published in 1866, he introduces a new question, a Darwinian question: Is sterility adaptive? He answers it with a firm 'No', indeed, three reasons for saying 'No', followed by a fourth reason in the sixth edition and several others in later works. Nevertheless, during the 1860s he was toying with a theory that he hoped would give the answer 'Yes', and in the fourth and fifth editions he sketches out this conjecture, albeit immediately repudiating it, and omitting it entirely in the last edition in 1872. (For Darwin's arguments for and against in the *Origin*, see Peckham 1959, pp. 443–7, 471; these discussions are repeated almost verbatim in both editions of

Variation under Domestication, in 1868 and 1875 (1868, ii, pp. 185–91, 2nd edn., pp. 211–18).) We'll look first at his four reasons for rejecting an adaptive explanation.

From the first edition of the *Origin* Darwin had mentioned in passing that sterility posed a problem for naturaì selection: 'On the theory of natural selection the case is especially important, inasmuch as the sterility of hybrids [and, he adds later, 'the sterility of species when first crossed'] could not possibly be of any advantage to them, and therefore could not have been acquired by the continued preservation of successive profitable degrees of sterility' (Peckham 1959, p. 424). But he had left the issue of natural selection at that. His question was not 'Incidental or selected?' but 'Incidental or specially endowed?' In the fourth edition, however, his question widens. To the summary at the beginning of the chapter, 'Sterility not a special endowment, but incidental on other differences', he adds the phrase 'not accumulated by natural selection' (Peckham 1959, p. 424).

'At one time it appeared to me probable', Darwin says, 'at it has to others, that the sterility of first crosses and of hybrids might have been acquired through the natural selection of slightly lessened degrees of fertility, which, like any other variation, spontaneously appeared in certain individuals of one variety when crossed with those of another' (Peckham 1959, p. 443; see also Darwin 1868, ii, p. 185). After all, sterility would preserve nascent differences: 'For it would clearly be advantageous to two varieties or incipient species, if they could be kept from blending, on the same principle that, when a man is selecting at the same time two varieties, it is necessary that he should keep them separate' (Peckham 1959, p. 443; see also Darwin 1868, ii, p. 185). (It is impossible to tell whether Darwin is thinking here of individuals or of a greater good; either way, it is clear what he is getting at.) But, as he puts it in *Variation*: 'when we endeavour to apply the principle of natural selection to the acquirement by distinct species of mutual sterility, we meet with great difficulties' (Darwin 1868, ii, p. 185). And Darwin eventually comes to the conclusion that the 'sterility of first crosses and of their hybrid progeny has not, as far as we can judge, been increased through natural selection' (Peckham 1959, p. 471).

His first argument (Peckham 1959, p. 443; see also Darwin 1868, ii, pp. 185–6) is that natural selection could not always have been the cause of interspecific sterility because in some cases the species are geographically separated from one another and could never have had the opportunity to cross. Selection would have had no occasion to stamp out cross-breeding because it was anyway prevented by geographical barriers.

Darwin's argument suggests that natural selection is not essential but it does not, of course, rule it out altogether. Indeed, as Darwin noted, if there had been selection for sterility between neighbouring species then sterility between geographically separated species could have arisen as an unintended side effect (Peckham 1959, p. 443). Wallace pointed out (Darwin, F. and Seward 1903, i, p. 294; Wallace 1889, p. 173) that, anyway, species that are now separated could still have been in contact when they became intersterile.

Darwin's second argument (Peckham 1959, pp. 443–4; see also Darwin 1868, ii, p. 186) takes the weapon that he had wielded against special creationism and turns it against natural selection. Infertility in reciprocal crosses is too unsystematic to be adaptive:

> it is as much opposed to the theory of natural selection as to that of special creation that in reciprocal crosses the male element of one form should be rendered utterly impotent on a second form, whilst at the same time the male element of this second form is enabled freely to fertilise the first form; for this peculiar state of the reproductive system could not possibly be advantageous to either species.
>
> (Peckham 1959, pp. 443–4)

(By 'either species' one would hope that he means 'individuals of either species', but his point applies no matter what level of selection he had in mind.)

Wallace rightly objected that natural selection doesn't have to do a complete adaptive job at a single stroke as long as a partial adaptation is useful; and even half-sterility, he said, was of some advantage in preventing interspecific crosses: 'the sterility of one cross would be advantageous even if the other cross was fertile: and just as characters now co-ordinated may have been separately accumulated by Natural Selection, so the reciprocal crosses may have become sterile one at a time' (Darwin, F. and Seward 1903, i, p. 293; see also p. 294). Darwin conceded Wallace's point as a principle but remained reluctant 'to admit [in practice the] probability of Natural Selection having done its work so queerly' (Darwin, F. and Seward 1903, i, p. 295). This is a rather arbitrary judgement coming from the foremost advocate of the idea that 'queer workings' were evidence for natural selection and against creationism.

Now we come to 'the greatest difficulty' (Peckham 1959, pp. 444–5; see also Darwin 1868, ii, p. 186). How could it ever be of advantage to an individual to be at all infertile? If, say, two interfertile varieties had been interbreeding, then no individual would have benefited from a loss of fertility: 'it could not have been of any direct advantage to an individual animal to breed poorly with another individual of a different variety, and thus to leave few offspring; consequently such individuals could not have been preserved or selected' (Peckham 1959, p. 444). Or take the case of two species that already show some infertility on crossing; again, says Darwin, no individual would benefit from further infertility:

> take the case of two species which in their present state when crossed, produce few and sterile offspring; now, what is there which could favour the survival of those individuals which happened to be endowed in a slightly higher degree with mutual infertility, and which thus approached by one small step towards absolute sterility? Yet an advance of this kind, if the theory of natural selection be brought to bear, must have incessantly occurred with many species ...   (Peckham 1959, p. 444)

What is there that could favour such survival? Darwin put that question rhetorically. But we have seen that modern Darwinism has a ready answer: opportunity costs. Darwin was right that natural selection cannot get interspecific sterility off the ground if there is full interfertility (and hybrid offspring are not inferior). But if two incipient species are partially sterile, there are likely to be enormous advantages to the individuals on both sides in becoming completely intersterile. Like some of his successors a century later, Darwin apparently fails to appreciate the full costs, above all the opportunity costs, of producing 'few and sterile offspring'. This point is crucial to understanding Darwin's position and we'll be returning to it.

It is in this argument that Darwin seems to be thinking most explicitly of altruism and to be acknowledging – but rejecting – the possibility of a higher-level explanation. Both Kottler and Ruse interpret him in this way (Kottler 1980, p. 406; Ruse 1979a, p. 217, 1980, pp. 623–4). Darwin certainly does seem to set up the problem as a contrast between group and individual: 'It may be admitted ... that it would profit an incipient species if it were rendered in some slight degree sterile when crossed with its parent-form or with some other variety; for thus fewer bastardised and deteriorated offspring would be produced to commingle their blood with the newly-forming variety' (Peckham 1959, p. 444). 'Bastardy' and the 'commingling of blood' sound like group- rather than individual-level worries (although 'deteriorated offspring' would not be welcome to an individual, either, if they got in the way of caring for better offspring). And his answer, as we have just seen, is that such sterility could not be the result of natural selection because it would be of no advantage to any individual. He then contrasts this case with sterility in social insects; there, he says, it can be favoured because they are in a social community:

> With sterile neuter insects we have reason to believe that modifications in their structure and fertility have been slowly accumulated by natural selection, from an advantage having been thus indirectly given to the community to which they belonged over other communities of the same species; but an individual animal not belonging to a social community, if rendered slightly sterile when crossed with

some other variety, would not thus itself gain any advantage or indirectly give any advantage to the other individuals of the same variety, thus leading to their preservation.   (Peckham 1959, p. 444–5; see also Darwin 1868, ii, pp. 186–7)

It's impossible to tell what level that is meant to be, but it's plain that Darwin is trying to establish some kind of contrast between the communal ways of insects and the solitary, wasted sterility of a non-social animal. (The greater the contrast, however, the less convincing it becomes to characterise Darwin as a thorough-going individual selectionist. Ruse, for example, gets into difficulties over this, ending up by declaring that Darwin 'saw a whole hive of bees as one individual' (Ruse 1979a, p. 217).)

In the sixth edition of the *Origin*, Darwin adds a fourth consideration (Peckham 1959, p. 447; see also Darwin 1868, 2nd edn., ii, p. 214). Natural selection definitely could not have been responsible for interspecific sterility in the case of plants. And yet the rules of sterility are much the same for plants as for animals. So it is unlikely that it is adaptive in animals either:

> with plants we have conclusive evidence that the sterility of crossed species must be due to some principle, quite independent of natural selection ... and from the laws governing the various grades of sterility being so uniform throughout the animal and vegetable kingdoms, we may infer that the cause, whatever it may be, is the same or nearly the same in all cases.   (Peckham 1959, p. 447)

The evidence from plants is this. In certain genera there is a gradation in interspecific sterility from species that produce seeds in abundance on some crosses, through those that produce fewer seeds, right down to those that produce no seeds at all but swell on contact with the foreign pollen. It was one of Darwin's standard tactics to take a series of intermediate stages in different species or genera as plausible evidence that natural selection had been at work; it was how he dealt, for example, with the evolution of that notoriously problematic organ, the eye (Darwin 1859, pp. 187–8). A gradation in sterility through a range of species, then, might at first seem to hint at adaptation. But in this case, Darwin argues, the appearance is misleading. There is a hiatus at the crucial point: the extreme of the characteristic – producing swelling but no seeds – is obviously not being passed on and therefore could not be the result of selection: 'It is here manifestly impossible to select the more sterile individuals, which have already ceased to yield seeds; so that this acme of sterility, when the germen alone is affected, cannot have been gained through selection' (Peckham 1959, p. 447).

It is not clear why Darwin finds this evidence more 'conclusive' than any other gradation of fertility that ends in sterility. Presumably no other cases had been investigated in such detail; as he pointed out elsewhere, it was exceedingly difficult to obtain similar data from animals (Darwin, F. and Seward 1903, i, p. 231) (particularly from species with internal fertilisation). Perhaps, too, this evidence seemed especially significant to him because it was the example of plants that had given him hope of explaining interspecific sterility adaptively.

Darwin, then, goes out of his way to insist that sterility is not an adaptation. But, curiously, at the same time, in the fourth and fifth editions of the *Origin* (but excised from the sixth) he also adds a conjecture about why sterility might, after all, be adaptive in plants (though not in animals) (Peckham 1959, pp. 445–7; see also Darwin 1868, pp. 187–8). What is this all about? Why does Darwin decide to bring in natural selection in the first place? And why does he both firmly repudiate and tentatively endorse the idea that sterility can be adaptive?

Darwin's adaptive interlude

What happened was this. In the early 1860s Darwin started some experimental crossbreeding in plants. His findings led to a heady, exciting interlude. It seemed to him that self-sterility in plants was an adaptation. And it seemed, too, that the means by which natural selection had achieved this within-species sterility could also have been used to evolve sterility between species. Eventually he decided that sterility within species was probably only rarely, if ever, adaptive, and interspecific sterility never so. But for a few years his publications reflected the conjectures and counter-arguments, the experimental results for and against, the changing fortunes of his ideas.

The story began in 1861, with Darwin's work on the genus *Primula* (Darwin 1862a). In some species, such as cowslips and primroses, the flowers occur in two forms, with only one kind on a plant and the plants in about equal proportions. In one form, the long-styled or pin, the style is long and the stamen is short (the style is the part that terminates in the stigma, which receives the pollen, and the stamen is the part that sheds pollen grains); in the other, the short-styled or thrum, the style is short and the stamen long. Darwin discovered that this was an adaptation, a structural device for achieving cross-pollination by insects. Naturalists had long noticed that offspring from self-pollinated plants lacked vigour and Darwin had earlier concluded that natural selection favoured cross-pollination (e.g. Darwin 1862). He showed that, in most of the dimorphic species of *Primula*, the heights of stamen and stigma in the two forms were cunningly matched, so that when an insect in search of nectar inserted its proboscis into a long-styled flower, the pollen from the stamen stuck to that part of the proboscis that would later touch the stigma of the short-styled – and vice versa. Between 1861 and 1863 Darwin found the same kind of structural provision in other groups: in flax and other species of *Linum*, and in purple loosestrife and other species of *Lythrum* (some of which are trimorphic, with one style and two stamens in each form) (Darwin 1864, 1865).

Whilst working on *Primula*, he noticed that the design was not fail-safe. Flowers could still become self-pollinated inadvertently. He wondered whether, as a back-up to promoting cross-pollination by structural arrangements, natural selection had developed some physiological barrier to self-fertilisation, so that even if the flower's own pollen did accidentally reach its stigma, the union would not be fertile. Certainly the stigma and pollen-grain were different in the two forms, perhaps a sign of such a barrier. He could not test this hypothesis by self-fertilising flowers because the ill effects of inbreeding (on his hypothesis, the very reason for such barriers) would influence the results. But he did find that when he crossed flowers of the same form (homomorphic crosses – short-styled with short, or long with long) fertility was much lower than in heteromorphic crosses. He assumed that this failure of a plant's fertility with half the members of its species was a by-product of selection for self-infertility. Because natural selection had to prevent flowers from being fertilised not only by themselves but by any flowers on the same plant, it had resorted to a blanket rule: Render infertile any unions between flowers of the same form. And, as an unintended by-product, even unions between different plants of the same form had become infertile.

What is more, Darwin apparently found striking independent evidence for this idea. Short-styled flowers were structurally more vulnerable to self-pollination. So natural selection should have taken greater precautions to make their inadvertent unions physiologically sterile. And he did indeed find that short-styled homomorphic crosses were far less fertile than long-styled homomorphic crosses:

> the chance of self-fertilization is much stronger in this [short-styled] than in the other form. On this view [that the purpose of dimorphism is to promote intercrossing] we can at once understand the good of the pollen of the short-styled form, relatively to its own stigma, being the most sterile; for this sterility would be the most requisite to check self-fertilization, or to favour intercrossing.
>
> (Darwin 1862a: Barrett 1977, ii, p. 60)

In *Linum*, too, he found that whilst one form was fertile with its own pollen, in plants of the other form 'their own pollen produces no more effect than the pollen of a plant of a different order, or than so much inorganic dust' (Darwin

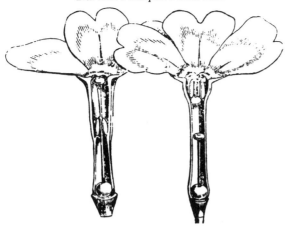

Long-styled form   Short-styled form

The loves of the plants

"No little discovery of mine ever gave me so much pleasure as the making out of the meaning of hetero-styled flowers." (Darwin: Autobiography)

Long-styled and short-styled *Primula veris* (pin and thrum cowslips) (from Darwin's *The Different Forms of Flowers*): Some *Primula* species produce two forms of flowers. One form has long styles but short stamens placed low in the floral tube; the other has short styles but long stamens borne near the mouth of the tube. (The style terminates in the stigma, which receives pollen, and the stamen is the part that sheds pollen.) Any one plant's flowers are all of the same form. Darwin discovered that this dimorphism is a structural device for promoting cross-pollination by insects. The heights of stamen and stigma in the two forms are cunningly matched, so that when a nectar-seeking insect inserts its proboscis into a long-styled flower, the pollen from the stamen sticks to that part of the proboscis that will later touch the stigma of the short-styled – and vice versa.

1862a: Barrett 1977, ii, p. 63; see also Darwin 1864). (He must have been dismayed to find that in *Lythrum* sterility could not be an adaptation because it was *inversely* related to the structural danger of 'illegitimate' union: 'If the rule be true, we must look at it as an incidental and useless result of the gradational changes through which this species has passed in arriving at its present condition' (Darwin 1865: Barrett 1977, ii, p. 120). But he persisted with his conjecture.) It seemed, then, that self-infertility was the work of natural selection, with homomorphic infertility as its unintended side effect.

Darwin conjectured that the physiological mechanism by which natural selection had evolved sterility was what he called 'prepotency'. Heteromorphic pollen had been made so much more potent than homomorphic pollen that if, by accident, a flower's stigma was exposed to its own pollen, then any heteromorphic pollen that reached the stigma 'would obliterate the action of the homomorphic pollen' (Darwin 1862a: Barrett

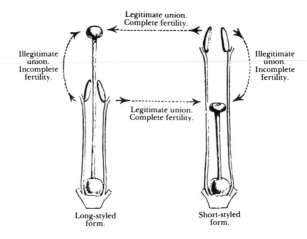

Legitimate and illegitimate unions in cowslips (from Darwin's *The Different Forms of Flowers*): Darwin found that, in cowslips, 'illegitimate' unions (short-styled with short or long with long) are less fertile than 'legitimate' ones. He thought at one time that this might be an adaptation: a physiological barrier to self-fertilisation to back up the structural adaptations for promoting cross-fertilisation. Such a barrier would ensure that, even if a flower's own pollen did accidentally reach its stigma, the marriage would be infertile. And it would also prevent fertile marriages among all flowers on the same plant. Darwin found that, in addition, unions between any flowers of the same form, even from different plants, were less fertile than 'legitimate' unions. He assumed that this last case – infertility between different plants – was an unintended by-product of selection for self-infertility, an inadvertent consequence of a blanket ban on fertility between flowers of the same form. It seemed, then, that natural selection had achieved self-infertility at the enormous cost of infertility with half the members of the flower's own species: a cowslip was 'sterile with half its brethren'.

1977, ii, p. 59) and take over the fertilisation (even, in some species, when it arrived as much as a day later). This 'prepotency' is what we now know to be sexual incompatibility – in this case, self-incompatibility. The plant can recognise and reject its own or similar pollen. As Darwin thought, it is a biochemical device for avoiding self-fertilisation, a back-up to mechanical devices for avoiding self-pollination. It generally acts as a contraceptive rather than by abortion, intervening early enough for the plant's eggs to be conserved for another attempt.

But what does all this have to do with sterility between species? For Darwin, the moral was apparently twofold. Sterility could be an adaptation. And prepotency – in the interspecific case, prepotency of own-species pollen over that of other species – could be natural selection's means of achieving that sterility. This is the model that Darwin seems to have had in mind, although he was cautious about drawing the analogy.

In his *Primula* paper, he hints at a parallel between sterility within a species and interspecific sterility. Homomorphic crosses, he says, are as sterile as interspecific crosses, and there is enough individual variation in sterility within species for natural selection to work on; this puts us in mind of interspecific sterility:

> Seeing that we thus have a groundwork of variability in sexual power, and seeing that sterility of a peculiar kind has been acquired by the species of *Primula* to favour intercrossing, those who believe in the slow modification of specific forms will naturally ask themselves whether sterility may not have been acquired for a distinct object, namely, to prevent two forms, whilst being fitted for distinct lines of life, becoming blended by marriage, and thus less well adapted for their new habits of life. (Darwin 1862a: Barrett 1977, ii, p. 61)

('Forms' here seems to mean individual- rather than higher-level selection; he is, after all, drawing an analogy from individual-level selection within species. Sterility in *Primula* is 'peculiar', by the way, because it is between likes.) But he brings his speculation to an abrupt halt: 'many great difficulties would remain, even if this view could be maintained' (Darwin 1862a: Barrett 1977, ii, p. 61). Nevertheless, writing to Huxley some months later, early in 1862, he is less cautious: 'the latter half of my Primula paper in the "Linn. Journal" ... leads me to suspect that sterility will hereafter have to be largely viewed as an acquired or *selected* character – a view which I wish I had had facts to maintain in the "Origin"' (Darwin, F. 1887, ii, p. 384). In December of that year he wrote to his great friend Joseph Hooker, the eminent botanist: 'my notions on hybridity are becoming considerably altered by my dimorphic work. I am now strongly inclined to believe that sterility is at first a selected quality to keep incipient species distinct' (Darwin, F. and Seward 1903, i, pp. 222–3).

And in the next two editions of the *Origin*, the fourth and fifth, in 1866 and 1869, he does indeed at least tentatively suggest how interspecific sterility in plants could have been acquired by natural selection:

> With many kinds [of plants], insects constantly carry pollen from neighbouring plants to the stigmas of each flower; and with some this is effected by the wind. Now, if the pollen of a variety, when deposited on the stigma of the same variety, should become by spontaneous variation in ever so slight a degree prepotent over the pollen of other varieties, this would certainly be an advantage to the variety; for its own pollen would thus obliterate the effects of the pollen of other varieties, and prevent deterioration of character. And the more prepotent the variety's own pollen could be rendered through natural selection, the greater the advantage would be. (Peckham 1959, p. 445)

Interspecific sterility, he says, is always accompanied by such prepotency. The great question is which came first, the prepotency or the sterility, for natural selection could not afford to promote sterility with foreign pollen unless self-fertilisation was assured: 'We know ... that, with species which are mutually sterile, the pollen of each is always prepotent on its own stigma over that of the other species; but we do not know whether this prepotency is a consequence of the mutual sterility, or the sterility a consequence of the prepotency' (Peckham 1959, pp. 445–6). If prepotency came before sterility, then we have a path that natural selection could have adopted: 'as the prepotency became stronger through natural selection, from being advantageous to a species in the process of formation, so the sterility consequent on prepotency would at the same time be augmented; and the final result would be various degrees of sterility, such as occurs with existing species' (Peckham 1959, p. 446). Clearly, Darwin sees prepotency as the key. If a flower sets out with an assured choice of pollens, then selection can gradually increase the power of its own species' pollen and diminish that of others.

Again, Darwin writes in higher-level language. In the fourth edition, he even says that prepotency of own-pollen would be an advantage to the variety because 'it would thus escape being bastardised and deteriorated in character' (Peckham 1959, p. 445) – though 'bastardy', surely a species-level concern, does get dropped in the fifth edition. And yet, presumably, he is thinking at an individual level, for his reason for latching on to prepotency is apparently because it provides a clue as to how individual costs and benefits could work.

Darwin has set out his conjecture. He then promptly withdraws it. The stumbling block, he says, is that there is no equivalent to 'pollen-choice' in animals. And yet the patterns of interspecific sterility in animals and plants are similar, so they are likely to have a common cause. Presumably, then, in neither case is natural selection that cause:

> This view might be extended to animals if the females before each birth received several males ... but ... most males and females pair for each birth, and some few for life ... [W]e may conclude that with animals the sterility of crossed species has not been slowly augmented, through natural selection; and as this sterility follows the same general laws in the vegetable as in the animal kingdom, it is improbable, though apparently possible, that with plants crossed species should have been rendered sterile by a different process.   (Peckham 1959, p. 446)

(It is ironic to find Darwin, of all people, claiming that females have no choice.) And so, he concludes, 'we must give up the belief that natural selection has come into play; and we are driven to our former proposition, that the sterility of first crosses, and indirectly of hybrids, is simply incidental on unknown differences in the reproductive systems of the parent-species' (Peckham 1959, p. 446; see also Darwin 1868, ii, pp. 228–9).

Darwin became increasingly convinced that interspecific sterility could not be adaptive. He dropped his speculation from the last editions of the *Origin* (the sixth, 1872) and *Variation under Domestication* (the second, 1875). And in *Effects of Cross and Self Fertilisation* (1876) and *Different Forms of Flowers* (1877) he listed several more reasons why sterility in plants even within species was unlikely to be adaptive. One argument was that the degree of self-sterility did not correspond to the degree of inferiority of offspring from self-fertilisation, so it was unlikely that this inferiority had been a selective pressure (Darwin 1876, pp. 345–6). Another was that the degree of self-sterility differed greatly in different offspring of the same parents (Darwin 1876, p. 346). (Darwin raises the intriguing possibility that some individuals have been selected for intercrossing and some for selfing 'to ensure the propagation of the species' (Darwin 1876, p. 346), but rejects the idea because self-sterile individuals are too rare. Was it a higher-level, greater-good explanation or, more interestingly, some kind of frequency-dependent concept that was uppermost in his mind?) He also argued that the degrees of sterility in the various homomorphic crosses differ 'capriciously' (Darwin 1877, p. 265). And that the degree of sterility is strongly affected by temperature, nutrients and other environmental factors (Darwin 1876, pp. 345–6). He stressed that self-sterility cannot be selected unless cross-fertilisation is ensured (Darwin 1876, pp. 381–2); but, he said, there are so many devices for achieving cross-fertilisation that 'sterility seems an almost superfluous acquirement for this purpose' (Darwin 1876, p. 346). Finally, natural selection would not have bought self-sterility at the enormous cost of homomorphic sterility, which cuts plants off from union with half their species (two-thirds in trimorphic species). Even in the *Primula* paper, when he still thought that self-sterility was an adaptation, he had said: 'it is not a little remarkable that the end [cross-fertilisation] has been gained ... [at this] expense' (Peckham 1959, p. 459). Now he argued:

It is incredible that so peculiar a form of mutual infertility should have been specially acquired unless it were highly beneficial to the species; and although it may be beneficial to an individual plant to be sterile with its own pollen, cross-fertilisation being thus ensured, how can it be of any advantage to a plant to be sterile with half its brethren, that is, with all individuals belonging to the same form?
(Darwin 1877, pp. 264–5)

(If Darwin took this argument seriously, he should have held that sex was not adaptive either; he certainly recognised that sex had much the same cost (e.g. Darwin 1865: Barrett 1977, ii, p. 126).) Armed with this barrage of arguments, Darwin dealt swiftly with interspecific sterility (Darwin 1877, pp. 466–9): the results of crossing and hybridisation follow much the same pattern within and between species, so there is no reason to assume that the causes are different. Thus, he concluded, no sterility is adaptive; it is all

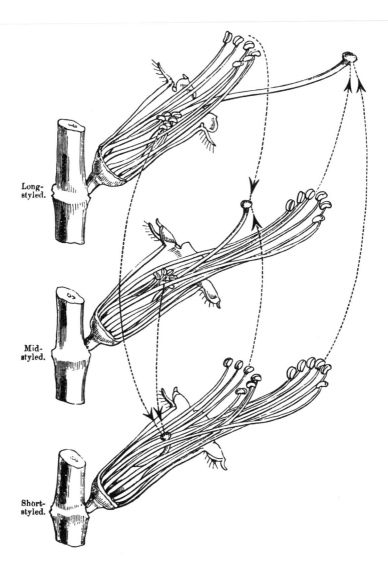

Long-
styled.

Mid-
styled.

Short-
styled.

incidental. And although he occasionally allowed that natural selection could, after all, give rise to self-sterility in some cases (e.g. Darwin 1876, p. 442, 1877, p. 258), he completely relinquished the view that this was true of interspecific sterility.

It is hard to understand why Darwin made such heavy weather of all this unless we try to put ourselves in his position. Remember that he had misgivings about the idea of sterility being adaptive because it seemed to him that natural selection would have to break what he thought was surely its fundamental rule of promoting individual reproductive success. He underestimated the opportunity costs involved in being only partially intersterile or in having inferior or sterile hybrid offspring.

It seems that prepotency liberated him from this way of thinking, allowing him to think instead in terms of opportunity costs, of selection not depriving an organism of reproductive opportunities but freeing it to take advantage of better ones. Sterility no longer emerged as an inevitable loss of reproductive powers but as the release of superior reproductive opportunities. Prepotency provided a model of the evolution of reproductive choice, a choice between one pollen and another, between fertilisation by likes or by unlikes. It enabled Darwin to envisage an evolutionary pathway along which selection first ensured that the better option was well established and only then withdrew the less attractive one. It is the guarantee of alternative fertilisation that turns sterility into an option. As Darwin put it, referring to self-sterility (this was when he had decided that sterility was not adaptive): 'The means for favouring cross-fertilisation must have been acquired before those which prevent self-fertilisation; as it would manifestly be injurious to a plant that its stigma should fail to receive its own pollen, unless it had already become well

'A most complex marriage-arrangement'

"In the manner of their fertilisation these plants offer a more remarkable case than can be found in any other plant or animal ... nature has ordained a most complex marriage-arrangement, namely a triple union between three hermaphrodites." (Darwin: The Different Forms of Flowers)

The three forms of purple loosestrife (*Lythrum salicaria*) (from Darwin's *The Different Forms of Flowers*): Purple loosestrife produces flowers with three arrangements of styles and stamens. (The style receives pollen where it terminates at the stigma; the stamen is the part that sheds pollen grains.) The stamens come in three lengths – long, medium and short; each flower has stamens of two lengths only. The style, too, can be long, medium or short; each flower has only one style and it is always of a different length from the stamens. Each plant bears flowers of only one type. Darwin realised that these structures are ingeniously devised to promote cross-pollination. As an insect flies from flower to flower, it will generally carry pollen from a stamen of one flower to a stigma of corresponding length – which will be on a flower of a different form.

adapted for receiving pollen from another individual' (Darwin 1876, p. 382).

What is more, Darwin felt that when he observed pollen prepotency (an increase in pollen fertility) he was probably seeing the very channel that selection could have used to promote pollen sterility, too. As he wrote to Joseph Hooker in 1862: 'If you have looked at *Lythrum* you will see how pollen can be modified merely to favour crossing; with equal readiness it could be modified to prevent crossing. It is this which makes me so much interested with dimorphism, etc' (Darwin, F. and Seward 1903, i, pp. 222–3).

Remember, also, that Darwin was thinking of sympatric speciation (having underestimated the importance of geographical isolation in speciation). He had in mind not two groups that have undergone dramatic adaptive divergence during geographical separation but two varieties that are initially very alike. He felt that, under these conditions, the costs of interbreeding were not very severe (unlike the case of sterility within species, where the costs of self-fertilisation are an enormous incentive to cross-pollination and self-sterility). Indeed, Darwin seemed more impressed by the risks of intersterility, of an individual being cut off from so many potential mates, than by the advantages to that individual of breeding only within its own group. Once again, he presumably saw prepotency as smoothing the evolutionary path. At first, when there was some advantage to breeding within the variety but the balance of risk was also in favour of continuing to breed with other varieties, then selection in favour of prepotency of own-variety pollen would exploit the advantage without completely precluding cross-breeding. Once the advantage of fertilisation within the variety had increased sufficiently (because of adaptive divergence between the two groups) and the risks had decreased sufficiently (because selection had promoted fertility within the group), selection would be able to favour sterility with other varieties.

So much for Darwin's brief encounter. Predictably, Wallace's attempt to explain interspecific sterility adaptively took a very different course.

Wallace: The power of natural selection

'I am deeply interested in all that concerns the powers of Natural Selection', Wallace wrote to Darwin, 'but, though I admit there are a few things it cannot do, I do not yet believe sterility to be one of them' (Marchant 1916, i, p. 203). Wallace was unwilling to accept that natural selection could not explain the evolution of a characteristic so useful, so widespread, so uniform. He was also worried that anti-Darwinians would attempt to exploit Darwinism's failure to provide an adaptive explanation of something so essential to the origin of species (Marchant 1916, i, p. 210). (Rightly worried, as it turned out; Vernon Kellogg's turn-of-the-century survey of criticisms of Darwinism cites this as one out of 'a rather formidable category of objections', mentioning in particular Thomas Hunt Morgan's attack (Kellogg 1907, p. 76).) Wallace tried to fill the adaptive gap. And he felt he had succeeded – though, with some justice, not many have agreed. (Wallace's final views on this question appeared after Darwin had died, in his *Darwinism* (Wallace 1889, ch. 7, particularly pp. 168–80, 184–6). This was a simplified version of the solution that he had proposed twenty years earlier in correspondence with Darwin in 1868 (Darwin, F. and Seward 1903, i, pp. 287–99; Marchant 1916, i, pp. 195–210 largely covers the same letters but with a few additions (pp. 203, 207)). The version in Wallace's book is much the same as the first part of his earlier text. The second part of the earlier version, which involved some more complicated assumptions about initial conditions, added nothing of value (as Wallace later recognised (Darwin, F. and Seward 1903, i, pp. 292–3, n1)) and we shall not examine it.)

At one point in his correspondence with Wallace about interspecific sterility Darwin remarked: 'I do not feel that I shall grapple with the sterility argument till my return home; I have tried once or twice, and it has made my stomach feel as if it had been placed in a vice' (Darwin, F. and Seward 1903, i, p. 293). Having tried more than once or twice, I can sympathise with Darwin's feelings – and with those of his family: 'Your paper has driven three of my children half mad – one sat up till 12 o'clock over it' (Darwin, F. and Seward 1903, i, p. 293). Wallace's discussion does not show him at his best. It is hard to decide what he is getting at and sometimes hard to make Darwinian sense of what he seems to be saying. In his book, Wallace himself admitted that 'this argument is a rather difficult one to follow' (Wallace 1889, p. 179) and went so far as to provide a summary, suspecting that some readers would be unable to cope with the full text (Wallace 1889, pp. 179–80). Wallace's argument seems to be made up of several independent strands. One way to do him justice is to examine these strands separately and then fit them back together for a general assessment.

Wallace's basic idea is that the evolution of interspecific and hybrid sterility proceeds in two stages. In the first stage, sterility is acquired incidentally. In the second stage, natural selection steps in and shapes these accidents into a systematic reproductive barrier between species. He stresses that his argument applies only when sterility arises at first incidentally, but that this is likely to happen very commonly (Marchant 1916, ii, p. 41; Wallace 1889, p. 179).

Consider, says Wallace, two varieties, in the same area, that are adapting to different environmental niches, such as damp and dry places, woods and open grounds. The two forms might be completely interfertile, in which case, if the hybrids are more vigorous, they will do better than the pure strains. ('Hybrid vigour' was a commonly-observed and well-documented effect.) But it is more likely that the hybrids would be somewhat infertile. Darwin has shown, Wallace reminds us, that successful reproduction requires a close compatibility between the sexes. Indeed, from a physiological point of view, it is fertility that is the precarious condition: 'It appears as if fertility depended on such a delicate adjustment of the male and female elements to each other, that, unless constantly kept up by the preservation of the most fertile individuals, sterility is always liable to arise' (Wallace 1889, p. 184). In particular, fertility is highly susceptible to change. We are, of course, assuming that the varieties are undergoing constitutional change (because of their new adaptations) and experiencing environmental change (the new niches). 'Let us suppose', then, 'that a partial sterility of the hybrids between the two forms arises, in correlation with the different modes of life and the slight external or internal peculiarities that exist between them, both of which we have seen to be real causes of infertility' (Wallace 1889, p. 175). To assume this partial sterility of hybrids is to go no further than Darwin. But Wallace now argues that natural selection can augment and systematise this incidental tendency to sterility:

we ... have obtained ... a starting-point ... All we need, now, is some means of increasing or accumulating this initial tendency ... Ample causes of infertility have been shown to exist, in the nature of the organism and the laws of correlation; the agency of natural selection is only needed to accumulate the effects produced by these causes, and to render their final results more uniform and more in accordance with the facts that exist.   (Wallace 1889, pp. 173–4)

Natural selection, Wallace argues, will select against hybrid offspring; and by doing so, it will promote interspecific sterility as a by-product. Hybrids are inferior to pure offspring partly because they are less fertile and partly because they are less well adapted (which will tell against them, in spite of hybrid vigour, if conditions become severe). As an automatic side effect of this selective force, natural selection will favour a tendency not to produce hybrids; it will therefore favour interspecific sterility. It is crucial to note, Wallace stresses, that natural selection works by selecting against hybrid offspring (because they are inferior), not by selecting in favour of intersterility:

It must particularly be noted that [selection for increased intersterility] ... would result, not by the preservation of the infertile variations on account of their infertility, but by the inferiority of the hybrid offspring, both as being fewer in numbers, less able to continue their race, and less adapted to the conditions of existence than either of the pure forms. It is this inferiority of the hybrid offspring that is the essential point ...   (Wallace 1889, p. 175)

The reason why this is so important, he says, is that selection will automatically stamp out sterility unless it is linked with some advantage. It is always the advantage, not the sterility, that selection favours; sterility is promoted as a by-product: 'no form of infertility or sterility between the individuals of a species, can be increased by natural selection unless correlated with some useful variation, while all infertility not so correlated has a constant tendency to effect its own elimination' (Wallace 1889, p. 183). Wallace feels that he has found such a correlation: selection favours non-hybrid offspring and this 'useful variation' is correlated with intersterility.

And so, Wallace concludes, he has carried the argument from the point at which Darwin left it, carried it from the merely incidental to the adaptive:

> Mr Darwin arrived at the conclusion that the sterility or infertility of species with each other ... is not a constant or necessary result of specific difference, but is incidental on unknown peculiarities of the reproductive system ... Here the problem was left by Mr Darwin; but we have shown that its solution may be carried a step further. (Wallace 1889, pp. 185–6)

To understand Wallace's argument, think of the position that he is in. He wants to explain sterility adaptively. But, like Darwin himself, he does not appreciate the costs to parents of having sterile or low-fertility hybrid offspring. He can't, then, envisage how selection could act on the parents. So he has to cast around for something that selection can act on. And his answer is that it's the hybrids themselves. Wallace believes that he must insist, as an orthodox adaptationist, that selection is on hybrid inferiority, not on the sterility of parents that have these inferior offspring, because he fails to see the opportunity costs that are involved. His argument is hard to understand (perhaps not even coherent) in detail; but it is clear what kind of case he is trying to make.

One of Darwin's main objections to Wallace's argument tellingly reflects the same oversight. There is no reason, Darwin says, why natural selection should favour greater infertility between individuals that are already somewhat intersterile; after all, their pure offspring will not benefit if the sterility of their parents' hybrid unions increases:

> take two species A and B, and assume that they are (by any means) half-sterile, *i.e.*, produce half the full number of offspring. Now try and make (by Natural Selection) A and B absolutely sterile when crossed, and you will find how difficult it is ... [Any] extra-sterile individuals of, we will say A, if they should hereafter breed with other individuals of A, will bequeath no advantage to their progeny, by which these families will tend to increase in number over other families of A, which are not more sterile when crossed with B. (Darwin, F. and Seward 1903, i, p. 289)

But this ignores opportunity costs, the reproductive capacities tied up in a sterile union that could perhaps have been employed more fruitfully elsewhere.

Darwin also objected to Wallace's argument on the grounds that Wallace sometimes assumed that hybrids were at an advantage from greater vigour and sometimes that they were at a disadvantage from being less well adapted (Marchant 1916, i, p. 207). Wallace maintained that this was not inconsistent because adaptive advantage depended on the conditions, the advantages of hybrid vigour being outweighed by the superior adaptation of the pure form during severe struggle.

The next strand of Wallace's argument is to do with mating preferences. Speciation and increasing intersterility, at least in animals, he says, would be 'greatly assisted by ... a disinclination of the two forms to pair together' (Wallace 1889, p. 176). This is because animals generally have a strong preference for breeding with likes, and constitutional differences are often correlated with external differences that animals would recognise, in particular peculiarities of colour:

> there is ... a very powerful cause of isolation in the mental nature – the likes and dislikes – of animals ... This constant preference of animals for their like ... is evidently a fact of great importance in considering the origin of species by natural selection, since it shows us that, so soon as a slight differentiation of form or colour has been effected, isolation will at once arise by the selective association of the animals themselves ...   (Wallace 1889, pp. 172–3)

This sounds very like the modern idea of natural selection reinforcing mating preferences to act as isolating mechanisms. And that would seem an obvious next step for Wallace to suggest, exactly the kind of adaptive stage building on the incidental that he was looking for. But, curiously, Wallace introduces the preference for mating with likes as a fortunate gift of nature rather than an adaptation. Certainly, these preferences reinforce the tendency to intersterility. But they just happen to do so. They are not an adaptive response for preventing unsuccessful unions and promoting successful ones. Strangely, this keen adaptationist, of all people, fails to point out here, of all places, that mate preference could be shaped by the advantages of avoiding hybrid offspring.

We know, however, from his theories of coloration for recognition that he did take reinforcement into account. Indeed, as we have seen, he seized on recognition as an alternative explanation to sexual selection: 'Some means of easy recognition ... enables the sexes to recognise their kind and thus avoid the evils of infertile crosses ... The wonderful diversity of colour and of marking that prevails, especially in birds and insects, may be due to the fact that one of the first needs of a new species would be, to keep separate from its nearest allies, and this could be most readily done by some easily seen external mark of difference' (Wallace 1889, pp. 217–18; see also pp. 217–28,

1891, pp. 367–8). Incidentally, this was an idea that he developed after his correspondence of 1868 with Darwin (e.g. Wallace 1891, pp. 349, 354); by 1889, when he published *Darwinism*, he had come to think that 'recognition [in general, not specifically for mating] has had a more widespread influence in determining the diversities of animal coloration than any other cause whatever' (Wallace 1889, p. 217). We have here, then, a confluence of some of Wallace's most cherished notions: adaptationism, natural selection in place of sexual selection, and the importance of coloration. Coloration for recognition was an alternative to Darwin's theory of sexual selection. It was also an adaptive explanation for species-specific differences in coloration, differences that some critics claimed were non-adaptive. And, in addition, it was an adaptation for avoiding the 'evils' of hybridisation.

Mate recognition was only one of several reproductive barriers that Wallace realised could isolate incipient species from one another. Again, although he does not bring them into his argument on interspecific sterility, he does discuss them elsewhere. In a paper on zoological distribution, for example, he pointed out that species that were externally similar could be separated by 'mode of life and habits' and species that 'agree closely in habits' were likely to differ in 'colour, form, or constitution' (Wallace 1879, pp. 257–8). Darwin, by the way, paid less attention to such barriers. This was partly because he preferred to explain coloration as far as possible by sexual selection. Perhaps it was also because he took plants as his paradigm case, and there sterility is the most obvious isolating mechanism (though perhaps not the most important (Willson and Burley 1983)); the most obvious ethological barriers occur at one remove, in the behaviour of insects.

To put it in the Darwinian language of today, Wallace can be credited with having emphasised the importance of ethological and other reproductive barriers (structure, colour and so on) in speciation, even though he perhaps did not appreciate how far natural selection could be responsible for reinforcing them. In Wallace's time this was unusual. For years, most Darwinians paid little attention to any barriers apart from two: accidents of geography and sterility. One notable exception was Karl Pearson: 'Natural selection', he argued, 'requires selective mating ... to produce that barrier to intercrossing on which the origin of species depends' (Pearson 1892, p. 423). But it was not until the publication of Dobzhansky's *Genetics and the Origin of Species* (1937), followed by Mayr's *Systematics and the Origin of Species* (1942), that the majority of Darwinians began to take seriously the kinds of reproductive isolating mechanisms that Wallace considered.

In recognition of Wallace's pioneering insight (and taking at face value Wallace's claim to have explained sterility adaptively) Verne Grant called the process of selection for reproductive isolation the 'Wallace effect' (Grant

1966, p. 99, 1971, p. 188): 'Wallace ... presented a model whereby natural selection could build up barriers of hybrid sterility and mating behaviour between diverging sympatric species. He argued that if the hybrids were adaptively inferior to the races or species, selection would favour sterility and ethological barriers between them' (Grant 1963, p. 503). Some critics have disputed whether Wallace deserves his eponymous effect, arguing that his theory was about the selection of post-mating mechanisms, which, they say, selection cannot influence (e.g. Kottler 1985, pp. 416–17, 430–1; Littlejohn 1981, p. 320):

> Wallace was not proposing the selective origin of reproductive isolation mechanisms in general, but rather the selective origin of the particular *post*-mating mechanisms of cross- and hybrid sterility. Since, according to current theory, these forms of sterility are precisely the types of reproductive isolation that cannot be produced by selection, the Darwin–Wallace debate provides little historical justification for the term 'Wallace effect'.   (Kottler 1985, p. 416)

As we have seen, this criticism rests on a spurious distinction. And, as we have also seen, Wallace did, anyway, stress the importance of selection for pre-mating barriers, mating preferences in particular – although admittedly he failed to integrate it into his theory of interspecific sterility, where it was most needed (see also Kottler 1985, pp. 430–1, n31).

Now to the third strand of Wallace's argument. He next suggests that the various factors that he has mentioned would be mutually reinforcing. It is possible, he says, that the degree of intersterility would correlate with, perhaps partly depend on, the degree of difference between the incipient species. In that case, intersterility would increase in proportion to the divergence of the two forms. All these isolating factors, then, would work in tandem: adaptive divergence, differences in external appearance, disinclination to mate with unlikes, and hybrid infertility 'would all proceed pari passu, and would ultimately lead to the production of two distinct forms having all the characteristics, physiological [by which he means intersterility] as well as structural, of true species' (Wallace 1889, p. 176).

Darwin and other critics disagreed with the idea that infertility tended to coincide with disinclination to cross or with structural dissimilarities; and anyway, said Darwin, it 'cannot hold with plants, or the lower fixed aquatic animals' (Darwin, F. and Seward 1903, i, p. 295; Marchant 1916, ii, p. 42). Wallace was convinced that correlations between infertility and other changes would arise because infertility was so readily induced by any disturbances to the organism. Darwin, of course, broadly agreed about the effect of disturbances; Wallace was, after all, drawing on his work. But were changes in colour, for example, likely to be accompanied by changes in mating preferences? There was no particular reason to believe that they would and no empirical evidence either way (even twenty years later, as Wallace regretfully pointed out (Marchant 1916, ii, p. 42)). Darwin was surely right to be suspicious if Wallace was assuming that all the complementary changes – colours, mating preferences and so on – had to coincide for selection to take off and that they were hereditary. As we've noted, that would 'require a miracle'. But we saw, too, that sympatric models of speciation can provide the kind of happy coincidence that Wallace needed without resort to miracles.

Finally, at several points, Wallace weaves in what seems to be a group-selectionist argument (e.g. Wallace 1889, p. 178). It is important for his theory, he says, that the proportion of hybrids within a particular area that are partially sterile should be fairly high, otherwise the incipient sterility between the two forms would be swamped. Once selection on these hybrids has increased the intersterility between the two forms, then the forms that are the most intersterile will take over from forms in other areas that have greater interfertility. So eventually the whole area will be taken over by the two forms with the greatest intersterility. And that sterility will be increased by natural selection.

Wallace is here at his most obscure and it is hard to know whether he is really appealing to group selection or merely using higher-level language. Darwin found the reasoning at this point of the theory extremely tortuous and their disagreement became a dispute over the mathematics, the details of which they did not bequeath to history. Darwin left the calculation to his mathematician son, then at Cambridge, and that year's Second Wrangler; but even he was driven to distraction by the task. All this reinforced Darwin's view that the idea that natural selection could promote sterility, however plausible it seemed at first, could not be made to work in detail (Darwin, F. and Seward 1903, i, p. 294). Nowadays the kinds of questions that they set up (about the effects of selection on neighbouring populations) could probably be settled quite readily by computer simulation.

Wallace has been widely criticised for being group-selectionist, often in contrast with Darwin's individual selectionism (e.g. Bowler 1984, p. 201; Kottler 1974, p. 190, 1985, pp. 387, 388, 408–10, 414–15; Ruse 1979, p. 14, 1979a, pp. 214–19 particularly p. 217, 1980, p. 624; Sober 1984, pp. 217–18, 1985, pp. 896–7; Vorzimmer 1972, pp. 203–9 particularly p. 207; for Darwin's individual selectionism see also Ghiselin 1969, pp. 149–50; Ruse 1982, pp. 191–2). Kottler (1985, pp. 407–10) gives a careful analysis of Wallace's argument, in which it certainly does emerge as being about 'selection between groups' (p. 410); but Kottler does not make clear how this ultimately involves anything that is not individual-selectionist. The evidence that is standardly cited against Wallace (e.g. Kottler 1985, p. 408; Ruse 1980,

p. 624; Sober 1984, pp. 217–18, 1985, p. 896; Vorzimmer 1972, p. 206) is a comment that he made in a letter to Darwin:

> I do not see your objection to sterility between allied species having been aided by Natural Selection. It appears to me that, given a differentiation of a species into two forms, each of which was adapted to a special sphere of existence, every slight degree of sterility would be a positive advantage, not to the individuals who were sterile, but to each form.   (Darwin, F. and Seward 1903, i, p. 288)

Wallace's words do, admittedly, seem to clinch the case for group selectionism. But then, taken alone, so would Darwin's statement, quoted earlier, on interspecific sterility: 'Natural Selection cannot effect what is not good for the individual, including in this term a social community' (Darwin, F. and Seward 1903, i, p. 294). The passage from Wallace was a single statement made at the beginning of his correspondence with Darwin. When he came to set out the details of his theory, he was never as unequivocal (or as clear!). Like Darwin, Wallace makes cavalier use of higher-level language perhaps without higher-level intentions. At one point, for example, he wrote to Darwin: 'is it not probable that Natural Selection can accumulate these variations [in degree of sterility] and thus *save the species*?' (Darwin, F. and Seward 1903, i, p. 294; my emphasis); but just a few months later he was writing: 'If "natural selection" could not accumulate varying degrees of sterility *for the plant's benefit*, then how did sterility ever come to be associated with one cross of a trimorphic plant rather than another?' (Darwin, F. and Seward 1903, i, p. 298; my emphasis). Unfortunately, with both Darwin and Wallace, their liberal use of higher-level language – species, varieties, types, forms – generally makes it impossible to decide definitively whether they really did have some kind of altruism in mind.

Wallace has also been accused of hyper-adaptationism, again in unfavourable contrast with Darwin (e.g. Ghiselin 1969, pp. 150–1; Kottler 1985, p. 388; Mayr 1959, 1976, pp. 129–34; see also Gillespie 1979, p. 72). But he was not intrinsically misguided to struggle for an adaptive explanation where Darwin had abandoned all hope. Indeed, we have seen that Darwin himself held out such a hope for several years. To put it in modern terms, Wallace was trying to develop a sympatric theory of speciation, assuming that natural selection could be sufficiently powerful to bring about divergent adaptation even when counteracted by intercrossing. If the defects of his undoubtedly defective theory are to be attributed to over-zealous adaptationism, this needs to be demonstrated not merely assumed.

Whatever the judgements of others, Wallace was proud of his explanation, so proud that in his autobiography he mentions it (along with animal coloration) as one of the two topics on which he managed to take Darwinism further than did Darwin himself:

> in several directions I believe that I have extended and strengthened [the theory of natural selection] ... The principle of 'utility', which is one of its chief foundation stones, I have always advocated unreservedly; while in extending this principle to almost every kind and degree of coloration, and in maintaining the power of natural selection to increase the infertility of hybrid unions, I have considerably extended its range. Hence it is that some of my critics declare that I am more Darwinian than Darwin himself, and in this, I admit, they are not far wrong.
>
> (Wallace 1905, ii, p. 22)

Origins elusive

To a modern Darwinian, it is puzzling that Darwin and Wallace so vastly underestimated geographical isolation, the crucial factor that could have resolved so many of their difficulties. In their earlier years, both had assumed that it played a prominent, even indispensable, role in speciation (on Darwin: Bowler 1984, pp. 160, 170–1, 200–1; Kottler 1978, pp. 284–8; Lesch 1975, pp. 484–5; Sulloway 1979, pp. 23–33; Vorzimmer 1972, pp. 168–9; on Wallace: Fichman 1981, pp. 34, 94–5; McKinney 1972, ch. 2). Why did they come to play down its importance? One reason – probably among several others (see e.g. Ghiselin 1969, pp. 148–9; Mayr 1959, pp. 221–3, 1976, pp. 120–3; Sulloway 1979, pp. 33–45) – was that the issue of geographical isolation and natural selection in speciation came to be curiously polarised, being seen almost as one versus the other rather than as the respective roles of each.

Most influentially, the nineteenth-century German naturalist and explorer Moritz Wagner came to attribute only a very minor role to selection (Wagner 1873; see also Sulloway 1979, pp. 49–58). He eventually argued that geographical isolation could bring about speciation almost without its help. Wagner first published on the importance of geographical isolation in the 1840s, elaborating his ideas during the 1860s and 70s. His work had its greatest impact in the following decades, once the issue had been widely taken up. As we noted when we looked at adaptive explanations, one of the leading figures in this later period was Romanes. He was particularly impressed with the writings of Gulick. To his essays, he said, he attributed 'a higher value than to any other work in the field of Darwinian thought since the date of Darwin's death' (adding, in a footnote, that he regarded 'Weismann's theory of heredity ... as still *sub judice*') (Romanes 1892–7, iii, p. 1). Romanes was convinced of the overwhelming importance of isolation:

I believe ... that in the principle of Isolation we have a principle so fundamental and so universal, that even the great principle of Natural Selection lies less deep, and pervades a region of smaller extent. Equalled only in its importance by the two basal principles of Heredity and Variation, this principle of Isolation constitutes the third pillar of a tripod on which is reared the whole superstructure of organic evolution.

(Romanes 1892–7, iii, pp. 1–2)

(Romanes included isolation through mate preference – 'discriminate selection' – in this principle; but in his writings he took a more idiosyncratic line on this behavioural aspect of isolation – a theory that he called 'physiological selection' (Romanes 1892–7, iii, pp. 41–100).) In the early 1900s, Vernon Kellogg reported that 'by some the species-forming influence of isolation is held to be as effective as selection itself – some deem it more effective'; both of these 'somes' were to be found particularly among 'systematists, students of distribution, and so-called field naturalists' (Kellogg 1907, p. 232). In less extreme vein, there was an increasingly popular view that most of the characteristic differences between closely allied species were not the result of adaptation but resulted merely from geographical isolation combined with chance or 'orthogenetic trends' (again, a view that we met when we discussed adaptationism – remember the land snails). According to Kellogg, it was the belief that many species-specific characteristics were non-adaptive that paved the way for the triumph of isolation over selection: 'It is indeed the general recognition by naturalists of the fact of the triviality or indifference of a majority of specific characters that has led to the recent renewal of the importance of isolation theories, particularly of geographical isolation' (Kellogg 1907, p. 43).

Against such claims, Darwin and Wallace wanted to emphasise that allied species were not only divergent but adaptively so. Natural selection, they stressed, deserved much of the tribute for differences between species. As Darwin put it: 'neither migration nor isolation in themselves can do anything' (Darwin 1859, p. 351). When he read Wagner's views, he scrawled on his copy: 'Most Wretched Rubbish ... There does not appear the least explanation how e.g. a woodpecker could be formed in an isolated region' (quoted in Vorzimmer 1972, p. 182). Or, as he more moderately wrote to Wagner himself in 1876:

my strongest objection to your theory is that it does not explain the manifold adaptations in structure in every organic being – for instance in a Picus [woodpecker] for climbing trees and catching insects – or in a Strix [owl] for catching animals at night, and so on *ad infinitum*. No theory is in the least satisfactory to me unless it clearly explains such adaptations.

(Darwin, F. 1887, iii, pp. 158–9; see also e.g. Darwin, F. 1887, iii, pp. 157–62; Darwin, F. and Seward 1903, i, p. 311; Peckham 1959, p. 196; for Wallace, see e.g. Wallace 1889, pp. 144–51)

At the beginning of this chapter I cited copious examples of geographical barriers from the *Origin*. What were these examples, many of them the results of Darwin's own detailed experiments, if not part of his argument about speciation? The answer is that they are part of his discussion of geographical distribution. His concern is to demonstrate that 'the individuals of the same species, and likewise of allied species, have proceeded from some one source [and that] ... all the grand leading facts of geographical distribution are explicable on the theory of migration ... together with subsequent modification' (Darwin 1859, p. 408). His concern is to demonstrate, in other words, that evolution, not a grand designer, placed living things where we find them now. Darwin showed a fine appreciation of how the accidents of geography could shape the history of life. But it was not the appreciation that we would expect.

Although Darwin and Wallace accepted sympatric theories of speciation, among the majority of evolutionists such theories have long been out of favour – and not merely unfashionable but derided. Ernst Mayr, in particular, has argued weightily and influentially for several decades that, although natural selection can reinforce reproductive isolating mechanisms, it cannot establish them from the beginning, entirely under its own steam: 'the same old arguments are cited again and again in favor of sympatric speciation, no matter how decisively they have been disproved previously ... Sympatric speciation is like the Lernaean Hydra which grew two new heads whenever one of its old heads was cut off' (Mayr 1963, p. 451). The problem with sympatric theories, he says, is that 'In the last analysis, [they] all ... make arbitrary postulates that at once endow the speciating individuals with all the attributes of a full species' (Mayr 1963, p. 451) – the main attribute being reproductive isolation.

I don't know why sympatric speciation has met with quite such acrimonious opposition. According to the eminent Australian cytologist M. J. D. White, vertebrate zoologists have been the least willing to entertain the idea, and plant evolutionists have also on the whole been unreceptive (except in the case of allopolyploidy). Entomologists have been more readily persuaded – perhaps, he suggests, because small animals are more able to speciate without geographical divides (White 1978, p. 229). (But are entomologists perhaps just taking a Gulliverian view of a Lilliputian barrier?)

Perhaps, also, we are yet again witnessing the familiar story of adaptationist versus non-adaptationist inclinations. We have seen how firmly

Wallace stuck to the idea that natural selection was sufficiently powerful to split species asunder without the help of geographical barriers. And Darwin, too, felt that slight differences between incipient species could accumulate in successive generations without help coming from the hills (or streams or whatever). (Darwin's disagreement with Wallace, remember, was not about geographical isolation but about whether natural selection would favour intersterility during speciation, particularly at the beginning, or whether intersterility would arise only as an incidental side effect of divergence.) So, for example, Darwin stated in the *Origin*: 'within the same area, varieties of the same animal can long remain distinct, from haunting different stations, from breeding at slightly different seasons, or from varieties of the same kind preferring to pair together' (Darwin 1859, p. 103). And he added the following comment to the fifth and sixth editions: 'Moritz Wagner has lately ... shown that the service rendered by isolation in preventing crosses between newly formed varieties is probably greater than I supposed. But ... I can by no means agree with this naturalist, that migration and isolation are necessary elements for the formation of new species' (Peckham 1959, p. 196). By contrast, those Darwinians who have been less convinced than Darwin and Wallace of selection's competence have assumed that, on the contrary, migration and isolation are crucial.

A few twentieth-century Darwinians have felt that even to allow reinforcement after geographical isolation is to concede too much to the power of selection. John Moore, the embryologist whom we met earlier in this chapter, for instance, claimed in the 1950s that Mayr's model of allopatric speciation showed that divergence during geographical separation was sufficient for turning out proper species, replete with isolating mechanisms; a final stage of reinforcement was possible but would be superfluous (Moore 1957, pp. 325–6, 332). In recent years, H. E. H. Paterson has taken Moore's claims further (Paterson 1978, 1982), arguing strenuously that Darwinians have clung to the idea of reinforcement only because:

> it provides a direct role for natural selection in the production of new species ... Dobzhansky believed species to be 'adaptive devices through which the living world has deployed itself to master a progressively greater range of environments and ways of living'. This view imposes on its holder the obligation to accept that species are the *direct* product of selection, which, in turn, requires that the reinforcement model of speciation be accepted. (Paterson 1978, pp. 369, 371)

Mate recognition, Paterson argues, has been overloaded with adaptive significance. Its only evolutionary function is to enable sex cells to get together. Any reproductive isolation that happens to come about is purely incidental and not an adaptation – and the same goes for mate preference within a species (Paterson 1982, p. 53). He even sees the standard view of species as making so free with adaptations that it attributes them to the species as a whole, thereby, he says, ending up (inconsistently) group-selectionist; as evidence, he points to the widespread use of terms like Dobzhansky's 'adaptive devices' or the 'integrity of the species' (an inconsistency, he suggests, because reinforcement mechanisms are individual-selectionist) (Paterson 1982, pp. 53–4).

But let's return to sympatric speciation. During several periods in the history of Darwinian theory, mainly towards the end of the last century and again since about the 1940s, geographical isolation has absorbed an enormous amount of attention, particularly among evolutionists whose prime interest is speciation rather than adaptation. It has been seen by some as a central tenet of Darwinian theory:

the development of physiological isolating mechanisms is preceded by a geographical isolation of parts of the original population ... Since Darwin, and especially since Wagner, it is regarded as probable that the formation of geographical races is an antecedent of species formation ... *Some systematists regard it as one of the greatest generalizations that has resulted from their work.*
(Dobzhansky 1937, 1st edn., pp. 256–7; my emphasis)

(– although, as we have seen, Dobzhansky is wrong about Darwin). Important as geographical isolation undoubtedly is in practice, it seems curious to hold it in such theoretical esteem as those systematists apparently did. Perhaps it is partly because sympatric speciation dispenses with what some Darwinians took to be a 'great generalisation' that it has seemed to them not merely wrong but thoroughly uncongenial?

Incidentally, if it's numerical generalisations that are the issue, then sympatric speciation probably wins hand down. As Guy Bush has pointed out, when it comes to sheer numbers, insects, which account for about 75% of named species, could well tip the balance, rendering sympatric speciation more common than allopatric:

Sympatric speciation appears to be limited to special kinds of animals, namely phytophagous and zoophagous parasites and parasitoids. However, this group encompasses a huge number of species (well over 500,000 described insects alone)
...
In the light of the fact that parasites are probably the most abundant of all eukaryotes, sympatric divergence seems an equally probable, and possibly even the normal, mode of speciation in many groups. The number of zoophagous and phytophagous parasites is staggering ... [According to one estimate] about 72.1% of the British insects (among the best known in the world) are parasitic on plants or animals ... If we consider that there are already 750,000 described species of insects worldwide, over 525,000 of these are parasites, a conservative figure as at least three

times this number remains undescribed. This amounts to more than all other plant and animal species combined.   (Bush 1975, pp. 352, 354)

Nevertheless, even if insects can come to the rescue quantitatively, Darwin and Wallace undoubtedly grossly underestimated the potential importance of geographical barriers and allopatric speciation.

The two fundamental problems that Darwin's theory was designed to solve were adaptation and diversity. The riddle of adaptation he solved superbly. As for diversity, on certain aspects he was equally successful. The patterns of geographical distribution, the fossil record, the taxonomic hierarchy, and comparative embryology all fell into place under his incisive analysis. But, in the midst of such success, there was one problem that remained just outside his grasp. It was – poignantly – the problem of the origin of species.

EPILOGUE

Darwinism is amongst the most comprehensively successful achievements of the human intellect. It gathers up and explains a vast, diverse collection of important and otherwise baffling facts. Like any scientific theory, it generates problems as well as solutions. We have looked at two of those problems: altruism and sexual selection. Problems once. Triumphs now. Other difficulties, however, remain for Darwinism: Why sex? How did the mind and other emergent properties evolve? What is the relationship between cultural and genetic evolution? These questions are as troubling to modern Darwinians as were the ant and the peacock to Darwin and Wallace. Those earlier anomalies were resolved in the Darwinian revolution of recent decades. Do we need another revolution to deal with these further difficulties? Or, more intriguingly, are the answers already staring us in the face?

NOTE ON THE LETTERS OF DARWIN AND WALLACE

In references to Darwin's *Life and Letters*, I have cited the first edition. The following list will help to identify these references in the numerous subsequent editions. It gives the dates of all letters cited from *The Life and Letters of Charles Darwin* (Darwin, F. 1887), and also from *More Letters of Charles Darwin* (Darwin, F. and Seward 1903) and *Alfred Russel Wallace: Letters and reminiscences* (Marchant 1916).

Chapter 2 A world without Darwin
Darwin, F. 1887, i, p. 314: Darwin to Julia Wedgwood, 11 July [1861]
Darwin, F. 1887, ii, p. 241: Darwin to Charles Lyell, [12 December 1859]
Darwin, F. 1887, ii, p. 373: Darwin to Asa Gray, 5 June [1861]
Darwin, F. 1887, ii, p. 378: Darwin to Asa Gray, 17 September [1861?]
Darwin, F. 1887, ii, p. 382: Darwin to Asa Gray, 11 December [1861]
Darwin, F. 1887, iii, pp. 61–2: Darwin to Joseph Dalton Hooker, 8 February [1867]
Darwin, F. 1887, iii, p. 266: Darwin to John Murray, 21 September [1861]
Darwin, F. 1887, iii, pp. 274–5: Editor's note
Darwin, F. and Seward 1903, i, pp. 190–2, n2: Editors' note
Darwin, F. and Seward 1903, i, pp. 191–3: Darwin to Charles Lyell, [2 August 1861]; Darwin to Charles Lyell, [13 August 1861]
Darwin, F. and Seward 1903, i, p. 202: Darwin to Asa Gray, 23 July [1862]
Darwin, F. and Seward 1903, i, p. 203: Darwin to Asa Gray, 23 July [1862]
Darwin, F. and Seward 1903, i, pp. 330–1, n1, n2: Editors' note
Darwin, F. and Seward 1903, i, p. 455: Darwin to Hugh Falconer, 17 December [1859]
Marchant 1916, i, p. 170: Wallace to Darwin, 2 July 1866

Chapter 3 Darwinism old and new
Darwin, F. 1887, ii, p. 273: Darwin to Asa Gray, [? February 1860]
Darwin, F. 1887, ii, p. 296: Darwin to Asa Gray, 3 April [1860]
Darwin, F. 1887, iii, p. 96: Darwin to Wallace, March [1867]

Chapter 5 The sting in the peacock's tail
Darwin, F. 1887, ii, p. 296: Darwin to Asa Gray, 3 April [1860]
Darwin, F. 1887, iii, pp. 90–1: Darwin to Wallace, 28 [May?] [1864]
Darwin, F. 1887, iii, pp. 90–6: Darwin to Wallace, 28 [May?] [1864]; Darwin to Wallace, 22 February [1867?]; Darwin to Wallace, 23 February [1867]; Darwin to Wallace, 26 February [1867]; Darwin to Wallace, March [1867]
Darwin, F. 1887, iii, pp. 95–6: Darwin to Wallace, March [1867]
Darwin, F. 1887, iii, pp. 111–12: Darwin to F. Müller, 22 February [1869?]
Darwin, F. 1887, iii, p. 135: Darwin to Wallace, 30 January [1871]

Darwin, F. 1887, iii, pp. 137–8: Darwin to Wallace, 16 March 1871
Darwin, F. 1887, iii, pp. 150–1: Darwin to F. Müller, 2 August [1871]
Darwin, F. 1887, iii, pp. 156–7: Darwin to August Weismann, 5 April 1872
Darwin, F. and Seward 1903, i, pp. 182–3: Darwin to Henry Walter Bates, 4 April
    [1861]
Darwin, F. and Seward 1903, i, p. 283: Darwin to Wallace, 12 and 13 October
    [1867]
Darwin, F. and Seward 1903, i, pp. 303–4: Darwin to Joseph Dalton Hooker, 21
    May [1868]
Darwin, F. and Seward 1903, i, p. 316: Darwin to Joseph Dalton Hooker, 13
    November [1869]
Darwin, F. and Seward 1903, i, pp. 324–7: Darwin to John Morley, 24 March 1871
Darwin, F. and Seward 1903, ii, pp. 35–6: Wallace to Darwin, 29 May [1864]
Darwin, F. and Seward 1903, ii, pp. 56–97: Darwin to James Shaw, 11 February
    [1866]; Darwin to James Shaw, April 1866; Darwin to Abraham Dee
    Bartlett, 16 February [1867?]; Darwin to William Bernhard Tegetmeier, 5
    March [1867]; Darwin to William Bernhard Tegetmeier, 30 March [1867];
    Darwin to Wallace, 29 April [1867]; Darwin to Wallace, 5 May [1867];
    Darwin to Wallace, 19 March 1868; Darwin to F. Müller, 28 March [1868];
    Darwin to John Jenner Weir, 27 February [1868]; Darwin to John Jenner
    Weir, 29 February [1868]; Darwin to John Jenner Weir, [6 March 1868];
    Darwin to John Jenner Weir, 13 March [1868]; Darwin to John Jenner Weir,
    22 March [1868]; Darwin to John Jenner Weir, 27 March [1868]; Darwin to
    John Jenner Weir, 4 April [1868]; Darwin to Wallace, 15 April [1868];
    Darwin to John Jenner Weir, 18 April [1868]; Darwin to Wallace, 30 April
    [1868]; Darwin to Wallace, 5 May [1868?]; Darwin to John Jenner Weir, 7
    May [1868]; Darwin to John Jenner Weir, 30 May [1868]; Darwin to F.
    Müller, 3 June [1868]; Darwin to John Jenner Weir, 18 June [1868]; Darwin
    to Wallace, 19 August [1868]; Darwin to Wallace, 23 September [1868];
    Wallace to Darwin, 27 September 1868; Wallace to Darwin, 4 October 1868;
    Darwin to Wallace, 6 October [1868]; Darwin to Benjamin Dann Walsh, 31
    October 1868; Darwin to Wallace, 15 June [1869?]; Darwin to George
    Henry Kendrick Thwaites, 13 February [N.D.]; Darwin to F. Müller, 28
    August [1870]; Wallace to Darwin, 27 January 1871; Darwin to G. B.
    Murdoch, 13 March 1871; Darwin to George Fraser, 14 April [1871];
    Darwin to Edward Sylvester Morse, 3 December 1871; Darwin to August
    Weismann, 29 February 1872; Darwin to H. Müller, [May 1872]
Darwin, F. and Seward 1903, ii, p. 59: Darwin to Wallace, 29 April [1867]
Darwin, F. and Seward 1903, ii, p. 62: Entry in Darwin's diary, 4 February 1868
Darwin, F. and Seward 1903, ii, p. 76: Darwin to Wallace, 30 April [1868]
Marchant 1916, i, p. 157: Wallace to Darwin, 29 May [1864]
Marchant 1916, i, p. 159: Wallace to Darwin, 15 June 1864
Marchant 1916, i, pp. 177–87: Darwin to Wallace, January 1867; Darwin to
    Wallace, 23 February 1867; note by Wallace; Darwin to Wallace, 26
    February 1867; Wallace to Darwin, 11 March 1867; Darwin to Wallace,
    March 1867; Darwin to Wallace, 29 April 1867; Darwin to Wallace, 5 May
    1867; Darwin to Wallace, 6 July 1867
Marchant 1916, i, pp. 190–5: Darwin to Wallace, 12 and 13 October 1867; Wallace
    to Darwin, 22 October; Darwin to Wallace, 22 February [1868?]
Marchant 1916, i, p. 199: Darwin to Wallace, 27 February 1868
Marchant 1916, i, pp. 202–5: Darwin to Wallace, 17 March 1868; Wallace to
    Darwin, 19 March; Darwin to Wallace, 19–24 March 1868
Marchant 1916, i, pp. 212–17: Darwin to Wallace, 15 April 1868; Darwin to
    Wallace, 30 April 1868; Darwin to Wallace, 5 May 1868
Marchant 1916, i, pp. 220–31: Darwin to Wallace, 19 August 1868; Wallace to
    Darwin, 30 August [1868?]; Darwin to Wallace, 16 September 1868;
    Wallace to Darwin, 18 September 1868; Darwin to Wallace, 23 September

1868; Wallace to Darwin, 27 September 1868; Wallace to Darwin, 4 October 1868; Wallace to Darwin, 6 October 1868

Marchant 1916, i, pp. 256–61: Wallace to Darwin, 27 January 1871; Darwin to Wallace, 30 January 1871; Wallace to Darwin, 11 March 1871; Darwin to Wallace, 16 March 1871

Marchant 1916, i, p. 270: Darwin to Wallace, 1 August 1871

Marchant 1916, i, p. 292: Darwin to Wallace, 17 June 1876

Marchant 1916, i, pp. 298–302: Wallace to Darwin, 23 July 1877; Darwin to Wallace, 31 August 1877; Wallace to Darwin, 3 September 1877; Darwin to Wallace, 5 September [1877]

Chapter 6 Nothing but natural selection?

Darwin, F. 1887, iii, p. 93: Darwin to Wallace, 22 February [1867?]

Darwin, F. 1887, iii, pp. 93–4: Darwin to Wallace, 23 February [1867]; Darwin to Wallace, 26 February [1867]

Darwin, F. 1887, iii, p. 94: Darwin to Wallace, 26 February [1867]

Darwin, F. 1887, iii, p. 138: Darwin to Wallace, 16 March 1871

Darwin, F. and Seward 1903, ii, p. 60: Darwin to Wallace, 29 April [1867]

Darwin, F. and Seward 1903, ii, p. 67: Darwin to John Jenner Weir, [6 March 1868]

Darwin, F. and Seward 1903, ii, p. 71: Darwin to John Jenner Weir, 4 April [1868]

Darwin, F. and Seward 1903, ii, p. 73: Darwin to Wallace, 15 April [1868]

Darwin, F. and Seward 1903, ii, p. 74: Darwin to Wallace, 15 April [1868]

Darwin, F. and Seward 1903, ii, p. 84: Darwin to Wallace, 19 August [1868]

Darwin, F. and Seward 1903, ii, p. 86: Wallace to Darwin, 27 September 1868

Darwin, F. and Seward 1903, ii, pp. 86–8: Wallace to Darwin, 27 September 1868

Darwin, F. and Seward 1903, ii, p. 87: Wallace to Darwin, 27 September 1868

Darwin, F. and Seward 1903, ii, pp. 91–2: Darwin to F. Müller, 28 August [1870]

Darwin, F. and Seward 1903, ii, p. 93: Darwin to G. B. Murdoch, 10 March 1871

Darwin, F. and Seward 1903, ii, p. 94: Darwin to G. B. Murdoch, 10 March 1871; B. T. Lowne, 1871

Marchant 1916, i, p. 177: Darwin to Wallace, January 1867

Marchant 1916, i, p. 217: Darwin to Wallace, 5 May 1868

Marchant 1916, i, p. 225: Wallace to Darwin, 18 September 1868

Marchant 1916, i, pp. 235–6: Wallace to Darwin, 10 March 1869

Marchant 1916, i, p. 298: Wallace to Darwin, 23 July 1877

Marchant 1916, i, p. 302: Darwin to Wallace, 5 September [1877]

Chapter 7 Can females shape males?

Darwin, F. 1887, iii, p. 138: Darwin to Wallace, 16 March 1871

Darwin, F. 1887, iii, p. 151: Darwin to F. Müller, 2 August [1871]

Darwin, F. 1887, iii, p. 157: Darwin to August Weismann, 5 April 1872

Darwin, F. and Seward 1903, i, pp. 324–5, n3: Editors' note

Darwin, F. and Seward 1903, i, p. 325: Darwin to John Morley, 24 March 1871

Darwin, F. and Seward 1903, i, pp. 325–6: Darwin to John Morley, 24 March 1871

Darwin, F. and Seward 1903, ii, pp. 62–3: Darwin to Wallace, 19 March 1868

Darwin, F. and Seward 1903, ii, p. 63: Darwin to Wallace, 19 March 1868

Chapter 9 'Until careful experiments are made ...'

Darwin, F. 1887, iii, pp. 94–5: Darwin to Wallace, 26 February [1867]

Darwin, F. 1887, iii, p. 151: Darwin to F. Müller, 2 August [1871]

Darwin, F. 1887, iii, p. 157: Darwin to August Weismann, 5 April 1872

Darwin, F. and Seward 1903, ii, pp. 57–9: Darwin to William Bernhard Tegetmeier, 5 March [1867]; Darwin to William Bernhard Tegetmeier, 30 March [1867]

Darwin, F. and Seward 1903, ii, pp. 64–5: Darwin to John Jenner Weir, 27 February [1868]; Darwin to John Jenner Weir, 29 February [1868]

Marchant 1916, i, p. 270: Darwin to Wallace, 1 August 1871

Chapter 10 Ghosts of Darwinism surpassed
Darwin, F. and Seward 1903, ii, p. 90: Darwin to Wallace, 15 June [1869?]

Chapter 15 Human altruism: a natural kind?
Darwin, F. 1887, ii, pp. 141–2: Darwin to Herbert Spencer, 25 November [1858]
Darwin, F. 1887, iii, pp. 55–6: Darwin to Joseph Dalton Hooker, 10 December [1866]
Darwin, F. 1887, iii, p. 99: Darwin to Alphonse de Candolle, 6 July 1868
Darwin, F. 1887, iii, p. 120: Darwin to E. Ray Lankester, 15 March [1870]
Darwin, F. 1887, iii, pp. 165–6: Darwin to Herbert Spencer, 10 June [1872]
Darwin, F. 1887, iii, p. 193: Darwin to John Fiske, 8 December 1874
Darwin, F. and Seward 1903, i, p. 271: Darwin to Wallace, 5 July [1866]
Darwin, F. and Seward 1903, ii, p. 235: Darwin to Joseph Dalton Hooker, 30 June [1866]
Darwin, F. and Seward 1903, ii, pp. 424–5: Darwin to Francis Maitland Balfour, 4 September 1880
Darwin, F. and Seward 1903, ii, p. 442: Darwin to Herbert Spencer, 9 December [1867]

Chapter 16 Breeding between the lines
Darwin, F. 1887, ii, p. 384: Darwin to T. H. Huxley, 14 [January?] [1862]
Darwin, F. 1887, iii, pp. 157–62: Darwin to Moritz Wagner, [1868?]; Darwin to Moritz Wagner, 13 October 1876; Darwin to Karl Semper, 26 November 1878; Darwin to Karl Semper, 30 November 1878
Darwin, F. 1887, iii, pp. 158–9: Darwin to Moritz Wagner, 13 October 1876
Darwin, F. and Seward 1903, i, pp. 137–8: Darwin to T. H. Huxley, 11 January [1860?]
Darwin, F. and Seward 1903, i, pp. 222–3: Darwin to Joseph Dalton Hooker, 12 [December 1862]
Darwin, F. and Seward 1903, i, pp. 225–6: Darwin to T. H. Huxley, 28 December [1862]
Darwin, F. and Seward 1903, i, pp. 230–2: Darwin to T. H. Huxley, 18 December [1862]; Darwin to T. H. Huxley, 10 [January] [1863]
Darwin, F. and Seward 1903, i, p. 231: Darwin to T. H. Huxley, 10 [January] [1863]
Darwin, F. and Seward 1903, i, p. 274: Darwin to T. H. Huxley, 22 December [1866?]
Darwin, F. and Seward 1903, i, p. 277: Darwin to T. H. Huxley, 7 January [1867]
Darwin, F. and Seward 1903, i, p. 287: Darwin to T. H. Huxley, 30 January [1868]
Darwin, F. and Seward 1903, i, pp. 287–99: Wallace to Darwin, February 1868; Darwin to Wallace, 27 February [1868]; Wallace to Darwin, 1 March 1868; Darwin to Wallace, 17 March 1868; Wallace to Darwin, 24 March [1868]; Darwin to Wallace, 6 April [1868]; Wallace to Darwin, 8 [April?] 1868; Wallace to Darwin, 16 August [1868]
Darwin, F. and Seward 1903, i, p. 288: Wallace to Darwin, February 1868
Darwin, F. and Seward 1903, i, p. 289: Darwin to Wallace, 27 February [1868]
Darwin, F. and Seward 1903, i, p. 293: Darwin to Wallace, 17 March 1868
Darwin, F. and Seward 1903, i, pp. 292–3, n1: Note by Wallace, 1899
Darwin, F. and Seward 1903, i, p. 293: Wallace to Darwin, 1 March 1868
Darwin, F. and Seward 1903, i, p. 294: Wallace to Darwin, 24 March [1868]
Darwin, F. and Seward 1903, i, p. 295: Darwin to Wallace, 6 April [1868]
Darwin, F. and Seward 1903, i, p. 298: Wallace to Darwin, 16 August [1868]
Darwin, F. and Seward 1903, i, p. 311: Darwin to August Weismann, 22 October 1868

Marchant 1916, i, pp. 195–210: Wallace to Darwin, [February 1868?]; Darwin to Wallace, 27 February 1868; Wallace to Darwin, 1 March 1868; Wallace to Darwin, 8 March 1868; Darwin to Wallace, 17 March 1868; Wallace to Darwin, 19 March 1868; Darwin to Wallace, 19–24 March 1868; Wallace to Darwin, 24 March 1868; Darwin to Wallace, 27 March 1868; Darwin to Wallace, 6 April 1868; Wallace to Darwin, 8 [April?] 1868

Marchant 1916, i, p. 203: Wallace to Darwin, 19 March 1868

Marchant 1916, i, p. 207: Darwin to Wallace, 27 March 1868

Marchant 1916, i, p. 210: Wallace to Darwin, 8 [April?] 1868

Marchant 1916, ii, p. 41: Wallace to Raphael Meldola, 20 March 1888

Marchant 1916, ii, p. 42: Wallace to Raphael Meldola, 12 April, 1888

BIBLIOGRAPHY

Normally, citations are to first editions. Where they are not, this is indicated either in the text or by an asterisk in the bibliography.

Alexander, R. D. (1979) *Darwinism and Human Affairs*, University of Washington Press, Washington

Alexander, R. D. (1987) *The Biology of Moral Systems*, Aldine de Gruyter, New York

Alexander, R. D. and Tinkle, D. W. (eds.) (1981) *Natural Selection and Social Behaviour: Recent research and new theory*, Blackwell, Oxford

Allee, W. C. (1938) *The Social Life of Animals*, William Heinemann, London

Allee, W. C. (1951) *Cooperation Among Animals with Human Implications*; a revised and amplified edition of *The Social Life of Animals*, Henry Schuman, New York

Allee, W. C., Emerson, A. E., Park, O., Park, T. and Schmidt, K. P. (1949) *Principles of Animal Ecology*, W. B. Saunders, Philadelphia

Allen, E. *et al.* [fifteen other signatories] (1975) 'Against "Sociobiology"', *New York Review of Books 13 November* ; *reprinted in Caplan 1978, pp. 259–64

Allen, G. (1879) *The Colour-Sense: Its origin and development. An essay in comparative psychology*, Trübner, London

Allen, G. E. (1978) *Thomas Hunt Morgan: The man and his science*, Princeton University Press, Princeton, New Jersey

Andersson, M. (1982) 'Female choice selects for extreme tail length in a widowbird', *Nature 299*, 818–20

Andersson, M. (1982a) 'Sexual selection, natural selection and quality advertisement', *Biological Journal of the Linnean Society 17*, 375–93

Andersson, M. (1983) 'Female choice in widowbirds', *Nature 302*, 456

Andersson, M. (1983a) 'On the function of conspicuous seasonal plumages in birds', *Animal Behaviour 31*, 1262–4

Andersson, M. (1986) 'Evolution of condition-dependent sex ornaments and mating preferences: sexual selection based on viability differences', *Evolution 40*, 804–16

Andersson, M. B. and Bradbury, J. W. (1987) 'Introduction' in Bradbury and Andersson 1987, pp. 1–8

Anon (1871) 'Artistic feeling of the lower animals', *The Spectator 11 March*, 280–1

Anon (1871a) 'Mr Darwin's *Descent of Man*', *The Spectator 18 March*, 319–20

Anon (1917) *Geoffrey Watkins Smith*, Printed for private circulation, Oxford

Arak, A. (1983) 'Male–male competition and mate choice in anuran amphibians' in Bateson 1983a, pp. 181–210

Arak, A. (1988) 'Female mate choice in the natterjack toad: active choice or passive attraction?', *Behavioral Ecology and Sociobiology 22*, 317–27

[Argyll, Duke of (Campbell, G. D.)] (1862) Review of Darwin's 'On the Various Contrivances by which British and Foreign Orchids are Fertilised by Insects', *Edinburgh Review 116*, 378–97

Argyll, Duke of (Campbell, G. D.) (1867) *The Reign of Law*, Alexander Strahan, London

Arnold, S. J. (1983) 'Sexual selection: the interface of theory and empiricism' in Bateson 1983a, pp. 67–107

Atchley, W. R. and Woodruff, D. S. (eds.) (1981) *Evolution and Speciation: Essays in honor of M. J. D. White*, Cambridge University Press, Cambridge

Axelrod, R. (1984) *The Evolution of Cooperation*, Basic Books, New York

Axelrod, R. (1986) 'An evolutionary approach to norms', *American Political Science Review 80*, 1095–111

Axelrod, R. and Hamilton, W. D. (1981) 'The evolution of cooperation', *Science 211*, 1390–6

Baer, K. E. von (1873) 'Zum Streit über den Darwinismus', *Augsburger Allgemeine Zeitung 130*, 1986–8; *reprinted in translation in Hull 1973, pp. 416–25

Bajema, C. J. (ed.) (1984) *Evolution by Sexual Selection Theory: Prior to 1900*, Benchmark Papers in Systematic and Evolutionary Biology 6, Van Nostrand Reinhold, New York

Baker, R. R. (1985) 'Bird coloration: in defence of unprofitable prey', *Animal Behaviour 33*, 1387–8

Baker, R. R. and Bibby, C. J. (1987) 'Merlin *Falco columbarius* predation and theories of the evolution of bird coloration', *Ibis 129*, 259–63

Baker, R. R. and Hounsome, M. V. (1983) 'Bird coloration: unprofitable prey model supported by ringing data', *Animal Behaviour 31*, 614–15

Baker, R. R. and Parker, G. A. (1979) 'The evolution of bird coloration', *Philosophical Transactions of the Royal Society of London B 287*, 63–130

Baker, R. R. and Parker, G. A. (1983) 'Female choice in widowbirds', *Nature 302*, 456

Barkow, J. H. (1984) 'The distance betweeen genes and culture', *Journal of Anthropological Research 40*, 367–79

Barkow, J. H., Cosmides, L. and Tooby, J. (eds.) (1992) *The Adapted Mind: Evolutionary psychology and the generation of culture*, Oxford University Press, New York

Barlow, G. W. and Silverberg, J. (eds.) (1980) *Sociobiology: Beyond Nature/Nurture? Reports, definitions and debate*, AAAS Selected Symposium 35, Westview Press, Boulder, Colorado

Barnard, C. J. and Behnke, J. M. (eds.) (1990) *Parasitism and Host Behaviour*, Taylor and Francis, London

Barnett, S. A. (ed.) (1958) *A Century of Darwin*, Heinemann, London; *reprinted Mercury Books, London, 1962

Barrett, P. H. (ed.) (1977) *The Collected Papers of Charles Darwin*, University of Chicago Press, Chicago

Bartholomew, M. J. (1975) 'Huxley's defence of Darwin', *Annals of Science 32*, 525–35

Barton, N. H. and Hewitt, G. M. (1985) 'Analysis of hybrid zones', *Annual Review of Ecology and Systematics 16*, 113–48

Barton, N. H. and Hewitt, G. M. (1989) 'Adaptation, speciation and hybrid zones', *Nature 341*, 497–503

Bartz, S. H. (1979) 'Evolution of eusociality in termites', *Proceedings of the National Academy of Sciences USA 76 (11)*, 5764–8

Bartz, S. H. (1980) 'Correction' [to Bartz 1979], *Proceedings of the National Academy of Sciences USA 77 (6)*, 3070

Bates, H. W. (1862) 'Contributions to an insect fauna of the Amazon Valley, Lepidoptera: Heliconidae', *Transactions of the Linnean Society of London 23*, 495–566

Bates, H. W. (1863) *The Naturalist on the River Amazons*, John Murray, London

Bateson, P. P. G. (1983) 'Rules for changing the rules' in Bendall 1983, pp. 483–507

Bateson, P. P. G. (ed.) (1983a) *Mate Choice*, Cambridge University Press, Cambridge

Bateson, P. P. G. (1983b) 'Optimal outbreeding' in Bateson 1983a, pp. 257–77

Bateson, P. P. G. and Hinde, R. A. (eds.) (1976) *Growing Points in Ethology*, Cambridge University Press, Cambridge

Bateson, W. (1910) 'Heredity and variation in modern lights' in Seward 1910, pp. 85–101

Bateson, W. (1922) 'Evolutionary faith and modern doubts', *Science, new series 55*, 55–61

Beddard, F. E. (1892) *Animal Coloration*, Swan Sonnenschein, London

Bell, G. (1978) 'The handicap principle in sexual selection', *Evolution 32*, 872–85

Bell, P. R. (ed.) (1959) *Darwin's Biological Work: Some aspects reconsidered*, Cambridge University Press, Cambridge; *reprinted John Wiley, New York, 1964

Bell, R. W. and Bell, N. J. (eds.) (1989) *Sociobiology and the Social Sciences*, Texas Tech University Press, Lubbock

Belt, T. (1874) *The Naturalist in Nicaragua: A narrative of a residence at the gold mines of Chontales; journeys in the savannahs and forests; with observations on animals and plants in reference to the theory of evolution of living forms*, John Murray, London

Bendall, D. S. (ed.) (1983) *Evolution from Molecules to Men*, Cambridge University Press, Cambridge

Bernal, J. D. (1954) *Science in History*, Watts, London

Bertram, B. C. R. (1979) 'Ostriches recognise their own eggs and discard others', *Nature 279*, 233–4

Bertram, B. C. R. (1979a) 'Breeding system and strategies of ostriches', *Proceedings of the XVII International Ornithological Congress*, 890–4

Bethel, W. M. and Holmes, J. C. (1973) 'Altered evasive behavior and responses to light in amphipods harboring acanthocephalan cystacanths', *Journal of Parasitology 59*, 945–56

Bethel, W. M. and Holmes, J. C. (1974) 'Correlation of development of altered evasive behavior in *Gammarus lacustris* (Amphipoda) harboring cystacanths of *Polymorphus paradoxus* (Acanthocephala) with the infectivity to the definitive host', *Journal of Parasitology 60*, 272–4

Bethel, W. M. and Holmes, J. C. (1977) 'Increased vulnerability of amphipods to predation owing to altered behavior induced by larval acanthocephalans', *Canadian Journal of Zoology 55*, 110–15

Blair, W. F. (1955) 'Mating call and stage of speciation in the *Microhyla olivacea–M. carolinensis* complex', *Evolution 9*, 469–80

Blair, W. F. (ed.) (1961) *Vertebrate Speciation*, University of Texas Press, Austin

Blaisdell, M. (1982) 'Natural theology and nature's disguises', *Journal of the History of Biology 15*, 163–89

Blake, C. C. (1871) 'The life of Dr Knox', *Journal of Anthropology 3*, 332–8

Bloch, M. (1977) 'The past and the present in the present', *Man, new series 12*, 278–92

Blum, M. S. and Blum, N. A. (eds.) (1979) *Sexual Selection and Reproductive Competition in Insects*, Academic Press, New York

Bonavia, E. (1870) 'Man's bare back', *Nature 3*, 127

Borgia, G. (1979) 'Sexual selection and the evolution of mating systems' in Blum and Blum 1979, pp. 19–80

Borgia, G. (1985) 'Bower quality, number of decorations and mating success of male satin bowerbirds (*Ptilonorhynchus violaceus*): an experimental analysis', *Animal Behaviour 33*, 266–71

Borgia, G. (1985a) 'Bower destruction and sexual competition in the satin bowerbird (*Ptilonorhynchus violaceus*)', *Behavioral Ecology and Sociobiology 18*, 91–100

Borgia, G. (1986) 'Sexual selection in bowerbirds', *Scientific American 254 (6)*, 70–9

Borgia, G. (1986a) 'Satin bowerbird parasites: a test of the bright male hypothesis', *Behavioral Ecology and Sociobiology 19*, 355–8

Borgia, G. and Collis, K. (1989) 'Female choice for parasite-free male satin bowerbirds and the evolution of bright male plumage', *Behavioral Ecology and Sociobiology 25*, 445–53

Borgia, G. and Gore, M. A. (1986) 'Feather stealing in the satin bowerbird (*Ptilonorhynchus violaceus*): male competition and the quality of display', *Animal Behaviour 34*, 727–38

Borgia, G., Kaatz, I. and Condit, R. (1987) 'Female choice and bower decoration in the satin bowerbird *Ptilonorhynchus violaceus*: a test of hypotheses for the evaluation of male display', *Animal Behaviour 35*, 1129–39

Bowler, P. J. (1976) 'Malthus, Darwin and the concept of struggle', *Journal of the History of Ideas 37*, 631–50

Bowler, P. J. (1977) 'Darwinism and the argument from design: suggestions for a reevaluation', *Journal of the History of Biology 10*, 29–43

Bowler, P. J. (1983) *The Eclipse of Darwinism: Anti-Darwinian evolution theories in the decades around 1900*, Johns Hopkins University Press, Baltimore

Bowler, P. J. (1984) *Evolution: The history of an idea*, University of California Press, Berkeley

Boyd, R. and Richerson, P. J. (1985) *Culture and the Evolutionary Process*, University of Chicago Press, Chicago

Bradbury, J. W. (1981) 'The evolution of leks' in Alexander and Tinkle 1981, pp. 138–69

Bradbury, J. W. and Gibson, R. M. (1983) 'Leks and mate choice' in Bateson 1983a, pp. 109–38

Bradbury, W. and Andersson, M. B. (eds.) (1987) *Sexual Selection: Testing the alternatives*, Report of the Dahlem Workshop on Sexual Selection, Life Sciences Research Report 39, Dahlem Konferenzen, Berlin, John Wiley, Chichester

Brandon, R. N. and Burian, R. M. (eds.) (1984) *Genes, Organisms, Populations: Controversies over the units of selection*, MIT Press, Cambridge, Mass

Brockmann, H. J. (1984) 'The evolution of social behaviour in insects' in Krebs and Davies 1978, second edition, pp. 340–61

Brooke, M. de L. and Davies, N. B. (1988) 'Egg mimicry by cuckoos *Cuculus canorus* in relation to discrimination by hosts', *Nature 335*, 630–2

Brooks, J. L. (1984) *Just Before the Origin: Alfred Russel Wallace's theory of evolution*, Columbia University Press, New York

Brown, J. L. (1978) 'Avian communal breeding systems', *Annual Review of Ecology and Systematics 9*, 123–55

Brown, J. L. (1983) 'Intersexual selection', *Nature 302*, 472

Brown, L. (1981) 'Patterns of female choice in mottled sculpins (Cottidae, Teleostei)', *Animal Behaviour 29*, 375–82

Burley, N. (1981) 'Sex ratio manipulation and selection for attractiveness', *Science 211*, 721–2

Burley, N. (1985) 'Leg-band color and mortality patterns in captive breeding populations of zebra finches', *Auk 102*, 647–51

Burley, N. (1986) 'Sexual selection for aesthetic traits in species with biparental care', *American Naturalist 127*, 415–45

Burley, N. (1986a) 'Comparison of the band-colour preferences of two species of estrildid finches', *Animal Behaviour 34*, 1732–41

Burley, N. (1986b) 'Sex-ratio manipulation in color-banded populations of zebra finches', *Evolution 40*, 1191–206

Burley, N. (1988) 'Wild zebra finches have band-colour preferences', *Animal Behaviour 36*, 1235–7

Burley, N. (1988a) 'The differential-allocation hypothesis: an experimental test', *American Naturalist 132*, 611–28

Bush, G. L. (1975) 'Modes of animal speciation', *Annual Review of Ecology and Systematics 6*, 339–64

Butler, S. (1879) *Evolution, Old and New; Or the theories of Buffon, Dr Erasmus Darwin, and Lamarck, as compared with that of Mr Charles Darwin*, Hardwicke and Bogue, London

Cade, W. (1979) 'The evolution of alternative male reproductive strategies in field crickets' in Blum and Blum 1979, pp. 343–79

Cade, W. (1980) 'Alternative male reproductive behaviors', *Florida Entomologist 63*, 30–45

Cain, A. J. (1964) 'The perfection of animals' in Carthy and Duddington 1964, pp. 36–63

Cameron, R. A. D., Carter, M. A and Palles-Clark, M. A. (1980) '*Cepaea* on Salisbury Plain: patterns of variation, landscape history and habitat stability', *Biological Journal of the Linnean Society 14*, 335–58

Campbell, B. (ed.) (1972) *Sexual Selection and the Descent of Man 1871–1971*, Heinemann, London

Canning, E. U. and Wright, C. A. (eds.) (1972) *Behavioural Aspects of Parasite Transmission*, Zoological Journal of Linnean Society 51, supplement 1, Academic Press, London

Caplan, A. L. (ed.) (1978) *The Sociobiology Debate: Readings on ethical and scientific issues*, Harper and Row, New York

Caplan, A. L. (1981) 'Popper's philosophy', *Nature 290*, 623–4

Carneiro, R. L. (ed.) (1967) *The Evolution of Society: Selections from Herbert Spencer's 'Principles of Sociology'*, University of Chicago Press, Chicago

[Carpenter, W. B.] (1847) Review of Owen's 'Lectures on the Comparative Anatomy and Physiology of the Vertebrate Animals, delivered at the Royal College of Surgeons of England, in 1844 and 1846', *British and Foreign Medical Review 23*, 472–92

Carthy, J. D. and Duddington, C. L. (eds.) (1964) *Viewpoints in Biology 3*, Butterworth, London

Catchpole, C. K. (1980) 'Sexual selection and the evolution of complex songs among European warblers of the genus *Acrocephalus*', *Behaviour 74*, 149–66

Catchpole, C. K. (1987) 'Bird song, sexual selection and female choice', *Trends in Ecology and Evolution 2*, 94–7

Catchpole, C. K. (1988) 'Sexual selection and the evolution of animal behaviour', *Science Progress 72*, 281–95

Catchpole, C. K., Dittani, J. and Leisler, B. (1984) 'Differential responses to male song repertoires in female songbirds implanted with oestradiol', *Nature 312*, 563–4

Cavalli-Sforza, L. L. and Feldman, M. W. (1981) *Cultural Transmission and Evolution: A quantitative approach*, Princeton University Press, Princeton, New Jersey

Chalmers, T. (1835) *The Adaptation of External Nature to the Moral and Intellectual Constitution of Man, The Bridgewater Treatises on the Power, Wisdom and Goodness of God, as Manifested in the Creation*, i, William Pickering, London

Charlesworth, W. R. and Kreutzer, M. A. (1973) 'Facial expressions of infants and children' in Ekman 1973, pp. 91–168

Charnov, E. L. and Krebs, J. R. (1975) 'The evolution of alarm calls: altruism or manipulation?', *American Naturalist 109*, 107–12

Cheng, P. W. and Holyoak, K. J. (1989) 'On the natural selection of reasoning theories', *Cognition 33*, 285–313

Clark, R. W. (1985) *The Survival of Charles Darwin: A biography of a man and an idea*, Weidenfeld and Nicolson, London

Clayton, D. H. (1990) 'Mate choice in experimentally parasitized rock doves: lousy males lose', *American Zoologist 30*, 251–62

Clutton-Brock, T. H. (1982) 'The functions of antlers', *Behaviour 79*, 108–25

Clutton-Brock, T. H. and Albon, S. D. (1979) 'The roaring of deer and the evolution of honest advertisement', *Behaviour 69*, 145–70

Clutton-Brock, T. H. and Harvey, P. H. (1979) 'Comparison and adaptation', *Proceedings of the Royal Society of London B 205*, 547–65

Clutton-Brock, T. H. and Harvey, P. H. (1980) 'Primates, brains and ecology', *Journal of Zoology 190*, 309–23

Clutton-Brock, T. H. and Harvey, P. H. (1984) 'Comparative approaches to investigating adaptation' in Krebs and Davies 1978, second edition, pp. 7–29

Clutton-Brock, T. H., Albon, S. D., Gibson, R. M. and Guinness, F. E. (1979) 'The logical stag: adaptive aspects of fighting in red deer (*Cervus elaphus* L.)', *Animal Behaviour 27*, 211–25

Clutton-Brock, T. H., Albon, S. D. and Guinness, F. E. (1981) 'Parental investment in male and female offspring in polygynous mammals', *Nature 289*, 487–9

Clutton-Brock, T. H., Albon, S. D. and Harvey, P. H. (1980) 'Antlers, body size and breeding group size in the Cervidae', *Nature 285*, 565–7

Clutton-Brock, T. H., Guinness, F. E. and Albon, S. D. (1982) *Red Deer: Behaviour and ecology of two sexes*, Edinburgh University Press, Edinburgh

Cohen, J. (1984) 'Sexual selection and the psychophysics of female choice', *Zeitschrift für Tierpsychologie 64*, 1–8

Coleman, W. (1971) *Biology in the Nineteenth Century: Problems of form, function, and transformation*, John Wiley, New York

Collins, J. P. (1986) 'Evolutionary ecology and the use of natural selection in ecological theory', *Journal of the History of Biology 19*, 257–88

Cooke, F. and Buckley, P. A. (eds.) (1987) *Avian Genetics: A population and ecological approach*, Academic Press, London

Cooke, F. and Davies, J. C. (1983) 'Assortative mating, mate choice and reproductive fitness in Snow Geese' in Bateson 1983a, pp. 279–95

Cosmides, L. (1989) 'The logic of social exchange: Has natural selection shaped how humans reason? Studies with the Wason selection task', *Cognition 31*, 187–276

Cosmides, L. and Tooby, J. (1987) 'From evolution to behavior: evolutionary psychology as the missing link' in Dupré 1987, pp. 277–306

Cosmides, L. and Tooby, J. (1989) 'Evolutionary psychology and the generation of culture, Part 2: Case study: a computational theory of social exchange', *Ethology and Sociobiology 10*, 51–97

Cott, H. B. (1940) *Adaptive Coloration in Animals*, Methuen, London

Cott, H. B. (1946) 'The edibility of birds: Illustrated by five years' experiments and observations (1941–1946) on the food preferences of the hornet, cat and man; and considered with special reference to the theories of adaptive coloration', *Proceedings of the Zoological Society of London 116*, 371–524

Cox, C. R. and Le Boeuf, B. J. (1977) 'Female incitation of male competition: a mechanism in mate selection', *American Naturalist 111*, 317–35

Cox, F. E. G. (1989) 'Parasites and sexual selection', *Nature 341*, 289

Coyne, J. (1974) 'The evolutionary origin of hybrid inviability', *Evolution 28*, 505–6

Crampton, H. E. (1916) *Studies on the Variation, Distribution, and Evolution of the Genus Partula: The species inhabiting Tahiti*, Carnegie Institution of Washington Publication 228, Washington

Crampton, H. E. (1925) *Studies on the Variation, Distribution, and Evolution of the Genus Partula: The species of the Mariana Islands, Guam and Saipan*, Carnegie Institution of Washington Publication 228a, Washington

Crampton, H. E. (1932) *Studies on the Variation, Distribution, and Evolution of the Genus Partula: The species inhabiting Moorea*, Carnegie Institution of Washington Publication 410, Washington

Crawford, C., Smith, M. and Krebs, D. (eds.) (1987) *Sociobiology and Psychology: Ideas, issues and applications*, Lawrence Erlbaum, Hillsdale, New Jersey

Crozier, R. H. and Luykx, P. (1985) 'The evolution of termite eusociality is unlikely to have been based on a male-haploid analogy', *American Naturalist 126*, 867–9

Daly, M. (1989) 'Parent–offspring conflict and violence in evolutionary perspective', in Bell and Bell 1989, pp. 25–43

Daly, M. and Wilson, M. (1978) *Sex, Evolution, and Behavior*, Duxbury Press, North Scituate, Mass; *second edition Willard Grant Press, Boston, Mass, 1983

Daly, M. and Wilson, M. (1984) 'A sociobiological analysis of human infanticide' in Hausfater and Hrdy 1984, pp. 487–502

Daly, M. and Wilson, M. (1988) *Homicide*, Aldine de Gruyter, New York

Daly, M. and Wilson, M. (1988a) 'The Darwinian psychology of discriminative parental solicitude', *Nebraska Symposium on Motivation 1987 35*, 91–144

Daly, M. and Wilson, M. (1989) 'Homicide and cultural evolution', *Ethology and Sociobiology 10*, 99–110

Daly, M. and Wilson, M. (1990) 'Killing the competition', *Human Nature 1*, 83–109

Darwin, C. (1845) *Journal of Researches into the Natural History and Geology of the Countries Visited During the Voyage of H.M.S. 'Beagle' Round the World, under the Command of Capt. Fitz Roy, R.N.*, John Murray, London; *new edition with a biographical introduction by G. T. Bettany, Ward, Lock, London, 1891

Darwin, C. (1859) *On the Origin of Species by means of Natural Selection or the Preservation of Favoured Races in the Struggle for Life*, John Murray, London; facsimile reproduction with an introduction by Ernst Mayr, Atheneum, New York, 1967

Darwin, C. (1862) *On the Various Contrivances by which British and Foreign Orchids are Fertilised by Insects*, John Murray, London; second edition 1877

Darwin, C. (1862a) 'On the two forms, or dimorphic condition, in the species of *Primula*, and on their remarkable sexual relations', *Journal of the Proceedings of the Linnean Society (Botany) 6*, 77–96; *reprinted in Barrett 1977, ii, pp. 45–63

[Darwin, C.] (1863) Review of Bates' 'Contributions to an insect fauna', *Natural History Review: Quarterly Journal of Biological Science*, 219–24; *reprinted in Barrett 1977, ii, pp. 87–92

Darwin, C. (1864) 'On the existence of two forms, and on their reciprocal sexual relation, in several species of the genus *Linum*', *Journal of the Proceedings of the Linnean Society (Botany) 7*, 69–83; *reprinted in Barrett 1977, ii, pp. 93–105

Darwin, C. (1865) 'On the sexual relations of the three forms of *Lythrum salicaria*', *Journal of the Proceedings of the Linnean Society (Botany) 8*, 169–96; *reprinted in Barrett 1977, ii, pp. 106–31

Darwin, C. (1868) *The Variation of Animals and Plants under Domestication*, John Murray, London; second edition 1875

Darwin, C. (1869) 'Origin of species', *Athenaeum 2174*, 861; *reprinted in Barrett, 1977, ii, pp. 156–7

Darwin, C. (1869a) 'Origin of species', *Athenaeum 2177*, 82; *reprinted in Barrett 1977, ii, pp. 157–8

Darwin, C. (1871) *The Descent of Man and Selection in Relation to Sex*, John Murray, London; second edition 1874; facsimile reproduction of first edition with an introduction by John Tyler Bonner and Robert M. May, Princeton University Press, Princeton, 1981

Darwin, C. (1872) *The Expression of the Emotions in Man and Animals*, John Murray, London; facsimile reproduction with an introduction by Konrad Lorenz, University of Chicago Press, Chicago, 1965

Darwin, C. (1876) *The Effects of Cross and Self Fertilisation in the Vegetable Kingdom*, John Murray, London; second edition 1878

Darwin, C. (1876a) 'Sexual selection in relation to monkeys', *Nature 15*, 18–19; *reprinted in Barrett 1977, ii, pp. 207–11

Darwin, C. (1877) *The Different Forms of Flowers on Plants of the Same Species*, John Murray, London; second edition 1892

Darwin, C. (1880) 'The sexual colours of certain butterflies', *Nature 21*, 237; *reprinted in Barrett 1977, ii, pp. 220–2

Darwin, C. (1882) 'A preliminary notice: "On the modification of a race of Syrian street-dogs by means of sexual selection"', *Proceedings of the Zoological Society of London*, 367–9; *reprinted in Barrett 1977, ii, pp. 278–80

Darwin, C. and Wallace, A. R. W. (1858) 'On the tendency of species to form varieties; and on the perpetuation of varieties and species by natural means of selection', *Journal of the Linnean Society of London (Zoology) 3*, 45–62; *reprinted in Linnean Society 1908, pp. 87–107

Darwin, F. (ed.) (1887) *The Life and Letters of Charles Darwin*, John Murray, London

Darwin, F. (ed.) (1892) *The Autobiography of Charles Darwin and Selected Letters*, John Murray, London; facsimile reproduction Dover, New York, 1958

Darwin, F. and Seward, A. C. (eds.) (1903) *More Letters of Charles Darwin: A record of his work in a series of hitherto unpublished letters*, John Murray, London

Davies, N. B. (1978) 'Territorial defence in the speckled wood butterfly (*Pararge aegeria*): the resident always wins', *Animal Behaviour 26*, 138–47

Davies, N. B. (1982) 'Cooperation and conflict in breeding groups', *Nature 296*, 702–3

Davis, J. W. F. and O'Donald, P. (1976) 'Sexual selection for a handicap: a critical analysis of Zahavi's model', *Journal of Theoretical Biology 57*, 345–54

Davison, G. W. H. (1981) 'Sexual selection and the mating system of *Argusianus argus* (Aves: Phasianidae)', *Biological Journal of the Linnean Society 15*, 91–104

Dawkins, M. S. (1986) *Unravelling Animal Behaviour*, Longman, Essex

Dawkins, R. (1976) *The Selfish Gene*, Oxford University Press, Oxford; second edition 1989

Dawkins, R. (1978a) 'Reply to Fix and Greene', *Contemporary Sociology 7*, 709–12

Dawkins, R. (1979) 'Twelve misunderstandings of kin selection', *Zeitschrift für Tierpsychologie 51*, 184–200

Dawkins, R. (1980) 'Good strategy or evolutionarily stable strategy?' in Barlow and Silverberg 1980, pp. 331–67

Dawkins, R. (1981) 'In defence of selfish genes', *Philosophy 56*, 556–73

Dawkins, R. (1982) *The Extended Phenotype: The gene as the unit of selection*, W. H. Freeman, Oxford

Dawkins, R. (1982a) 'The necessity of Darwinism', *New Scientist 94*, 130–2

Dawkins, R. (1983) 'Universal Darwinism' in Bendall 1983, pp. 403–25

Dawkins, R. (1986) *The Blind Watchmaker*, Longman, Essex

Dawkins, R. (1986a) 'Sociobiology: the new storm in a teacup' in Rose and Appignanesi 1986, pp. 61–78

Dawkins, R. (1989) 'The evolution of evolvability' in Langton 1989, pp. 201–20

Dawkins, R. (1990) 'Parasites, desiderata lists and the paradox of the organism' in Keymer and Read 1990, pp. S63–73

Dawkins, R. and Krebs, J. R. (1978) 'Animal signals: information or manipulation?' in Krebs and Davies 1978, pp. 282–309

Dawkins, R. and Krebs, J. R. (1979) 'Arms races within and between species', *Proceedings of the Royal Society of London B 205*, 489–511

Dawkins, R. and Ridley, M. (eds.) (1985) *Oxford Surveys in Evolutionary Biology 2*, Oxford University Press, Oxford

de Beer, G. R. (ed.) (1938) *Evolution: Essays on aspects of evolutionary biology, presented to Professor E. S. Goodrich on his seventieth birthday*, Clarendon Press, Oxford

de Beer, G. R. (1963) *Charles Darwin: Evolution by natural selection*, Thomas Nelson, London

de Beer, G. R. (1971) 'Darwin, Charles Robert' in Gillispie 1971, iii, pp. 565–77

de Beer, G. R., Rowlands, M. J. and Skramovsky, B. M. (eds.) (1960–7) 'Darwin's Notebooks on Transmutation of Species', *Bulletin of The British Museum (Natural History) Historical Series 2 (2–6), 3 (5)*

Delfino, V. P. (ed.) (1987) *International Symposium on Biological Evolution, Bari 9–14 April 1985*, Adriatica Editrice, Bari

Dennett, D. C. (1984) *Elbow Room: The varieties of free will worth wanting*, Oxford University Press, Oxford

Dewar, D. and Finn, F. (1909) *The Making of Species*, John Lane, London

Dewey, J. (1909) 'The influence of Darwinism on philosophy', *Popular Science Monthly*; *reprinted in Dewey 1910, pp. 1–19

Dewey, J. (1910) *The Influence of Darwin on Philosophy and Other Essays in Contemporary Thought*, Henry Holt, New York; facsimile reproduction Peter Smith, New York, 1951

Diamond, J. (1981) 'Birds of paradise and the theory of sexual selection', *Nature 293*, 257–8

Diamond, J. (1982) 'Evolution of bowerbirds' bowers: animal origins of the aesthetic sense', *Nature 297*, 99–102

Diamond, J. (1987) 'Biology of birds of paradise and bowerbirds', *Annual Review of Ecology and Systematics 17*, 17–37

Diamond, J. (1988) 'Experimental study of bower decoration by the bowerbird *Amblyornis inornatus*, using colored poker chips', *American Naturalist 131*, 631–53

Diver, C. (1940) 'The problem of closely related species living in the same area' in Huxley 1940, pp. 303–28

Dobzhansky, Th. (1937) *Genetics and the Origin of Species*; *third edition Columbia University Press, New York, 1951

Dobzhansky, Th. (1940) 'Speciation as a stage in evolutionary divergence', *American Naturalist 74*, 312–21

Dobzhansky, Th. (1956) 'What is an adaptive trait?', *American Naturalist 90*, 337–47

Dobzhansky, Th. (1970) *Genetics of the Evolutionary Process*, Columbia University Press, New York

Dobzhansky, Th. (1975) 'Analysis of incipient reproductive isolation within a species of *Drosophila*', *Proceedings of the National Academy of Sciences USA 72 (9)*, 3638–41

Doherty, J. A. and Gerhardt, H. C. (1983) 'Acoustic communication in hybrid treefrogs: sound production by males and selective phonotaxis by females', *Journal of Comparative Physiology A 154*, 319–30

Dohrn, A. (1871) Review of Wallace's 'Contributions to the Theory of Natural Selection: A series of essays', *The Academy 2*, 159–60

Dominey, W. J. (1983) 'Sexual selection, additive genetic variance and the "phenotypic handicap"', *Journal of Theoretical Biology 101*, 495–502

Douglass, G. N. (1895) 'On the Darwinian hypothesis of sexual selection', *Natural Science 7*, 326–32, 398–406

Downhower, J. F. and Brown, L. (1980) 'Mate preferences of female mottled sculpins, *Cottus bairdi*', *Animal Behaviour 28*, 728–34

Downhower, J. F. and Brown, L. (1981) 'The timing of reproduction and its behavioural consequences for mottled sculpins, *Cottus bairdi*' in Alexander and Tinkle 1981, pp. 78–95

Duncan, D. (ed.) (1908) *The Life and Letters of Herbert Spencer*, Methuen, London

Dunford, C. (1977) 'Kin selection for ground squirrel alarm calls', *American Naturalist 111*, 782–5

Dupré, J. (ed.) (1987) *The Latest on the Best: Essays on evolution and optimality*, MIT Press, Cambridge, Mass

Durant, J. R. (1979) 'Scientific naturalism and social reform in the thought of Alfred Russel Wallace', *British Journal for the History of Science 12*, 31–58

Durant, J. R. (1981) 'Innate character in animals and man: a perspective on the origins of ethology' in Webster 1981, pp. 157–92

Durant, J. R. (1985) 'The ascent of nature in Darwin's *Descent of Man*' in Kohn 1985, pp. 283–306

Eberhard, W. G. (1979) 'The function of horns in *Podischnus agenor* (Dynastinae) and other beetles' in Blum and Blum 1979, pp. 231–58

Eberhard, W. G. (1980) 'Horned beetles', *Scientific American 242 (3)*, 124–31

Eberhard, W. G. (1985) *Sexual Selection and Animal Genitalia*, Harvard University Press, Cambridge, Mass

Ebling, F. J. and Stoddart, D. M. (eds.) (1978) *Population Control by Social Behaviour*, Proceedings of a symposium held at the Royal Geographical Society, London, on 20 and 21 September 1977, Symposia of the Institute of Biology 23, Institute of Biology, London

Egerton, F. N. (1973) 'Changing concepts of the balance of nature', *Quarterly Review of Biology 48*, 322–50

Eibl-Eibesfeldt, I. (1970) *Ethology: The biology of behavior*, Holt, Rinehart and Winston, New York

Eiseley, L. (1958) *Darwin's Century: Evolution and the men who discovered it*, Doubleday, New York; *reprinted Anchor Books, New York, 1961

Eisenberg, J. F. and Dillon, W. S. (eds.) (1971) *Man and Beast: Comparative social behavior*, Papers delivered at the Smithsonian Institution Annual Symposium 1969, Smithsonian Annual 3, Smithsonian Institution Press, Washington DC

Ekman, P. (ed.) (1973) *Darwin and Facial Expression: A century of research in review*, Academic Press, New York

Ellegård, A. (1958) *Darwin and the General Reader: The reception of Darwin's theory of evolution in the British periodical press, 1859–1872*, University of Göteborg, Göteborg

Elster, J. (1983) *Explaining Technical Change: A case study in the philosophy of science*, Cambridge University Press, Cambridge

Emerson, A. E. (1958) 'The evolution of behavior among social insects' in Roe and Simpson 1958, pp. 311–35

Emerson, A. E. (1960) 'The evolution of adaptation in population systems', in Tax 1960, i, *The Evolution of Life: Its origin, history and future*, pp. 307–48

Emlen, S. T. (1984) 'Cooperative breeding in birds' in Krebs and Davies 1978, second edition, pp. 305–39

Endler, J. A. (1977) *Geographic Variation, Speciation and Clines*, Princeton University Press, Princeton, New Jersey

Engelhard, G., Foster, S. P. and Day, T. H. (1989) 'Genetic differences in mating success and female choice in seaweed flies (*Coelopa frigida*)', *Heredity 62*, 123–31

Eriksson, D. and Wallin, L. (1986) 'Male bird song attracts females – a field experiment', *Behavioral Ecology and Sociobiology 19*, 297–9

Eshel, I. (1978) 'On the handicap principle – a critical defence', *Journal of Theoretical Biology 70*, 245–50

Evans, J. St. B. T. (ed.) (1983) *Thinking and Reasoning: Psychological approaches*, Routledge and Kegan Paul, London

Farley, J. (1982) *Gametes and Spores: Ideas about sexual reproduction 1750–1914*, Johns Hopkins University Press, Baltimore

Fichman, M. (1981) *Alfred Russel Wallace*, Twayne, Boston

Fischer, E. A. (1980) 'The relationship between mating system and simultaneous hermaphroditism in the coral reef fish, *Hypoplectrus nigricans* (Serranidae)', *Animal Behaviour 28*, 620–33

Fisher, R. A. (1915) 'The evolution of sexual preference', *Eugenics Review 7*, 184–92

Fisher, R. A. (1930) *The Genetical Theory of Natural Selection*, Clarendon Press, Oxford; *revised edition Dover, New York, 1958

Fodor, J. A. (1983) *The Modularity of Mind: An essay on faculty psychology*, MIT Press, Cambridge, Mass

Ford, E. B. (1964) *Ecological Genetics*, Methuen, London; *fourth edition Chapman and Hall, London, 1975

Foster, M. and Lankester, E. R. (eds.) (1898) *Scientific Memoirs of T. H. Huxley*, Macmillan, London

Fox, D. L. (1953) *Animal Biochromes and Structural Colours: Physical, chemical, distributional and physiological features of coloured bodies in the animal world*, Cambridge University Press, Cambridge; *second edition University of California Press, Berkeley, 1976

Fox, R. (ed.) (1975) *Biosocial Anthropology*, ASA Studies 1, Malaby Press, London

Fraser, G. (1871) 'Sexual selection', *Nature 3*, 489

Freeman, R. B. (1978) *Charles Darwin: A companion*, William Dawson, Folkestone, Kent

Futuyma, D. J. (1986) *Evolutionary Biology*, Sinauer Associates, Sunderland, Mass, second edition

Gadgil, M. (1981) 'Evolution of reproductive strategies' in Scudder and Reveal 1981, pp. 91–2

Gale, B. G. (1972) 'Darwin and the concept of a struggle for existence: a study in the extrascientific origins of scientific ideas', *Isis 63*, 321–44

Geddes, P. and Thomson, J. A. (1889) *The Evolution of Sex*, Walter Scott, London; *second edition 1901

Geist, V. (1974) 'On fighting strategies in animal combat', *Nature 250*, 354

George, W. (1982) *Darwin*, Fontana

Ghiselin, M. T. (1969) *The Triumph of the Darwinian Method*, University of California Press, Berkeley

Ghiselin, M. T. (1974) *The Economy of Nature and the Evolution of Sex*, University of California Press, Berkeley

Gibson, R. M. and Bradbury, J. W. (1985) 'Sexual selection in lekking sage grouse: phenotypic correlates of male mating success', *Behavioral Ecology and Sociobiology 18*, 117–23

Gillespie, N. C. (1979) *Charles Darwin and the Problem of Creation*, University of Chicago Press, Chicago

Gillispie, C. C. (1951) *Genesis and Geology*, Harvard University Press, Cambridge, Mass; *reprinted Harper and Row, New York, 1959

Gillispie, C. C. (ed.) (1971) *Dictionary of Scientific Biography*, Charles Scribner's Sons, New York

Gliserman, S. (1975) 'Early Victorian science writers and Tennyson's *In Memoriam*: a study in cultural exchange', *Victorian Studies 18*, 277–308, 437–59

Gould, S. J. (1978) *Ever Since Darwin: Reflections in natural history*, Burnett Books, London; *reprinted Penguin, Middlesex, 1980

Gould, S. J. (1978a) 'Sociobiology and human nature: a postpanglossian vision', *Human Nature 1*; *reprinted in Montagu 1980, pp. 283–90

Gould, S. J. (1980) *The Panda's Thumb: More reflections in natural history*, W. W. Norton, New York

Gould, S. J. (1980a) 'Sociobiology and the theory of natural selection' in Barlow and Silverberg 1980, pp. 257–69

Gould, S. J. (1983) *Hen's Teeth and Horse's Toes: Further reflections in natural history*, W. W. Norton, New York; *reprinted Penguin, Middlesex, 1984

Gould, S. J. and Lewontin, R. C. (1979) 'The spandrels of San Marco and the Panglossian paradigm: a critique of the adaptationist programme', *Proceedings of the Royal Society of London B 205*, 581–98

Gowaty, P. A. and Karlin, A. A. (1984) 'Multiple maternity and paternity in single broods of apparently monogamous eastern bluebirds (*Sialia sialis*)', *Behavioral Ecology and Sociobiology 15*, 91–5

Grafen, A. (1984) 'Natural selection, kin selection and group selection' in Krebs and Davies 1978, second edition, pp. 62–84

Grafen, A. (1985) 'A geometric view of relatedness' in Dawkins and Ridley 1985, pp. 28–89

Grafen, A. (1990) 'Sexual selection unhandicapped by the Fisher process', *Journal of Theoretical Biology 144*, 473–516

Grafen, A. (1990a) 'Biological signals as handicaps', *Journal of Theoretical Biology 144*, 517–46

Graham, W. (1881) *The Creed of Science: Religious, moral and social*, Kegan Paul, London

Grant, V. (1963) *The Origin of Adaptations*, Columbia University Press, New York

Grant, V. (1966) 'The selective origin of incompatibility barriers in the plant genus *Gilia*', *American Naturalist 100*, 99–118

Grant, V. (1971) *Plant Speciation*, Columbia University Press, New York; *second edition 1981

Gray, A. (1876) *Darwiniana: Essays and reviews pertaining to Darwinism*, Appleton, New York; *reprinted with an introduction by A. Hunter Dupree, Harvard University Press, Cambridge, Mass, 1963

Gray, R. D. (1988) 'Metaphors and methods: behavioural ecology, panbiogeography and the evolving synthesis' in Ho and Fox 1988, pp. 209–42

Greene, J. C. (1959) *The Death of Adam: Evolution and its impact on western thought*, Iowa State University Press, Ames

Gregorio, M. A. di (1982) 'The dinosaur connection: a reinterpretation of T. H. Huxley's evolutionary view', *Journal of the History of Biology 15*, 397–418

Grene, M. (ed.) (1983) *Dimensions of Darwinism: Themes and counterthemes in twentieth-century evolutionary theory*, Cambridge University Press, Cambridge

Grinnell, G. J. (1985) 'The rise and fall of Darwin's second theory', *Journal of the History of Biology 18*, 51–70

Groos, K. (1898) *The Play of Animals: A study of animal life and instinct*, Chapman and Hall, London

Gruber, H. E. (1974) *Darwin on Man: A psychological study of scientific creativity, together with Darwin's early and unpublished notebooks transcribed and annotated by Paul H. Barrett*, Wildwood House, London

Gulick, J. T. (1872) 'On the variation of species as related to their geographical distribution, illustrated by the *Achatinellinae*', *Nature 6*, 222–4

Gulick, J. T. (1873) 'On diversity of evolution under one set of external conditions', *Journal of the Linnean Society (Zoology) 11*, 496–505

Gulick, J. T. (1890) 'Divergent evolution through cumulative segregation', *Journal of the Linnean Society (Zoology) 20*, 189–274

Haas, R. (1978) 'Sexual selection in *Notobranchius guentheri* (Pisces: Cyprinodontidae)', *Evolution 30*, 614–22

Haldane, J. B. S. (1932) *The Causes of Evolution*, Longmans, Green, London; *reprinted Cornell University Press, Ithaca, 1966

Haldane, J. B. S. (1939) *Science and Everyday Life*, Lawrence and Wishart, London; reprinted in part in Maynard Smith 1985b

Haldane, J. B. S. (1955) 'Population genetics', *New Biology 18*, 34–51

Halliday, T. R. (1983) 'Do frogs and toads choose their mates?', *Nature 306*, 226–7

Halliday, T. R. (1983a) 'The study of mate choice' in Bateson 1983a, pp. 3–32

Hamilton, W. D. (1963) 'The evolution of altruistic behavior', *American Naturalist 97*, 354–6

Hamilton, W. D. (1964) 'The genetical evolution of social behaviour', *Journal of Theoretical Biology 7*, 1–16, 17–52

Hamilton, W. D. (1971) 'Notes and addendum' in Williams 1971, pp. 62, 63, 87–9

Hamilton, W. D. (1971a) 'Selection of selfish and altruistic behavior in some extreme models' in Eisenberg and Dillon 1971, pp. 57–91

Hamilton, W. D. (1972) 'Altruism and related phenomena, mainly in social insects', *Annual Review of Ecology and Systematics 3*, 193–232

Hamilton, W. D. (1975) 'Innate social aptitudes of man: an approach from evolutionary genetics' in Fox 1975, pp. 133–55

Hamilton, W. D. (1979) 'Wingless and fighting males in fig wasps and other insects' in Blum and Blum 1979, pp. 167–220

Hamilton, W. D. and Zuk, M. (1982) 'Heritable true fitness and bright birds: a role for parasites?', *Science 218*, 384–7

Hamilton, W. D. and Zuk, M. (1989) 'Parasites and sexual selection', *Nature 341*, 289–90

Harcourt, A. H., Harvey, P. H., Larson, S. G. and Short, R. V. (1981) 'Testis weight, body weight and breeding system in primates', *Nature 293*, 55–7

Harris, M. (1968) *The Rise of Anthropological Theory: A history of theories of culture*, Thomas Y. Crowell, New York

Harris, M. (1974) *Cows, Pigs, Wars and Witches: The riddles of culture*, Random House, New York; *reprinted Hutchinson, London, 1975

Harris, M. (1986) *Good to Eat: Riddles of food and culture*, Allen and Unwin, London

Harvey, P. H. (1986) 'Birds, bands and better broods?', *Trends in Ecology and Evolution 1*, 8–9

Harvey, P. H. and Clutton-Brock, T. H. (1983) 'The survival of the theory', *New Scientist 98*, 312–15

Harvey, P. H. and Mace, G. M. (1982) 'Comparisons between taxa and adaptive trends: problems of methodology' in King's College Sociobiology Group 1982, pp. 343–61

Harvey, P. H. and Pagel, M. D. (1991) *The Comparative Method in Evolutionary Biology*, Oxford University Press, Oxford

Harvey, P. H. and Wilcove, D. S. (1985) 'Sex among the dunnocks', *Nature 313*, 180

Harvey, P. H., Kavanagh, M. and Clutton-Brock, T. H. (1978) 'Sexual dimorphism in primate teeth', *Journal of Zoology 186*, 475–85

Harvey, P. H., Kavanagh, M. and Clutton-Brock, T. H. (1978a) 'Canine tooth size in female primates', *Nature 276*, 817–18

Hausfater, G. and Hrdy, S. B. (eds.) (1984) *Infanticide: Comparative and evolutionary perspectives*, Aldine, New York

Hausfater, G., Gerhardt, H. C. and Klump, G. M. (1990) 'Parasites and mate choice in gray treefrogs, *Hyla versicolor*', *American Zoologist 30*, 299–311

Herbert, S. (1971) 'Darwin, Malthus, and selection', *Journal of the History of Biology 4*, 209–17

Herbert, S. (1977) 'The place of man in the development of Darwin's theory of transmutation', *Journal of the History of Biology 1*, 155–227

Herschel, J. F. W. (1861) *Physical Geography*, Adam and Charles Black, Edinburgh

Heslop-Harrison, J. (1958) 'Darwin as a botanist' in Barnett 1958, pp. 267–95

Heslop-Harrison, J. (1959) 'The origin of isolation', *New Biology 28*, 65–9

Hillgarth, N. (1990) 'Parasites and female choice in the ring-necked pheasant', *American Zoologist 30*, 227–33

Himmelfarb, G. (1959) *Darwin and the Darwinian Revolution*, Doubleday, New York; second edition 1962; *reprint of second edition W. W. Norton, New York, 1968

Hingston, R. W. G. (1933) *The Meaning of Animal Colour and Adornment*, Edward Arnold, London

Hirst, P. Q. (1976) *Social Evolution and Sociological Categories*, George Allen and Unwin, London

Hitching, F. (1982) *The Neck of the Giraffe: Or where Darwin went wrong*, Pan, London

Ho, M-W. (1988) 'On not holding nature still: evolution by process, not by consequence' in Ho and Fox 1988, pp. 117–45

Ho, M-W. and Fox, S. W. (eds.) (1988) *Evolutionary Processes and Metaphors*, John Wiley, Chichester

Hofstadter, D. R. (1982) 'Can inspiration be mechanized?', *Scientific American 247 (3)*, 18–31; *reprinted as 'On the seeming paradox of mechanizing creativity' in Hofstadter 1985, pp. 526–46

Hofstadter, D. R. (1985) *Metamagical Themas: Questing for the essence of mind and pattern*, Basic Books; *reprinted Penguin, Middlesex, 1986

Hogan-Warburg, A. J. (1966) 'Social behaviour of the ruff, *Philomachus pugnax* (L.)', *Ardea 54*, 109–229

Holmes, J. C. and Bethel, W. M. (1972) 'Modification of intermediate host behaviour by parasites' in Canning and Wright 1972, pp. 123–47

Horn, D. J., Stairs, G. R. and Mitchell, R. D. (eds.) (1979) *Analysis of Ecological Systems*, Ohio State University Press, Columbus

Howard, J. C. (1981) 'A tropical Volute shell and the Icarus syndrome', *Nature 290*, 441–2; *reprinted with Addendum in Maynard Smith 1982c, pp. 100–5

Howlett, R. (1988) 'Sexual selection by female choice in monogamous birds', *Nature 332*, 583–4

Hoy, R. H., Hahn, J. and Paul, R. C. (1977) 'Hybrid cricket auditory behavior: evidence for genetic coupling in animal communication', *Science 195*, 82–4

Hoyle, F. and Wickramasinghe, N. C. (1981) *Evolution from Space*, J. M. Dent, London

Hudson, W. H. (1892) *The Naturalist in La Plata*, Chapman and Hall, London

Hull, D. L. (1973) *Darwin and His Critics: The reception of Darwin's theory of evolution by the scientific community*, Harvard University Press, Cambridge, Mass

Hull, D. L. (1981) 'Units of evolution: a metaphysical essay' in Jensen and Harré 1981, pp. 23–44

Hull, D. L. (1983) 'Darwin and the nature of science' in Bendall 1983, pp. 63–80

Hull, D. L. (1984) 'Rich man, poor man', *Nature 308*, 798–9

Hume, D. (1779) *Dialogues Concerning Natural Religion*; *reprinted Hafner, New York, 1948

Humphrey, N. (1976) 'The social function of intellect' in Bateson and Hinde 1976, pp. 303–25

Humphrey, N. (1986) *The Inner Eye*, Faber and Faber, London

Huxley, J. S. (1914) 'The courtship-habits of the great crested grebe (*Podiceps cristatus*); with an addition to the theory of sexual selection', *Proceedings of the Zoological Society of London*, 491–562

Huxley, J. S. (1921) 'The accessory nature of many structures and habits associated with courtship', *Nature 108*, 565–6

Huxley, J. S. (1923) 'Courtship activities in the red-throated diver (*Colymbus Stellatus* pontopp.); together with a discussion of the evolution of courtship in birds', *Journal of the Linnean Society of London (Zoology) 35*, 253–92

Huxley, J. S. (1923a) *Essays of a Biologist*, Chatto and Windus, London; *second edition 1923

Huxley, J. S. (1931) 'The relative size of antlers in deer', *Proceedings of the Zoological Society of London*, 819–64

Huxley, J. S. (1932) *Problems of Relative Growth*, Methuen, London

Huxley, J. S. (1938) 'Darwin's theory of sexual selection and the data subsumed by it, in the light of recent research', *American Naturalist 72*, 416–33

Huxley, J. S. (1938a) 'The present standing of the theory of sexual selection' in de Beer 1938, pp. 11–42

Huxley, J. S. (ed.) (1940) *The New Systematics*, Oxford University Press, Oxford

Huxley, J. S. (1942) *Evolution: The modern synthesis*, George Allen and Unwin, London; *second edition 1963

Huxley, J. S. (1947) 'The vindication of Darwinism' in Huxley and Huxley 1947, pp. 153–76

Huxley, J. S. (1966) 'Introduction', *Philosophical Transactions of the Royal Society B 251*, 249–71

Huxley, L. (ed.) (1900) *The Life and Letters of Thomas Henry Huxley*, Macmillan, London

Huxley, T. H. (1856) 'On natural history, as knowledge, discipline, and power', *Proceedings of the Royal Institution 2 1854–8*, 187–95; reprinted in Foster and Lankester 1898, i, pp. 305–14

Huxley, T. H. (1860) 'The origin of species' in Huxley 1893–4, ii, pp. 22–79

Huxley, T. H. (1860a) 'On species and races, and their origin', *Proceedings of the Royal Institution 3 1858–62*, 195–200; reprinted in Foster and Lankester 1898, ii, pp. 388–94

Huxley, T. H. (1863) 'On the relations of man to the lower animals' in Huxley 1893–4, vii, pp. 76–156

Huxley, T. H. (1863a) *On Our Knowledge of the Causes of the Phenomena of Organic Nature*, Robert Hardwicke, London

Huxley, T. H. (1871) 'Mr Darwin's critics', *Contemporary Review 18*, 443–76

Huxley, T. H. (1887) 'On the reception of the *Origin of Species*' in Darwin, F. 1887, ii, pp. 179–204

Huxley, T. H. (1888) 'The struggle for existence in human society: a programme', *Nineteenth Century 23*, 161–80; *reprinted in Huxley 1893–4, ix, pp. 195–236

Huxley, T. H. (1892) 'Prologue [Controverted Questions]' in Huxley 1893–4, v, pp. 1–58

Huxley, T. H. (1893) 'Evolution and ethics', The Romanes Lecture 1893, Macmillan, London; *reprinted in Huxley 1893–4, ix, pp. 46–116

Huxley, T. H. (1893–4) *Collected Essays*, Macmillan, London

Huxley, T. H. (1894) 'Evolution and ethics: prolegomena' in Huxley 1893–4, ix, pp. 1–45

Huxley, T. H. (1894a) 'Preface' in Huxley 1893–4, ix, pp. v–xiii

Huxley, T. H. and Huxley, J. S. (1947) *Evolution and Ethics 1893–1943*, Pilot Press, London

Irvine, W. (1955) *Apes, Angels and Victorians: The story of Darwin, Huxley and evolution*, McGraw–Hill, New York

Jacob, F. (1977) 'Evolution and tinkering', *Science 196*, 1161–6

Jaenike, J. (1988) 'Parasitism and male mating success in *Drosophila testacea*', *American Naturalist 131*, 774–80

Jarvie, I. (1964) *The Revolution in Anthropology*, Routledge and Kegan Paul, London; *reprinted with corrections and additions 1967

Jensen, V. J. and Harré, R. (eds.) (1981) *The Philosophy of Evolution*, Harvester Press, Brighton

Johnson, L. L. and Boyce, M. S. (1991) 'Female choice of males with low parasite load in sage grouse' in Loye and Zuk 1991, pp. 377–88

Jones, J. S. (1982) 'Of cannibals and kin', *Nature 299*, 202–3

Jones, J. S. (1982a) 'Genetic differences in individual behaviour associated with shell polymorphism in the snail *Cepea nemoralis*', *Nature 298*, 749–50

Jones, J. S., Leith, B. H. and Rawlings, P. (1977) 'Polymorphism in *Cepea*: a problem with too many solutions?', *Annual Review of Ecology and Systematics 8*, 109–43

Kellogg, V. L. (1907) *Darwinism To–day: A discussion of present-day scientific criticism of the Darwinian selection theories, together with a brief account of the principal other proposed auxiliary and alternative theories of species-forming*, Henry Holt, New York

Kennedy, C. E. J., Endler, J. A., Poynton, S. L. and McMinn, H. (1987) 'Parasite load predicts mate choice in guppies', *Behavioral Ecology and Sociobiology 21*, 291–5

Keymer, A. E. and Read, A. F. (eds.) (1990) *The Evolutionary Biology of Parasitism*, Symposia of the British Society for Parasitology 27, Parasitology 100, supplement 1990, Cambridge University Press, Cambridge

Kimler, W. C. (1983) 'Mimicry: views of naturalists and ecologists before the modern synthesis' in Grene 1983, pp. 97–127

Kimura, M. (1983) *The Neutral Theory of Molecular Evolution*, Cambridge University Press, Cambridge

King's College Sociobiology Group (eds.) (1982) *Current Problems in Sociobiology*, Cambridge University Press, Cambridge

Kirkpatrick, M. (1982) 'Sexual selection and the evolution of female choice', *Evolution 36*, 1–12

Kirkpatrick, M. (1986) 'The handicap mechanism of sexual selection does not work', *American Naturalist 127*, 222–40

Kirkpatrick, M. (1987) 'Sexual selection by female choice in polygynous animals', *Annual Review of Ecology and Systematics 18*, 43–70

Kirkpatrick, M. (1989) 'Is bigger always better?', *Nature 337*, 116–17

Kirkpatrick, M., Price, T. and Arnold, S. J. (1990) 'The Darwin–Fisher theory of sexual selection in monogamous birds', *Evolution 44*, 180–93

Knox, R. (1831) 'Observations on the structure of the stomach of the Peruvian lama; to which are prefixed remarks on the analogical reasoning of anatomists, in the determination a priori of unknown species and unknown structures', *Transactions of the Royal Society of Edinburgh 11*, 479–98

Kodric-Brown, A. and Brown, J. H. (1984) 'Truth in advertising: the kinds of traits favoured by sexual selection', *American Naturalist 124*, 309–23

Koestler, A. (1971) *The Case of the Midwife Toad*, Hutchinson, London; *reprinted Pan Books, London 1974

Koestler, A. (1978) *Janus: A summing up*, Hutchinson, London

Kohn, D. (1980) 'Theories to work by: rejected theories, reproduction and Darwin's path to natural selection', *Studies in the History of Biology 4*, 67–170

Kohn, D. (ed.) (1985) *The Darwinian Heritage*, Princeton University Press, Princeton

Kottler, M. J. (1974) 'Alfred Russel Wallace, the origin of man and spiritualism', *Isis 65*, 144–92

Kottler, M. J. (1978) 'Charles Darwin's biological species concept and theory of geographic speciation: the Transmutation Notebooks', *Annals of Science 35*, 275–97

Kottler, M. J. (1980) 'Darwin, Wallace, and the origin of sexual dimorphism', *Proceedings of the American Philosophical Society 124*, 203–26

Kottler, M. J. (1985) 'Charles Darwin and Alfred Russel Wallace: Two decades of debate over natural selection' in Kohn 1985, pp. 367–432

Krebs, J. R. (1979) 'Bird colours', *Nature 282*, 14–16

Krebs, J. R. and Davies, N. B. (eds.) (1978) *Behavioural Ecology: An evolutionary approach*, Blackwell, Oxford; second edition 1984

Krebs, J. R. and Dawkins, R. (1984) 'Animal signals: mind-reading and manipulation' in Krebs and Davies 1978, second edition, pp. 380–402

Krebs, J. R. and Harvey, P. (1988) 'Lekking in Florence', *Nature 333*, 12–13

Kroodsma, D. E. (1976) 'Reproductive development in a female songbird: differential stimulation by quality of male song', *Science 192*, 574–5

Kropotkin, P. (1899) *Memoirs of a Revolutionist*, Smith, Elder, London

Kropotkin, P. (1902) *Mutual Aid: A factor of evolution*; *reprinted Penguin, Middlesex, 1939

Kummer, H. (1978) 'Analogs of morality among non-human primates' in Stent 1978, pp. 31–47

Lack, D. (1966) *Population Studies of Birds*, Clarendon Press, Oxford

Lack, D. (1968) *Ecological Adaptations for Breeding in Birds*, Methuen, London

Lacy, R. C. (1980) 'The evolution of eusociality in termites: a haplodiploid analogy?', *American Naturalist 116*, 449–451

Lacy, R. C. (1984) 'The evolution of eusociality: reply to Leinaas', *American Naturalist 123*, 876–8

Lande, R. (1981) 'Models of speciation by sexual selection on polygenic traits', *Proceedings of the National Academy of Sciences USA 78 (6)*, 3721–5

Langton, C. (ed.) (1989) *Artificial Life*, Addison-Wesley, Redwood City, California

Lankester, E. R. (1889) Review of Wallace's 'Darwinism', *Nature 40*, 566–70

Leinaas, H. P. (1983) 'A haploid analogy in the evolution of termite eusociality? Reply to Lacy', *American Naturalist 121*, 302–4

Lesch, J. E. (1975) 'The role of isolation in evolution: George J. Romanes and John T. Gulick', *Isis 66*, 483–503

Lewis, D. (1979) *Sexual Incompatibility in Plants*, Institute of Biology's Studies in Biology 110, Edward Arnold, London

Lewontin, R. C. (1978) 'Adaptation', *Scientific American 239 (3)*, 156–69

Lewontin, R. C. (1979) 'Sociobiology as an adaptationist program', *Behavioral Science 24*, 5–14

Lewontin, R. C. (1979a) 'Fitness, survival and optimality' in Horn, Stairs and Mitchell 1979, pp. 3–21

Lewontin, R. C. (1983) 'The organism as the subject and object of evolution', *Scientia 118*, 65–82

Ligon, J. D. and Ligon, S. H. (1978) 'Communal breeding in green woodhoopoes as a case for reciprocity', *Nature 276*, 496–8

Lill, A. (1976) 'Lek behavior in the golden-headed manakin, *Pipra erythrocephala* in Trinidad (West Indies)', *Fortschritte der Verhaltensforschung Zugleich Beiheft 18 zur Zeitschrift für Tierpsychologie*, Paul Parey Verlag, Berlin

Linnean Society (1908) *The Darwin–Wallace Celebration held on Thursday 1 July 1908 by the Linnean Society of London*, Linnean Society, London

Littlejohn, M. J. (1981) 'Reproductive isolation: a critical review' in Atchley and Woodruff 1981, pp. 298–334

Lloyd, J. E. (1979) 'Sexual selection in luminescent beetles' in Blum and Blum 1979, pp. 293–342

Lombardo, M. P. (1985) 'Mutual restraint in tree swallows: a test of the Tit for Tat model of reciprocity', *Science 227*, 1363–5

Lorenz, K. (1965) 'Preface' in Darwin 1872, pp. ix–xiii

Lorenz, K. (1966) *On Aggression*, Methuen, London

Lovelock, J. E. (1979) *Gaia*, Oxford University Press, Oxford

Loye, J. E., and Zuk, M. (eds.) (1991) *Bird–Parasite Interactions: Ecology, evolution and behaviour*, Oxford University Press, Oxford

Lumsden, C. J. and Wilson, E. O. (1981) *Genes, Mind, and Culture: The coevolutionary process*, Harvard University Press, Cambridge, Mass

Lumsden, C. J. and Wilson, E. O. (1983) *Promethean Fire: Reflections on the origin of mind*, Harvard University Press, Cambridge, Mass

Lyon, B. E. and Montgomerie, R. D. (1985) 'Conspicuous plumage of birds: sexual selection or unprofitable prey?', *Animal Behaviour 33*, 1038–40

MacBride, E. W. (1925) 'Zoology' in *Evolution in the Light of Modern Knowledge: A collective work*, Blackie, London; *1932 edition, pp. 211–61

Manier, E. (1978) *The Young Darwin and His Cultural Circle: A study of the influences which helped shape the language and logic of the first drafts of the theory of natural selection*, Studies in the History of Modern Science 2, Reidel, Dordrecht

Manier, E. (1980) 'Darwin's language and logic', *Studies in the History and Philosophy of Science 11*, 305–23

Marchant, J. (ed.) (1916) *Alfred Russel Wallace: Letters and reminiscences*, Cassell, London

Marler, P. (1985) 'Foreword' in Ryan 1985, pp. ix–xi

Marshall, J. (1980) 'The new organology', *Behavioral and Brain Sciences 3*, 23–5

Mayer, A. G. (1900) 'On the mating instinct in moths', *Psyche 9*, 15–20

Mayer, A. G. and Soule, C. G. (1906) 'Some reactions of caterpillars and moths', *Journal of Experimental Zoology 3*, 427–31

Maynard Smith, J. (1958) *The Theory of Evolution*, Penguin, Middlesex; *third edition 1975

Maynard Smith, J. (1958a) 'Sexual selection' in Barnett 1958, pp. 231–44

Maynard Smith, J. (1964) 'Group selection and kin selection', *Nature 201*, 1145–7

Maynard Smith, J. (1972) *On Evolution*, Edinburgh University Press, Edinburgh

Maynard Smith, J. (1974) 'The theory of games and the evolution of animal conflicts', *Journal of Theoretical Biology 47*, 209–21

Maynard Smith, J. (1976) 'Group selection', *Quarterly Review of Biology 51*, 277–83

Maynard Smith, J. (1976a) 'Sexual selection and the handicap principle', *Journal of Theoretical Biology 57*, 239–42

Maynard Smith, J. (1976b) 'Evolution and the theory of games', *American Scientist 64*, 41–5

Maynard Smith, J. (1978) *The Evolution of Sex*, Cambridge University Press, Cambridge

Maynard Smith, J. (1978a) 'The handicap principle – a comment', *Journal of Theoretical Biology 70*, 251–2

Maynard Smith, J. (1978b) 'The evolution of behavior', *Scientific American 239 (3)*, 136–45

Maynard Smith, J. (1978c) 'Optimization theory in evolution', *Annual Review of Ecology and Systematics 9*, 31–56

Maynard Smith, J. (1978d) 'Constraints on human nature', *Nature 276*, 120

Maynard Smith, J. (1982) *Evolution and the Theory of Games*, Cambridge University Press, Cambridge

Maynard Smith, J. (1982a) 'Storming the fortress', *New York Review of Books 29*, 41–2

Maynard Smith, J. (1982b) 'The evolution of social behaviour – a classification of models' in King's College Sociobiology Group 1982, pp. 29–44

Maynard Smith, J. (ed.) (1982c) *Evolution Now: A century after Darwin*, Macmillan, London

Maynard Smith, J. (1984) 'Game theory and the evolution of behaviour', *Behavioral and Brain Sciences 7*, 95–125

Maynard Smith, J. (1984a) 'Preface, August 1983' in Brandon and Burian 1984, pp. 238–9

Maynard Smith, J. (1985) 'Sexual selection, handicaps and true fitness', *Journal of Theoretical Biology 115*, 1–8

Maynard Smith, J. (1985a) 'Biology and the behaviour of man', *Nature 318*, 121–2

Maynard Smith, J. (ed.) (1985b) *On Being the Right Size and Other Essays by J. B. S. Haldane*, Oxford University Press, Oxford

Maynard Smith, J. (1986) *The Problems of Biology*, Oxford University Press, Oxford

Maynard Smith, J. (1987) 'Sexual selection – a classification of models' in Bradbury and Andersson 1987, pp. 9–20

Maynard Smith, J. and Parker, G. A. (1976) 'The logic of asymmetric contests', *Animal Behaviour 24*, 159–75

Maynard Smith, J. and Price, G. R. (1973) 'The logic of animal conflict', *Nature 246*, 15–18

Maynard Smith, J. and Ridpath, M. G. (1972) 'Wife sharing in the Tasmanian native hen, *Tribonyx mortierii*: a case of kin selection?', *American Naturalist 106*, 447–52

Maynard Smith, J. and Warren, N. (1982) 'Models of cultural and genetic change', *Evolution 36*, 620–7

Mayr, E. (1942) *Systematics and the Origin of Species from the Viewpoint of a Zoologist*, Columbia University Press, New York; *corrected edition with a new Preface by the author, Dover, New York, 1964

Mayr, E. (ed.) (1957) *The Species Problem*, A symposium presented at the Atlanta meeting of the American Association for the Advancement of Science, December 28–29 1955, Publication 50 of the AAAS, Washington DC

Mayr, E. (1959) 'Isolation as an evolutionary factor', *Proceedings of the American Philosophical Society 103*, 221–30

Mayr, E. (1963) *Animal Species and Evolution*, Harvard University Press, Cambridge, Mass

Mayr, E. (1972) 'Sexual selection and natural selection' in Campbell 1972, pp. 87–104

Mayr, E. (1976) *Evolution and the Diversity of Life: Selected essays*, Harvard University Press, Cambridge, Mass

Mayr, E. (1982) *The Growth of Biological Thought: Diversity, evolution, and inheritance*, Harvard University Press, Cambridge, Mass

Mayr, E. (1983) 'How to carry out the adaptationist programme?', *American Naturalist 121*, 324–34

Mayr, E. and Provine, W. B. (1980) *The Evolutionary Synthesis: Perspectives on the unification of biology*, Harvard University Press, Cambridge, Mass

McClelland, D. C., Atkinson, J. W., Clark, R. A. and Lowell, E. L. (eds.) (1953) *The Achievement Motive*, Appleton-Century-Crofts, New York

McClelland, J. L., Rumelhart, D. E. and Hinton, G. E. (1986) 'The appeal of parallel distributed processing' in Rumelhart, McClelland and the PDP Research Group 1986, pp. 3–44

McKinney, H. L. (1966) 'Alfred Russel Wallace and the discovery of natural selection', *Journal of the History of Medicine and Allied Sciences 21*, 333–57

McKinney, H. L. (1972) *Wallace and Natural Selection*, Yale University Press, New Haven

Mecham, J. S. (1961) 'Isolating mechanisms in anuran amphibians' in Blair 1961, pp. 24–61

Medawar, P. B. (1963) 'Onwards from Spencer: Evolution and evolutionism', *Encounter 21 (3) September*, 35–43; *reprinted as 'Herbert Spencer and the law of general evolution' (Spencer Lecture for 1963) in Medawar 1982, pp. 209–27

Medawar, P. B. (1967) *The Art of the Soluble*, Methuen, London

Medawar, P. B. (1982) *Pluto's Republic*, Oxford University Press, Oxford

Meek, R. L. (ed.) (1953) *Marx and Engels on Malthus: Selections from the writings of Marx and Engels dealing with the theories of Thomas Robert Malthus*, Lawrence and Wishart, London

Meeuse, B. and Morris, S. (1984) *The Sex Life of Flowers*, Faber and Faber, London

Midgley, M. (1979) *Beast and Man: The roots of human nature*, Harvester Press, Brighton; *reprinted Methuen, London, 1980

Midgley, M. (1979a) 'Gene-juggling', *Philosophy 54*, 439–58

Milinski, M. (1987) 'Tit for Tat in sticklebacks and the evolution of cooperation', *Nature 325*, 433–5

Miller, L. G. (1976) 'Fated genes: An essay review of E. O. Wilson, *Sociobiology: The new synthesis*', *Journal of the History of the Behavioral Sciences 12*, 183–90; *reprinted in Caplan 1978, pp. 269–79

[Mivart, St G.] (1871) Review of Darwin's 'Descent of Man', *Quarterly Review* *131*, 47–90

Mivart, St G. (1871a) *On the Genesis of Species*, Macmillan, London

Møller, A. P. (1988) 'Female choice selects for male sexual tail ornaments in the monogamous swallow', *Nature 332*, 640–2

Møller, A. P. (1990) 'Effects of a haematophagous mite on secondary sexual tail ornaments in the barn swallow (*Hirundo rustica*): a test of the Hamilton and Zuk hypothesis', *Evolution 44*, 771–84

Møller, A. P. (1991) 'Parasites, sexual ornaments and mate choice in the barn swallow' in Loye and Zuk 1991, pp. 328–43

Montagu, A. (1952) *Darwin: Competition and Cooperation*, Henry Schuman, New York

Montagu, A. (ed.) (1980) *Sociobiology Examined*, Oxford University Press, Oxford

Montagu, A. (1980a) 'Introduction' in Montagu 1980, pp. 3–14

Moodie, G. E. E. (1972) 'Predation, natural selection and adaptation in an unusual threespine stickleback', *Heredity 28*, 155–67

Moore, J. (1984) 'Parasites that change the behavior of their host', *Scientific American 250 (5)*, 82–9

Moore, J. (1984a) 'Altered behavioral responses in intermediate hosts – an acanthocephalan parasite strategy', *American Naturalist 123*, 572–7

Moore, J. and Gotelli, N. J. (1990) 'A phylogenetic perspective on the evolution of altered host behaviours: a critical look at the manipulation hypothesis' in Barnard and Behnke 1990, pp. 193–229

Moore, J. A. (1957) 'An embryologist's view of the species concept' in Mayr 1957, pp. 325–38

Morgan, C. L. (1890–1) *Animal Life and Intelligence*, Edward Arnold, London

Morgan, C. L. (1896) *Habit and Instinct*, Edward Arnold, London

Morgan, C. L. (1900) *Animal Behaviour*, Edward Arnold, London

Morgan, T. H. (1903) *Evolution and Adaptation*, Macmillan, New York

Mott, F. T. (1874) 'Insects and colour in flowers', *Nature 11*, 28

Mottram, J. C. (1914) *Controlled Natural Selection and Value Marking*, Longmans, Green, London

Mottram, J. C. (1915) 'The distribution of secondary sexual characters amongst birds, with relation to their liablity to the attack of enemies', *Proceedings of the Zoological Society of London*, 663–78

Munn, C. A. (1986) 'Birds that cry "wolf"', *Nature 319*, 143–5

Murray, J. (1972) *Genetic Diversity and Natural Selection*, Oliver and Boyd, Edinburgh

Myles, T. G. and Nutting, W. L. (1988) 'Termite eusocial evolution: a re-examination of Bartz's hypothesis and assumptions', *Quarterly Review of Biology 63*, 1–23

Nordenskiöld, E. (1929) *The History of Biology: A survey*, Kegan Paul, Trench, Trubner, London

Nur, N. and Hasson, O. (1984) 'Phenotypic plasticity and the handicap principle', *Journal of Theoretical Biology 110*, 275–97

O'Donald, P. (1962) 'The theory of sexual selection', *Heredity 17*, 541–52

O'Donald, P. (1980) *Genetic Models of Sexual Selection*, Cambridge University Press, Cambridge

O'Donald, P. (1987) 'Polymorphism and sexual selection in the Arctic skua' in Cooke and Buckley 1987, pp. 433–50

Ochman, H., Jones, J. S. and Selander, R. K. (1983) 'Molecular area effects in *Cepaea*', *Proceedings of the National Academy of Sciences USA 80 (3)*, 4189–93

Ochman, H., Jones, J. S. and Selander, R. K. (1987) 'Large scale patterns of genetic differentiation at enzyme loci in the land snails *Cepaea nemoralis* and *Cepaea hortensis*', *Heredity 58*, 127–38

Oldroyd, D. R. (1980) *Darwinian Impacts: An introduction to the Darwinian revolution*, Open University Press, Milton Keynes

Ospovat, D. (1978) 'Perfect adaptation and teleological explanation: approaches to the problem of the history of life in the mid-nineteenth century', *Studies in the History of Biology 2*, 33–56

Ospovat, D. (1980) 'God and natural selection: the Darwinian idea of design', *Journal of the History of Biology 13*, 169–94

Ospovat, D. (1981) *The Development of Darwin's Theory: Natural history, natural theology and natural selection 1838–1859*, Cambridge University Press, Cambridge

Otte, D. (1979) 'Historical developments of sexual selection theory' in Blum and Blum 1979, pp. 1–18

Owen, R. (1849) *On the Nature of Limbs*, John van Voorst, London

Packer, C. (1977) 'Reciprocal altruism in *Papio anubis*', *Nature 265*, 441–3; *reprinted in Maynard Smith 1982c, pp. 204–8

Pagel, M. D. and Harvey, P. H. (1988) 'Recent developments in the analysis of comparative data', *Quarterly Review of Biology 63*, 413–40

Pagel, M. D., Trevelyan, R. and Harvey, P. H. (1988) 'The evolution of bowerbuilding', *Trends in Ecology and Evolution 3*, 288–90

Paley, W. (1802) *Natural Theology; or evidences of the existence and attributes of the deity, collected from the appearances of nature*; *thirteenth edition Faulder, London, 1810

Paradis, J. G. (1978) *T. H. Huxley: Man's place in nature*, University of Nebraska Press, Lincoln

Parker, G. A. (1974) 'Assessment strategy and the evolution of fighting behaviour', *Journal of Theoretical Biology 47*, 223–43

Parker, G. A. (1979) 'Sexual selection and sexual conflict' in Blum and Blum 1979, pp. 123–66

Parker, G. A. (1983) 'Mate quality and mating decisions' in Bateson 1983a, pp. 141–66

Parker, G. A. (1984) 'Evolutionarily stable strategies' in Krebs and Davies 1978, second edition, pp. 30–61

Partridge, L. (1980) 'Mate choice increases a component of offspring fitness in fruit flies', *Nature 283*, 290–1; *reprinted in Maynard Smith 1982c, pp. 224–6

Partridge, L. and Halliday, T. (1984) 'Mating patterns and mate choice' in Krebs and Davies 1978, second edition, pp. 222–50

Paterson, H. E. H. (1978) 'More evidence against speciation by reinforcement', *South African Journal of Science 74*, 369–71

Paterson, H. E. H. (1982) 'Perspective on speciation by reinforcement', *South African Journal of Science 78*, 53–7

Payne, R. B. (1983) 'Bird songs, sexual selection, and female mating strategies' in Wasser 1983, pp. 55–90

Payne, R. B. and Payne, K. Z. (1977) 'Social organization and mating success in local song populations of village indigobirds, *Vidua chalybeata*', *Zeitschrift für Tierpsychologie 45*, 113–73

Pearson, K. (1892) *The Grammar of Science*, Adam and Charles Black, London; *second edition 1900

Peckham, G. W. and Peckham, E. G. (1889) 'Observations on sexual selection in spiders of the family Attidae', *Occasional Papers of the Natural History Society of Wisconsin 1*, 1–60

Peckham, G. W. and Peckham, E. G. (1890) 'Additional observations on sexual selection in spiders of the family Attidae, with some remarks on Mr Wallace's theory of sexual ornamentation', *Occasional Papers of the Natural History Society of Wisconsin 1*, 115–51

Peckham, M. (ed.) (1959) *The Origin of Species by Charles Darwin: A variorum text*, University of Pennsylvania Press, Philadelphia

Peek, F. W. (1972) 'An experimental study of the territorial function of vocal and visual display in the male red-winged blackbird (*Agelaius phoeniceus*)', *Animal Behaviour 20*, 112–18

Peel, J. D. Y. (1971) *Herbert Spencer: The evolution of a sociologist*, Heinemann, London

Peel, J. D. Y. (ed.) (1972) *Herbert Spencer: On social evolution – Selected writings*, University of Chicago Press, Chicago

Petrie, M. (1983) 'Female moorhens compete for small fat males', *Science 220*, 413–15

Petrie, M., Halliday, T. and Sanders, C. (1991) 'Peahens prefer peacocks with elaborate trains', *Animal Behaviour 41*, 323–31

Pittendrigh, C. S. (1958) 'Adaptation, natural selection and behavior' in Roe and Simpson 1958, pp. 390–416

Pocock, R. I. (1890) 'Sexual selection in spiders', *Nature 42*, 405–6

Pomiankowski, A. (1987) 'Sexual selection: the handicap principle does work – sometimes', *Proceedings of the Royal Society of London B 231*, 123–45

Pomiankowski, A. (1989) 'Mating success in male pheasants', *Nature 337*, 696

Popper, K. R. (1957) 'The aim of science', *Ratio 1*, 24–35; *reprinted with revisions in Popper 1972, pp. 191–205

Popper, K. R. (1972) *Objective Knowledge*, Oxford University Press, Oxford

Poulton, E. B. (1890) *The Colours of Animals: Their meaning and use, especially considered in the case of insects*, Kegan Paul, Trench, Trübner, London

Poulton, E. B. (1896) *Charles Darwin and the Theory of Natural Selection*, Cassell, London

Poulton, E. B. (1908) *Essays on Evolution 1889–1907*, Clarendon Press, Oxford

Poulton, E. B. (1909) *Charles Darwin and the Origin of Species*, Longmans, Green, London

Poulton, E. B. (1910) 'The value of colour in the struggle for life' in Seward 1910, pp. 271–97

Powell, B. (1857) *The Study of the Evidences of Natural Theology*, Oxford Essays, John W. Parker, London

Provine, W. B. (1985) 'Adaptation and mechanisms of evolution after Darwin: a study in persistent controversies' in Kohn 1985, pp. 825–66

Provine, W. B. (1985a) 'The R. A. Fisher–Sewall Wright controversy and its influence upon modern evolutionary theory' in Dawkins and Ridley 1985, pp. 197–219

Provine, W. B. (1986) *Sewall Wright and Evolutionary Biology*, University of Chicago Press, Chicago

Pruett-Jones, S. G., Pruett-Jones, M. A. and Jones, H. I. (1990) 'Parasites and sexual selection in birds of paradise', *American Zoologist 30*, 287–98

Pruett-Jones, S. G., Pruett-Jones, M. A. and Jones, H. I. (1991) 'Parasites and sexual selection in a New Guinea avifauna', *Current Ornithology 8*, 213–46

Rádl, E. (1930) *The History of Biological Theories* (translated and adapted by E. J. Hatfield), Oxford University Press, London

Radnitzky, G. and Andersson, G. (eds.) (1978) *Progress and Rationality in Science*, Boston Studies in the Philosophy of Science 125, Reidel, Dordrecht

Read, A. F. (1987) 'Comparative evidence supports the Hamilton and Zuk hypothesis on parasites and sexual selection', *Nature 328*, 68–70

Read, A. F. (1988) 'Sexual selection and the role of parasites', *Trends in Ecology and Evolution 3*, 97–101

Read, A. F. (1990) 'Parasites and the evolution of host sexual behaviour' in Barnard and Behnke 1990, pp. 117–57

Read, A. F. and Harvey, P. H. (1989) 'Reassessment of comparative evidence for Hamilton and Zuk theory on the evolution of secondary sexual characters', *Nature 339*, 618–20

Read, A. F. and Harvey, P. H. (1989a) 'Validity of sexual selection in birds', *Nature 340*, 104–5

Read, A. F. and Weary, D. M. (1990) 'Sexual selection and the evolution of bird song: a test of the Hamilton–Zuk hypothesis', *Behavioral Ecology and Sociobiology 26*, 47–56

Rehbock, P. F. (1983) *The Philosophical Naturalists: Themes in early nineteenth-century British biology*, University of Wisconsin Press, Madison

Reid, J. B. (1984) 'Bird coloration: predation, conspicuousness and the unprofitable-prey model', *Animal Behaviour 32*, 294

Reighard, J. (1908) 'An experimental field-study of warning coloration in coral-reef fishes', *Carnegie Institution of Washington Publications 103*, 257–325

Rensch, B. (1959) *Evolution Above the Species Level*, Methuen, London

Richards, O. W. (1927) 'Sexual selection and allied problems in the insects', *Biological Reviews 2*, 298–360

Richards, O. W. (1953) *The Social Insects*, Macdonald, London

Richards, R. J. (1979) 'The influence of the sensationalist tradition on early theories of the evolution of behavior', *Journal of the History of Ideas 40*, 85–105

Richards, R. J. (1981) 'Instinct and intelligence in British natural theology: some contributions to Darwin's theory of the evolution of behavior', *Journal of the History of Biology 14*, 193–230

Richards, R. J. (1982) 'The emergence of evolutionary biology of behaviour in the early nineteenth century', *British Journal for the History of Science 15*, 241–80

Ridley, M. (1983) *The Explanation of Organic Diversity: The comparative method and adaptations for mating*, Oxford University Press, Oxford

Ridley, M. (1985) *The Problems of Evolution*, Oxford University Press, Oxford

Ridley, M. (1985a) 'More Darwinian detractors', *Nature 318*, 124–5

Ridley, M. and Dawkins, R. (1981) 'The natural selection of altruism' in Rushton and Sorrentino 1981, pp. 19–39

Ridley, Matt (1981) 'How the peacock got his tail', *New Scientist 91*, 398–401

Robson, G. C. and Richards, O. W. (1936) *The Variation of Animals in Nature*, Longmans, Green, London

Roe, A. and Simpson, G. G. (eds.) (1958) *Behavior and Evolution*, Yale University Press, New Haven

Romanes, G. J. (1886) 'Physiological selection: an additional suggestion on the origin of species', *Journal of the Linnean Society (Zoology) 19*, 337–411

Romanes, G. J. (1886a) 'Physiological selection: an additional suggestion on the origin of species', *Nature 34*, 314–16, 336–40, 362–5

Romanes, G. J. (1890) 'Before and after Darwin', *Nature 41*, 524–5

Romanes, G. J. (1892–7) *Darwin, and After Darwin: An exposition of the Darwinian theory and a discussion of post-Darwinian questions*, Longmans, Green, London; *new edition 1900–5

Rood, J. P. (1978) 'Dwarf mongoose helpers at the den', *Zeitschrift für Tierpsychologie 48*, 277–87

Rose, M. and Charlesworth, B. (1980) 'A test of evolutionary theories of senescence', *Nature 287*, 141–2

Rose, S. and Appignanesi, L. (eds.) (1986) *Science and Beyond*, Basil Blackwell, Oxford

Rozin, P. (1976) 'The evolution of intelligence and access to the cognitive unconscious' in Sprague and Epstein 1976, pp. 245–80

Ruben, D-H. (1985) *The Metaphysics of the Social World*, Routledge and Kegan Paul, London

Rumelhart, D. E., McClelland, J. L. and the PDP Research Group (1986) *Parallel Distributed Processing: Explorations in the microstructures of cognition*, i, *Foundations*, MIT Press, Cambridge, Mass

Ruse, M. (1971) 'Natural selection in *The Origin of Species*', *Studies in the History and Philosophy of Science 1*, 311–51

Ruse, M. (1979) *Sociobiology: Sense or Nonsense*, Reidel, Dordrecht

Ruse, M. (1979a) *The Darwinian Revolution*, University of Chicago Press, Chicago

Ruse, M. (1980) 'Charles Darwin and group selection', *Annals of Science 37*, 615–30

Ruse, M. (1982) *Darwinism Defended: A guide to the evolution controversies*, Addison-Wesley, Reading, Mass

Ruse, M. (1986) *Taking Darwin Seriously: A naturalistic approach to philosophy*, Blackwell, Oxford

Rushton, J. P. and Sorrentino, R. M. (eds.) (1981) *Altruism and Helping Behavior: Social, personality and developmental perspectives*, Lawrence Erlbaum, Hillsdale, New Jersey

Ryan, M. J. (1985) *The Túngara Frog: A study in sexual selection and communication*, University of Chicago Press, Chicago

Ryan, M. J., Tuttle, M. D. and Rand, A. S. (1982) 'Bat predation and sexual advertisement in a neotropical anuran', *American Naturalist 119*, 136–9

Sahlins, M. (1976) *The Use and Abuse of Biology: An anthropological critique of sociobiology*, University of Michigan Press, Ann Arbor

Schuster, A. and Shipley, A. (1917) *Britain's Heritage of Science*, Constable, London

Schwartz, J. S. (1984) 'Darwin, Wallace, and the *Descent of Man*', *Journal of the History of Biology 17*, 271–89

Schweber, S. S. (1977) 'The origin of the *Origin* revisited', *Journal of the History of Biology 10*, 229–316

Schweber, S. S. (1980) 'Darwin and the political economists: divergence of character', *Journal of the History of Biology 13*, 195–289

Scudder, G. E. and Reveal, J. L. (1981) *Evolution Today*, Proceedings of the Second International Congress of Systematic and Evolutionary Biology, University of British Columbia, Vancouver, Canada, 1980, Hunt Institute for Botanical Documentation, Carnegie-Mellon University, Pittsburgh

Searcy, W. A. (1982) 'The evolutionary effects of mate selection', *Annual Review of Ecology and Systematics 13*, 57–85

Searcy, W. A. and Yasukawa, K. (1983) 'Sexual selection and red-winged blackbirds', *American Scientist 71*, 166–74

Seger, J. (1985) 'Unifying genetic models for the evolution of female choice', *Evolution 39*, 1185–93

Selander, R. K. (1972) 'Sexual selection and dimorphism in birds' in Campbell 1972, pp. 180–230

Selous, E. (1910) 'An observational diary on the nuptial habits of the blackcock (*Tetrao tetrix*) in Scandinavia and England', *The Zoologist, fourth series 14*, 51–6, 176–82, 248–65

Selous, E. (1913) 'The nuptial habits of the blackcock', *The Naturalist 673*, 96–8

Semler, D. E. (1971) 'Some aspects of adaptation in a polymorphism for breeding colours in the Threespine stickleback (*Gasterosteus aculeatus*)', *Journal of Zoology 165*, 291–302

Seward, A. C. (ed.) (1910) *Darwin and Modern Science: Essays in commemoration of the centenary of the birth of Charles Darwin and of the fiftieth anniversary of the publication of 'The Origin of Species'*, Cambridge University Press, Cambridge

Seyfarth, R. M. and Cheney, D. L. (1984) 'Grooming, alliances and reciprocal altruism in vervet monkeys', *Nature 308*, 541–3

Shepard, R. N. (1987) 'Evolution of a mesh between principles of the mind and regularities of the world' in Dupré 1987, pp. 251–75

Sheppard, P. M. (1958) *Natural Selection and Heredity*, Hutchison, London; *fourth edition 1975

Sherman, P. and Morton, M. L. (1988) 'Extra-pair fertilizations in mountain white-crowned sparrows', *Behavioral Ecology and Sociobiology 22*, 413–20

Sherman, P. W. (1977) 'Nepotism and the evolution of alarm calls', *Science 197*, 1246–53; *reprinted in Maynard Smith 1982c, pp. 186–203

Sherman, P. W. (1980) 'The meaning of nepotism', *American Naturalist 116*, 604–6

Sherman, P. W. (1980a) 'The limits of ground squirrel nepotism' in Barlow and Silverberg 1980, pp. 505–44

Shull, A. F. (1936) *Evolution*, McGraw–Hill, New York

Simpson, G. G. (1944) *Tempo and Mode in Evolution*, Columbia University Press, New York

Simpson, G. G. (1950) *The Meaning of Evolution: A study of the history of life and of its significance for man*, Oxford University Press, London

Simpson, G. G. (1953) *The Major Features of Evolution*, Columbia University Press, New York

Singer, C. (1931) *A Short History of Biology: A general introduction to the study of living things*, Oxford University Press, Oxford

Sloan, P. R. (1981) Review of Ruse's 'The Darwinian Revolution', *Philosophy of Science 48*, 623–7

Smith, D. G. (1972) 'The role of the epaulets in the red-winged blackbird, (*Agelaius phoeniceus*) social system', *Behaviour 41*, 251–68

Smith, R. (1972) 'Alfred Russel Wallace: philosophy of nature and man', *British Journal for the History of Science 6*, 177–99

Sober, E. (1984) *The Nature of Selection: Evolutionary theory in philosophical focus*, MIT Press, Cambridge, Mass

Sober, E. (1985) 'Darwin on natural selection: a philosophical perspective' in Kohn 1985, pp. 867–99

Spencer, H. (1863–7) *The Principles of Biology*, Williams and Norgate, London

Spencer, H. (1887) *Factors of Organic Evolution*, Williams and Norgate, London

Sprague, J. M. and Epstein, A. N. (eds.) (1976) *Progress in Psychobiology and Physiological Psychology 6*, Academic Press, New York

Spurrier, M. F., Boyce, M. S. and Manly, B. F. (1991) 'Effects of parasites on mate choice by captive sage grouse' in Loye and Zuk 1991, pp. 389–98

Stacey, P. B. and Koenig, W. D. (1984) 'Cooperative breeding in the acorn woodpecker', *Scientific American 251 (2)*, 100–7

Stacey, P. B. and Koenig, W. D. (eds.) (1990) *Cooperative Breeding in Birds*, Cambridge University Press, Cambridge

Stauffer, R. C. (ed.) (1975) *Charles Darwin's Natural Selection, Being the Second Part of His Big Species Book Written from 1856 to 1858*, Cambridge University Press, London

Stebbins, G. L. (1950) *Variation and Evolution in Plants*, Oxford University Press, Oxford

Steele, E. J. (1979) *Somatic Selection and Adaptive Evolution: On the inheritance of acquired characters*, Williams and Wallace, Toronto

Stent, G. S. (ed.) (1978) *Morality as a Biological Phenomenon*, Report of the Dahlem Workshop on Biology and Morals, Life Sciences Research Report 9, Dahlem Konferenzen, Berlin; *revised edition University of California Press, Berkeley and Los Angeles, California, 1980

Stolzmann, J. (1885) 'Quelques remarques sur le dimorphisme sexuel', *Proceedings of the Zoological Society of London*, 421–32

Stonehouse, B. and Perrins, C. (eds.) (1977) *Evolutionary Ecology*, Macmillan, London

Sulloway, F. J. (1979) 'Geographic isolation in Darwin's thinking: the vicissitudes of a crucial idea', *Studies in the History of Biology 3*, 23–65

Sulloway, F. J. (1982) 'Darwin and his finches: the evolution of a legend', *Journal of the History of Biology 15*, 1–53

Swift, J. (1726) *Travels into Several Remote Nations of the World. In Four parts. By Lemuel Gulliver*; *reprinted Penguin, Middlesex, 1976

Symondson, A. (ed.) (1970) *The Victorian Crisis of Faith*, SPCK and Victorian Society, London

Symons, D. (1979) *The Evolution of Human Sexuality*, Oxford University Press, New York

Symons, D. (1980) 'Précis of *The Evolution of Human Sexuality*' and 'Author's response', *Behavioral and Brain Sciences 3*, 171–81, 203–14

Symons, D. (1987) 'If we're all Darwinians, what's the fuss about?' in Crawford, Smith and Krebs 1987, pp. 121–46

Symons, D. (1989) 'A critique of Darwinian anthropology', *Ethology and Sociobiology 10*, 131–44

Symons, D. (1992) 'On the use and misuse of Darwinism in the study of human behavior' in Barkow, Cosmides and Tooby, 1992

Syren, R. M. and Luykx, P. (1977) 'Permanent segmental interchange complex in the termite *Incisitermes schwarzi*', *Nature 266*, 167–8

Tax, S. (ed.) (1960) *Evolution After Darwin, The University of Chicago Centennial*, University of Chicago Press, Chicago

Thomson, W. (Lord Kelvin) (1872) 'Presidential Address', *Report of the Forty-First Meeting of the British Association for the Advancement of Science, Edinburgh, 1871*, pp. lxxxiv–cv, John Murray, London

Thornhill, R. (1976) 'Sexual selection and nuptial feeding behavior in *Bittacus apicalis* (Insecta: Mecoptera)', *American Naturalist 110*, 529–48

Thornhill, R. (1979) 'Male and female sexual selection and the evolution of mating strategies in insects' in Blum and Blum 1979, pp. 81–121

Thornhill, R. (1980) 'Competitive, charming males and choosy females: was Darwin correct?', *Florida Entomologist 63*, 5–29

Thornhill, R. (1980a) 'Sexual selection in the black-tipped hangingfly', *Scientific American 242 (6)*, 138–45

Thornhill, R. (1980b) 'Mate choice in *Hylobittacus apicalis* (Insecta: Mecoptera) and its relation to some models of female choice', *Evolution 34*, 519–38

Thornhill, R. and Alcock, J. (1983) *The Evolution of Insect Mating Systems*, Harvard University Press, Cambridge, Mass

Tinbergen, N. (1963) 'On aims and methods of ethology', *Zeitschrift für Tierpsychologie 20*, 410–33

Tooby, J. and Cosmides, L. (1989) 'Evolutionary psychology and the generation of culture, Part 1: Theoretical considerations', *Ethology and Sociobiology 10*, 29–49

Tooby, J. and Cosmides, L. (1989a) 'The innate versus the manifest: How universal does universal have to be?', *Behavioral and Brain Sciences 12*, 36–7

Tooby, J. and Cosmides, L. (1989b) 'Evolutionary psychologists need to distinguish between the evolutionary process, ancestral selection pressures, and psychological mechanisms', *Behavioral and Brain Sciences 12*, 724–5

Trivers, R. L. (1971) 'The evolution of reciprocal altruism', *Quarterly Review of Biology 46*, 35–57

Trivers, R. L. (1972) 'Parental investment and sexual selection' in Campbell 1972, pp. 136–79

Trivers, R. L. (1974) 'Parent–offspring conflict', *American Zoologist 14*, 249–64

Trivers, R. L. (1983) 'The evolution of a sense of fairness', *Absolute Values and the Creation of the New World*, Proceedings of the Eleventh International Conference on the Unity of the Sciences, ii, International Cultural Foundation Press, New York, pp. 1189–208

Trivers, R. L. (1985) *Social Evolution*, Benjamin/Cummings, Menlo Park, California

Turner, F. M. (1974) *Between Science and Religion: The reaction to scientific naturalism in late Victorian England*, Yale University Press, New Haven

Turner, J. R. G. (1978) 'Why male butterflies are non-mimetic: natural selection, sexual selection, group selection, modification and sieving', *Biological Journal of the Linnean Society 10*, 385–432

Turner, J. R. G. (1983) '"The hypothesis that explains mimetic resemblance explains evolution": the gradualist–saltationist schism' in Grene 1983, pp. 129–69

Tylor, A. (1886) *Coloration in Animals and Plants*, Alabaster, Passmore, London

Urbach, P. (1987) *Francis Bacon's Philosophy of Science: An account and a reappraisal*, Open Court, La Salle, Illinois

Vehrencamp, S. L. and Bradbury, J. W. (1984) 'Mating systems and ecology' in Krebs and Davies 1978, second edition, pp. 251–78

von Schantz, T., Göransson, G., Andersson, G., Fröberg, I., Grahn, M., Helgée, A. and Wittzell, H. (1989) 'Female choice selects for a viability-based male trait in pheasants', *Nature 337*, 166–9

Vorzimmer, P. (1969) 'Darwin, Malthus and natural selection', *Journal of the History of Ideas 30*, 527–42

Vorzimmer, P. (1972) *Charles Darwin: The years of controversy – The 'Origin of Species' and its critics 1859–82*, University of London Press, London

Wagner, M. (1873) *The Darwinian Theory and the Law of the Migration of Organisms* (translated by J. L. Laird), Edward Stanford, London

Wallace, A. R. (1853) *A Narrative of Travels on the Amazon and Rio Negro*, Reeve, London; second edition Ward, Lock, London, 1889; *reprint of second edition with a new introduction by H. Lewis McKinney, Dover Publications, New York, 1972

Wallace, A. R. (1856) 'On the habits of the orang-utan of Borneo', *Annals and Magazine of Natural History, second series 18*, 26–32

Wallace, A. R. (1864) 'The origin of human races and the antiquity of man deduced from the theory of "natural selection"', *Anthropological Review and Journal of Anthropological Society of London 2*, 158–87; revised and reprinted as 'The development of human races under the law of natural selection' in Wallace 1870, pp. 303–31 and in Wallace 1891, pp. 167–85

[Wallace, A. R.] (1864a) '"Natural selection" as applied to man', *Natural History Review 4*, 328–36

[Wallace, A. R.] (1869) 'Geological climates and the origin of species', *Quarterly Review 126*, 359–94

Wallace, A. R. (1869a) *The Malay Archipelago: The land of the orang-utan, and the bird of paradise; a narrative of travel, with studies of man and nature*, Macmillan, London

Wallace, A. R. (1870) *Contributions to the Theory of Natural Selection: A series of essays*, Macmillan, London

Wallace, A. R. (1870a) 'Man and natural selection', *Nature 3*, 8–9

Wallace, A. R. (1871) Review of Darwin's Descent of Man, *The Academy 2*, 177–83

Wallace, A. R. (1877) 'Presidential Address', Biology Section, *Report of the Forty-Sixth Meeting of the British Association for the Advancement of Science, Glasgow, 1876*, Transactions of the Sections, pp. 110–19, John Murray, London; reprinted as 'By-paths in the domain of biology' in Wallace 1878, pp. 249–303 and reprinted in part as 'The antiquity and origin of man' in Wallace 1891, pp. 416–32

Wallace, A. R. (1878) *Tropical Nature and Other Essays*, Macmillan, London

Wallace, A. R. (1879) 'Animals and their countries', *Nineteenth Century 5*, 247–59

Wallace, A. R. (1889) *Darwinism*, Macmillan, London; *third edition 1901

Wallace, A. R. (1890) 'Human selection', *Fortnightly Review, new series 48*, 325–37; *reprinted in Wallace 1900, i, pp. 509–26

Wallace, A. R. (1890a) Review of Poulton's 'The Colours of Animals', *Nature 42*, 289–91

Wallace, A. R. (1891) *Natural Selection and Tropical Nature: Essays on descriptive and theoretical biology*, Macmillan, London

Wallace, A. R. (1892) 'Note on sexual selection', *Natural Science 1*, 749–50

Wallace, A. R. (1893) 'Are individually acquired characters inherited?', *Fortnightly Review, new series 53 (old series 59)*, 490–8, 655–68

Wallace, A. R. (1900) *Studies, Scientific and Social*, Macmillan, London

Wallace, A. R. (1905) *My Life: A record of events and opinions*, Chapman and Hall, London

Ward, P. I. (1988) 'Sexual dichromatism and parasitism in British and Irish freshwater fish', *Animal Behaviour 36*, 1210–15

Wason, P. C. (1983) 'Realism and rationality in the selection task' in Evans 1983, pp. 44–75

Wasser, S. K. (ed.) (1983) *Social Behavior of Female Vertebrates*, Academic Press, New York

Watkins, J. W. N. (1984) *Science and Scepticism*, Princeton University Press, Princeton

Watt, W. B., Carter, P. A. and Donohue, K. (1986) 'Females' choice of "good genotypes" as mates is promoted by an insect mating system', *Science 233*, 1187–90

Webster, C. (ed.) (1981) *Biology, Medicine and Society 1840–1940*, Cambridge University Press, Cambridge

Weismann, A. (1893) 'The all-sufficiency of natural selection: a reply to Herbert Spencer', *Contemporary Review 64*, 309–38, 596–610

West, M. J., King, A. P. and Eastzer, D. H. (1981) 'Validating the female bioassay of cowbird song: relating differences in song potency to mating success', *Animal Behaviour 29*, 490–501

West-Eberhard, M. J. (1979) 'Sexual selection, social competition, and evolution', *Proceedings of the American Philosophical Society 123*, 222–34

West-Eberhard, M. J. (1983) 'Sexual selection, social competition, and speciation', *Quarterly Review of Biology 58*, 155–83

Westermarck, E. (1891) *The History of Human Marriage*, Macmillan, London; *fifth edition 1921

Westneat, D. F. (1987) 'Extra-pair copulations in a predominantly monogamous bird: observations of behaviour', *Animal Behaviour 35*, 865–76

Westneat, D. F. (1987a) 'Extra-pair fertilizations in a predominantly monogamous bird: genetic evidence', *Animal Behaviour 35*, 877–86

Wheeler, W. M. (1911) 'The ant-colony as an organism', *Journal of Morphology 22*, 307–25

Wheeler, W. M. (1928) *The Social Insects: Their origin and evolution*, Kegan Paul, Trench, Trubner, London

White, M. J. D. (1978) *Modes of Speciation*, W. H. Freeman, San Francisco

Whitehouse, H. L. K. (1959) 'Cross- and self-fertilization in plants' in Bell 1959, pp. 207–61

Wiley, R. H. (1973) 'Territoriality and non-random mating in sage grouse, *Centrocerus urophasianus*', *Animal Behaviour Monographs 6*, 85–169

Wilkinson, G. S. (1984) 'Reciprocal food sharing in the vampire bat', *Nature 308*, 181–4

Wilkinson, G. S. (1985) 'The social organization of the common vampire bat', *Behavioral Ecology and Sociobiology 17*, 111–21

Williams, G. C. (1957) 'Pleiotropy, natural selection and the evolution of senescence', *Evolution 11*, 398–411

Williams, G. C. (1966) *Adaptation and Natural Selection: A critique of some current evolutionary thought*, Princeton University Press, Princeton, New Jersey

Williams, G. C. (ed.) (1971) *Group Selection*, Aldine Atherton, Chicago

Willson, M. F. and Burley, N. (1983) *Mate Choice in Plants: Tactics, mechanisms and consequences*, Princeton University Press, Princeton, New Jersey

Wilson, E. O. (1971) *The Insect Societies*, Harvard University Press, Cambridge, Mass

Wilson, E. O. (1975) *Sociobiology: The new synthesis*, Harvard University Press, Cambridge, Mass

Wilson, E. O. (1978) *On Human Nature*, Harvard University Press, Cambridge, Mass

Wilson, M. and Daly, M. (1985) 'Competitiveness, risk taking, and violence: the young male syndrome', *Ethology and Sociobiology 6*, 59–73

Worrall, J. (1978) 'The ways in which the methodology of scientific research programmes improves on Popper's methodology' in Radnitzky and Andersson 1978, pp. 45–70

Wrege, P. H. and Emlen, S. T. (1987) 'Biochemical determination of parental uncertainty in white-fronted bee-eaters', *Behavioral Ecology and Sociobiology 20*, 153–60

Wright, C. (1870) Review of Wallace's 'Contributions to the Theory of Natural Selection: A series of essays', *North American Review 111*, 282–311

Wright, S. (1932) 'The roles of mutation, inbreeding, crossbreeding and selection in evolution', *Proceedings of the Sixth International Congress of Genetics*, i, 356–66; *reprinted in Wright 1986, pp. 161–71

Wright, S. (1945) 'Tempo and mode in evolution: a critical review', *Ecology 26*, 415–19

Wright, S. (1951) 'Fisher and Ford on "the Sewall Wright effect"', *American Scientist 39*, 452–8, 479; *reprinted in Wright 1986, pp. 515–22

Wright, S. (1968–78) *Evolution and the Genetics of Populations: A treatise in four volumes*, University of Chicago Press, Chicago

Wright, S. (1986) *Evolution: Selected papers, edited by W. B. Provine*, University of Chicago Press, Chicago

Wynne-Edwards, V. C. (1959) 'The control of population-density through social behaviour: a hypothesis', *Ibis 101*, 436–41

Wynne-Edwards, V. C. (1962) *Animal Dispersion in Relation to Social Behaviour*, Oliver and Boyd, Edinburgh

Wynne-Edwards, V. C. (1963) 'Intergroup selection in the evolution of social systems', *Nature 200*, 623–6

Wynne-Edwards, V. C. (1964) 'Group selection and kin selection' and 'Survival of young swifts in relation to brood-size', *Nature 201*, 1147, 1148–9

Wynne-Edwards, V. C. (1977) 'Society versus the individual in animal evolution' in Stonehouse and Perrins 1977, pp. 5–17

Wynne-Edwards, V. C. (1978) 'Intrinsic population control: an introduction' in Ebling and Stoddart 1978, pp. 1–22

Wynne-Edwards, V. C. (1982) Review of Boorman and Levitt's 'The Genetics of Altruism', *Social Science and Medicine 16*, 1095–8

Wynne-Edwards, V. C. (1986) *Evolution Through Group Selection*, Blackwell Scientific Publications, Oxford

Yasukawa, K. (1981) 'Song repertoires in the red-winged blackbird (*Agelaius phoeniceus*): a test of the Beau Geste hypothesis', *Animal Behaviour 29*, 114–25

Yasukawa, K., Blank, J. L. and Patterson, C. B. (1980) 'Song repertoires and sexual selection in the red-winged blackbird', *Behavioural Ecology and Sociobiology 7*, 233–8

Yeo, R. (1979) 'William Whewell, natural theology and the philosophy of science in mid nineteenth century Britain', *Annals of Science 36*, 493–516

Young, J. Z. (1957) *The Life of Mammals*, Oxford University Press, London

Young, R. M. (1969) 'Malthus and the evolutionists: the common context of biological and social theory', *Past and Present 43*, 109–45

Young, R. M. (1970) 'The impact of Darwin on conventional thought' in Symondson 1970, pp. 13–35

Young, R. M. (1971) 'Darwin's metaphor: does nature select?', *Monist 55*, 442–503

Zahar, E. (1973) 'Why did Einstein's programme supersede Lorentz's?', *British Journal for the Philosophy of Science 24*, 95–123, 223–62

Zahavi, A. (1975) 'Mate selection – a selection for a handicap', *Journal of Theoretical Biology 53*, 205–14

Zahavi, A. (1977) 'Reliability in communication systems and the evolution of altruism' in Stonehouse and Perrins 1977, pp. 253–9

Zahavi, A. (1978) 'Decorative patterns and the evolution of art', *New Scientist 80*, 182–4

Zahavi, A. (1980) 'Ritualization and the evolution of movement signals', *Behaviour 72*, 77–81

Zahavi, A. (1981) 'Natural selection, sexual selection and the selection of signals' in Scudder and Reveal 1981, pp. 133–8

Zahavi, A. (1987) 'The theory of signal selection and some of its implications' in Delfino 1987, pp. 305–27

Zahavi, A. (1990) 'Arabian babblers: the quest for social status in a cooperative breeder' in Stacey and Koenig 1990, pp. 103–30

Zuk, M. (1987) 'The effects of gregarine parasites, body size, and time of day on spermatophore production and sexual selection in field crickets', *Behavioral Ecology and Sociobiology 21*, 65–72

Zuk, M. (1988) 'Parasite load, body size, and age of wild-caught male field crickets (Orthoptera: Gryllidae): effects on sexual selection', *Evolution 42*, 969–76

Zuk, M. (1989) 'Validity of sexual selection in birds', *Nature 340*, 104–5

Zuk, M. (1991) 'Parasites and bright birds: new data and a new prediction' in Loye and Zuk 1991, pp. 317–27

Zuk, M., Thornhill, R., Ligon, J. D. and Johnson, K. (1990) 'Parasites and mate choice in red jungle fowl', *American Zoologist 30*, 235–44

INDEX

Page numbers in italic refer to illustrations